Studies in Fuzziness and Soft Computing

Volume 411

Murat Diker

Texture Spaces

Springer

Murat Diker
Department of Mathematics
Hacettepe University
Ankara, Türkiye

ISSN 1434-9922 ISSN 1860-0808 (electronic)
Studies in Fuzziness and Soft Computing
ISBN 978-3-031-39747-9 ISBN 978-3-031-39748-6 (eBook)
https://doi.org/10.1007/978-3-031-39748-6

This Springer imprint is published by the registered company Springer Nature Switzerland AG
The registered company address is: Gewerbestrasse 11, 6330 Cham, Switzerland

To my wife Nilgün and our grandson Bora with love.

Preface

This book contains the texture theory and its relations to rough sets and fuzzy sets. It provides the intelligibility of the notions due to textures and presents concrete examples with easy explanations for the basic connections among the rough sets, fuzzy sets and textures, and the recent investigations. Many observations related to textures are scattered in the literature, and this book aims to combine the basic tools and theoretical arguments of textures in this respect. Prospective readers are not only mathematicians who interest in purely mathematical theories related to textures, but also engineers of information sciences who need theoretical models for their interdisciplinary studies due to rough sets and fuzzy sets. Therefore, I tried to make self-contained and reader friendly this source as much as possible. The theory of texture spaces was produced by Lawrence M. Brown and presented first by him at the 2nd International Conference of the Balcanic Union for Fuzzy Systems and Artificial Intelligence in 1992. The first formal paper about textures was published in the Special Issue of the Journal of Fuzzy Sets and Systems entitled "Topics of the Mathematics of Fuzzy Objects" by L. M. Brown and M. Diker in 1998 [1].

A texture is a family of subsets of a domain of discourse which is an alternative point-set-based setting for fuzzy sets. The notions of approximation operator and definable set of rough set theory are also key concepts in the theory of texture spaces. Recent investigations show that the theory of texture is one of the remarkable mathematical models for fuzzy sets and rough sets. However, the peculiar structure of textures makes difficult to understand the basic concepts and results of the theory of textures for the researchers who study real-life applications using fuzzy sets and rough sets. Texture space, as one of the main subjects of this book, is essentially motivated by fuzzy sets. Therefore, most of the basic arguments of fuzzy sets can be considered in terms of textures. Rough set theory whose creator is Zdzislaw Pawlak [2] arosed from the need for the solutions of the problems related to imperfect data with respect to information systems. The theory of fuzzy sets was created by Lotfi A. Zadeh [6], and it is also subjected to imperfect or imprecise data. However, the rough set theory uses indiscernibility relations while fuzzy set theory considers the membership degrees of the objects. It is remarkable to say that both of the theories are complementary to each other as it was discussed in detail by Pawlak and Skowron in

[3]. Texture, as a purely mathematical structure, combines fuzzy set theory and rough set theory in various directions. I believe that this book will be a useful source for the researchers (computer scientists, engineers, mathematicians) who need development of the theoretical background for their works due to information sciences.

The contents of the book consist of 7 chapters.

Chapter 1 is devoted to some basic facts from lattice theory. Complete distributivity plays an important role in texture spaces and hence, we give a characterization for the complete distributivity in textures using the Raney's theorem facilitating the proofs with respect to complete distributivity.

In Chap. 2, Sect. 2.1 starts with one of the canonic examples of fuzzy lattices providing the motivation of textures. Section 2.2 presents the main tools of a texture called a p-set and q-set assuring a strong duality in almost all textural results. Section 2.3 focuses on the equivalences of completely distributivity, in particular in terms of p-set and q-set. Chapter 2 is ended with the important subclasses of textures called simple and plain textures.

Chapter 3 presents Brown's Representation Theorem for textures. In Sects. 3.1 and 3.2, textural isomorphism and the product of textures are discussed. Sections 3.3 and 3.4 discuss two important examples of textures called L-fuzzy texture and orthopair texture, respectively.

In Chap. 4, the concept of a direlation is defined as a morphism between textures. A direlation is a generalization of the ordinary relation between sets, and it is compatible with the product of textures. The notions of image and preimage due to a direlation called section and presection, respectively, are the remarkable generalizations of Pawlak approximations. This unexpected connection makes these concepts even more valuable and so, in Sects. 4.2–4.4, together with seriality and injectivity classes, almost all properties of direlations are given in detail.

A summary about Pawlak's rough sets and generalizations presented by Yiyu Yao [4, 5] are given in Sects. 5.1, 5.2 and 5.3 taking into account the theoretical development of approximation spaces, respectively. Sections 5.4 and 5.5 discuss the textural rough sets and the connections to Pawlak's rough set approximations and generalizations with a single or two domains of discourse. A further discussion is also devoted to textural definability in Sect. 5.5.

Section 6.1 considers the formal connection between textures and rough sets. Textural definability will be used in Sect. 6.2 to prove the results due to definability in generalized approximation spaces. Chapter 6 is essentially devoted to the basic properties of rough set theory. In Sect. 6.3, the revised rough set approximations are improved using the textural arguments given in Sect. 6.4.

Finally, Chap. 7 presents a detailed study on fuzzy rough sets. The combinations of fuzzy sets and rough sets can be realized using fuzzy logical connectives. Therefore, Sect. 7.1 reviews the information about the fuzzy logical connectives. The continuity of fuzzy logical connectives is widely used in the literature. It is known that some algebraic properties such as adjointness can be obtained from continuity condition. However, in most cases, continuity is also a necessary and significant condition for the basic results of fuzzy rough set approximations. Therefore, for the sake of the readers,

Sect. 7.2 provides some basic topological arguments due to continuity of the binary relations defined on the cartesian product of the unit interval. The fuzzy version of a direlation, called t-fuzzy direlation, is a key concept for a general approach to fuzzy rough set models and hence, textural fuzzy direlations are discussed in Sect. 7.3. Section 7.4 argues the propositional fuzzy logic construction for a bridge between fuzzy direlations and fuzzy relations. Sections 7.5 and 7.6 are devoted to adjointness, duality and the well-known fuzzy rough set models. This makes it possible to express the concepts of seriality, injectivity, symmetricity and transitivity of a fuzzy relation in terms of continuous t-norms and implicators. These arguments can be found in Sect. 7.7. Further, Sects. 7.8 and 7.9 are devoted to t-fuzzy rough set approximations and definability. The results given here are used for the basic arguments due to fuzzy rough set approximations with two domains of discourse in Sects. 7.10 and 7.11. Finally, Sects. 7.12 and 7.13 argue the revised fuzzy rough set approximations.

Ankara, Türkiye
May 2023

Murat Diker

References

1. L.M. Brown, M. Diker, Ditopological texture spaces and intuitionistic sets. Fuzzy Sets Syst. **98**, 217224 (1998)
2. Z. Pawlak, Rough Sets. Int. J. Comput. Info. Sci. 341–356 (1982)
3. Z. Pawlak, A. Skowron. Rudiments of rough sets. Info. Sci. **177** (2007) 3âĂŞ27.
4. Y. Y. Yao, Relational interpretations of neigborhood operators and rough set approximations operators. Inf. Sci. **111**(1-4), 239–259 (1998)
5. Y.Y. Yao, Combination of rough and fuzzy sets based on alpha-level sets, in *Rough Sets and Data Mining: Analysis for Imprecise Data*, eds. by T.Y. Lin, N. Cercone (Kluwer Academic Publishers, Boston, 1997) pp. 301–321
6. L.A. Zadeh. Inf. Control. **8**(3), 338–353 (1965)

Acknowledgements

I would like to express my gratitude to Professor Janusz Kacprzyk, the Editor in-Chief of the book series, for his valuable encouragement. I am pleased to extend my appreciation to Professors Ayşegül Altay Uğur and Sadık Bayhan, who thoroughly reviewed the book and provided significant recommendations and feedback. I am also grateful to production staffs of Springer Nature, especially to Jayarani Premkumar and Femina Joshi A. I am indebted to journals of Fuzzy Sets and Systems, Information Sciences and International Journal of Approximate Reasoning.

Contents

Chapter 1
Introduction

A Texturing is a lattice of ordinary sets satisfying certain conditions which are compatible with the structure of a lattice of fuzzy subsets. In this chapter, we give some basic properties of lattices which are related to textures. For further information about lattices, we refer to [1].

1.1 Lattices

As is known that the notion of (complete) distributivity is an algebraic property providing remarkable representations of lattices in terms of sets of a family. In other words, distributivity gives more concrete information about the structure of a lattice through the sets and the basic set operations. For instance, Birkhoff's theorem says that any finite distributive lattice can be represented with the lattice of lower sets of the partial order of the join-irreducible elements [1]. *A complete rings of sets* is a family which is closed under intersections and unions. Raney showed that a complete lattice is completely distributive if and only if it is homeomorphic to a complete ring of sets [2]. The unit interval $[0, 1]$ is a complete and completely distributive lattice with the ordinary order "$\leq$" of real numbers. Moreover, the family of all fuzzy subsets of a given domain of discourse U, that is, the family $\mathcal{F}(U) = \{f \mid f : U \to [0, 1]\}$ is also a complete and completely distributive lattice where f denotes a function from U to $[0, 1]$. A basic motivation of a definition of a texture notion is to find an alternative point-set-based setting for such lattices. A *texturing* is a complete lattice of a family of some ordinary sets with respect to set inclusion satisfying the completely distributivity. From this point of view, Brown's representation theorem provides a family of some crisp subsets of U as a point-set-based setting for $\mathcal{F}(U)$ which will be called *a texturing*. Further, (complete) distributivity not only supply with a kind of ring of sets for $\mathcal{F}(U)$, but also allows to perform operations in the classical sense

M. Diker, *Texture Spaces*, Studies in Fuzziness and Soft Computing 411,
https://doi.org/10.1007/978-3-031-39748-6_1

related to points and sets in view of Raney's theorem [3] given in the sequel (see, p-set and q-set).

Now, let us start with a non-empty set L. *A partially order* $\leq$ is a binary relation on L satisfying the following conditions:

For all $r, s, t \in L$,
$r \leq r$ (reflexivity),
$r \leq s$ and $s \leq r \implies r = s$ (anti-symmetry) and
$r \leq s$ and $s \leq t \implies r \leq t$ (transitivity).

Then we call the pair $(L, \leq)$ *a partially ordered set*. Let $A \subseteq L$ with $A \neq \emptyset$. Then $r \in L$ is called
(i) *a lower bound* of A if $r \leq s$ for all $s \in A$, and
(ii) *an upper bound* of A if $s \leq r$ for all $s \in A$,
(iii) *the greatest lower bound* of A, if
  r is a lower bound of A, and
  $s \leq r$ for all lower bound s of A,
(iv) *the least upper bound* of A if
  r is an upper bound of A, and
  $r \leq s$ for all upper bound s of A.

In this case, we write $r = \bigwedge A$ or $r = \bigvee A$, for the greatest lower bound or least upper bound of A, respectively. A partially ordered set $(L, \leq)$ is called *a lattice* if for any two elements $r, s \in L$, the set $\{r, s\}$ has the least upper bound and the greatest lower bound. Note that a mapping ϕ between two lattices L and L' is *a lattice isomorphism* if it is a one-to-one and onto mapping preserving arbitrary meets and joins, that is,

$$\phi(\bigvee_{j \in J} r_j) = \bigvee_{j \in J} \phi(r_j) \text{ and } \phi(\bigwedge_{j \in J} r_j) = \bigwedge_{j \in J} \phi(r_j)$$

for all $\{r_j \mid j \in J\} \subseteq L$. A lattice L is said to be *complete* if for each non-empty subset A of L,

the least upper bound $\bigvee A$ (join), and
the greatest lower bound $\bigwedge A$ (meet)

exist. If a lattice L has smallest and largest elements denoted by $\mathbf{0}$ and $\mathbf{1}$ respectively, then it is called *a bounded lattice*.

The partially ordered set $(\mathcal{P}(U), \subseteq)$ is clearly a complete and bounded lattice with

$$\bigvee_{j \in J} A_j = \bigcup_{j \in J} A_j, \ \bigwedge_{j \in J} A_j = \bigcap_{j \in J} A_j$$

where $\{A_j \mid j \in J\} \subseteq \mathcal{P}(U)$, $\mathbf{0} = \emptyset$ and $\mathbf{1} = U$.

The following result will be useful in the sequel:

Theorem 1.1.1 *Let U be set and $\mathcal{A} \subseteq \mathcal{P}(U)$ where $\mathcal{P}(U)$ is the family of all subsets of U. If $U \in \mathcal{A}$ and $\mathcal{A}$ is closed under arbitrary intersections, then $(\mathcal{A}, \subseteq)$ is a complete lattice.*

Proof Clearly, $(\mathcal{A}, \subseteq)$ is a partially ordered set. Let $\mathcal{B} \subseteq \mathcal{A}$ and $\mathcal{B} \neq \emptyset$. Note that $\bigcap \mathcal{B} \in \mathcal{P}(U)$ and by the assumption, we also have that $\bigcap \mathcal{B} \in \mathcal{A}$. Hence $\bigcap B$ is the greatest lower bound of $\mathcal{B}$ in $\mathcal{A}$. Let us consider the family

$$C = \bigcap \{A \in \mathcal{A} \mid \bigcup \mathcal{B} \subset A\}.$$

Since $C \in \mathcal{A}$ and $U \in \mathcal{A}$, $C \neq \emptyset$. Note that for all $A \in \mathcal{B}$, $A \subseteq C$, that is C is an upper bound for $\mathcal{B}$. Moreover, C is the least upper bound of B.

Let L be a lattice and $m \in L$ with $m \neq 0$. Then m is called
(i) *a molecule* or *join prime*, if
$\forall r, s \in L,\ m \leq r \vee s \Rightarrow m \leq r \text{ or } m \leq s,$
(ii) *join irreducible*, if
$\forall r, s \in L,\ m = r \vee s \Rightarrow m = r \text{ or } m = s.$
Clearly, if m is a molecule, then it is also join irreducible. Indeed, if $m = r \vee s$, then $m \leq r \vee s$ and so $m \leq r$ or $m \leq s$. However, $r \leq m$ and $s \leq m$ yields that $m = r$ and $m = s$.

A lattice L is called *distributive* if for all $r, s, t \in L$, the equalities
$r \wedge (s \vee t) = (r \wedge s) \vee (r \wedge t)$, and
$r \vee (s \wedge t) = (r \vee s) \wedge (r \vee t)$
hold. If L is distributive, then the notions of molecule and join-irreducible element are coincide [4]. That is, if m is join irreducible, then it is also a molecule. Let $m \leq r \vee s$. Then by distributivity, we have

$$m = (r \vee s) \wedge m = (r \wedge m) \vee (s \wedge m).$$

Since m is join irreducible, $m = r \wedge m$ or $m = s \wedge m$, that is, $m \leq r$ or $m \leq s$.

For any $r \in L$, if there exists a $s \in L$ such that

$$r \wedge s = 0 \text{ and } r \vee s = 1,$$

then r is called *a complement* of s. *An order reversing involution* $' : L \to L$ is a mapping satisfying the conditions
$(r')' = r$ and $r \leq s \implies s' \leq r'$
for all $r, s \in L$. If for all $r \in L$, r' is a complement of r, then the order reversing involution $' : L \to L$ is called *a complementation mapping*. The unit interval $[0, 1]$ is a complemented and bounded complete lattice with the ordinary relation $\leq$.

A complementation mapping $' : L \to L$ has the following properties:

Theorem 1.1.2 *Let* $\{r_j \mid j \in J\} \subseteq L$. *Then*
(i) $(\bigwedge_{j \in J} r_j)' = \bigvee_{j \in J} r_j'$.
(ii) $(\bigvee_{j \in J} r_j)' = \bigwedge_{j \in J} r_j'$.

Proof (i) Note that for all $j \in J$, $\bigwedge_{j\in J} r_j \leq r_j$ and hence, we get $r_j' \leq (\bigwedge_{j\in J} r_j)'$ and so $\bigvee_{j\in J} r_j' \leq (\bigwedge_{j\in J} r_j)'$. Further, for all $j \in J, r_j' \leq \bigvee_{j\in J} r_j'$ and then we have $(\bigvee_{j\in J} r_j')' \leq r_j$. This implies that $(\bigvee_{j\in J} r_j')' \leq \bigwedge_{j\in J} r_j$. Finally, we obtain that $(\bigwedge_{j\in J} r_j)' \leq \bigvee_{j\in J} r_j'$.
(ii) The proof is similar.

Theorem 1.1.3 *For a bounded and distributive lattice L, we have the following propositions:*
(i) *If an element $r \in L$ has a complement s, then s is unique.*
(ii) *If every element of L has a complement, then the complementation mapping $' : L \to L$ is also unique.*

Proof (i) Suppose that t is also a complement of r where $t \neq s$. Then

$$s = 1 \wedge s = (r \vee t) \wedge s = (r \wedge s) \vee (t \wedge s) = 0 \vee (t \wedge s) = (t \wedge s)$$

and

$$t = 1 \wedge t = (r \vee s) \wedge t = (r \wedge t) \vee (s \wedge t) = 0 \vee (s \wedge t) = (s \wedge t).$$

This yields that $s = t$.
(ii) The proof is immediate by (i).

1.2 Completely Distributive Lattices

Recall that a lattice L is called *completely distributive* if for all index set K, we have

$$\bigwedge_{k\in K} \bigvee_{j\in J_k} a_j^k = \bigvee_{\gamma\in C} \bigwedge_{k\in K} a_{\gamma(k)}^k$$

where C is the set of all choice functions $\gamma : K \to \bigcup_{k\in K} J_k$ such that $\gamma(k) \in J_k \subseteq K$ and $a_j^k \in L$.

Trivially, for a finite lattice, completely distributivity is equivalent to distributivity.

If L is completely distributive, then every element of L can be written as a join of irreducible elements [5]. The following characterization of Raney [3] makes easy the proof of complete distributivity of a given lattice.

Theorem 1.2.1 *A complete lattice $(L, \preceq)$ is completely distributive if and only if for all $r, s \in L$, there exist $x, y \in L$ such that*
(i) $r \npreceq s \implies r \npreceq x$ *and* $y \npreceq s$, *and*
(ii) $\forall t \in L$, *either* $t \preceq x$ *or* $y \preceq t$.

Using the theorem of Raney, we can easily prove the following theorem:

Theorem 1.2.2 *Let U be a non-empty domain of discourse. Then the family of all subsets of U, that is, the family $\mathcal{P}(U) = \{A \mid A \subseteq U\}$ is completely distributive with the order of ordinary set inclusion.*

Proof $(\mathcal{P}(U), \subseteq)$ has the following properties:
If $A, B \in \mathcal{P}(U)$ and $A \nsubseteq B$, then there exist $C, D \in \mathcal{P}(U)$ such that $A \nsubseteq C$ and $D \nsubseteq B$, and
If $E \in \mathcal{P}(U)$, $E \subseteq C$ or $D \subseteq E$.

Note that if $A \nsubseteq B$, then $u \in A$ and $u \notin B$ for some $u \in U$. If we get $C = U \setminus \{u\}$ and $D = \{u\}$, we easily see that the first condition of Raney's theorem holds. For the second condition, if $E \subseteq U$ and $E \nsubseteq U \setminus \{u\}$, then we clearly have that $D = \{u\} \subseteq E$. As a result, the complete lattice $(\mathcal{P}(U), \subseteq)$ is completely distributive.

L-Fuzzy Sets

Let U be a non-empty domain of discourse and $(L, \leq, ')$ be a complemented lattice. A mapping $\alpha : U \to L$ is called *a L-fuzzy subset of U* [6]. We denote the family of all L-fuzzy subsets of U by $\mathcal{F}_L(U)$. The family $\mathcal{F}_L(U)$ is a complete lattice with the pointwise ordering

$$\forall u \in U,\ \alpha \leq \beta \Longleftrightarrow \alpha(u) \leq \beta(u)$$

for all $\alpha, \beta \in \mathcal{F}_L(U)$. We denote the top and bottom elements of $\mathcal{F}_L(U)$ by $\mathbf{1}$ and $\mathbf{0}$, respectively where

$\mathbf{1} : U \to L, \mathbf{1}(u) = 1$ for all $u \in L$,
$\mathbf{0} : U \to L, \mathbf{0}(u) = 0$ for all $u \in L$

where 1 and 0 are the top and bottom elements of L. *The complement* α' of an L-fuzzy subset α is defined by $\alpha'(u) = (\alpha(u))'$ for all $u \in U$. Arbitrary joins and meets of a family of L-fuzzy subsets are defined as follows:

For any $\mathcal{A} \subseteq \mathcal{F}_L(U)$,
$(\bigvee \mathcal{A})(u) = \bigvee\{\alpha(u) \mid \alpha \in \mathcal{A}\}$,
$(\bigwedge \mathcal{A})(u) = \bigwedge\{\alpha(u) \mid \alpha \in \mathcal{A}\}$
where $\bigvee$ and $\bigwedge$ are the least upper bound and greatest lower bound for the subset $\{\alpha(u) \mid \alpha \in \mathcal{A}\}$ in L where $u \in U$.

One of the important fuzzifications of a crisp point in fuzzy set theory is the fuzzy point defined by Wong in [7]. This concept provides many fundamental results due to local properties in fuzzy topological spaces [8]. As we will see in the sequel, in textures, we have special sets corresponding to fuzzy points and fuzzy co-points (see, Lemma 3.3.2).

For $\lambda \in L$, with $\lambda \neq 0$, *a fuzzy point* u_λ of $\mathcal{F}_L(U)$ is defined by

$$u_\lambda(z) = \begin{cases} \lambda, & \text{if } z = u \\ 0, & \text{if } z \neq u \end{cases}$$

and for $\lambda \neq 1$, *a fuzzy co-point* u^λ is defined by

$$u^\lambda(z) = \begin{cases} \lambda, & \text{if } z = u \\ 1, & \text{if } z \neq u \end{cases}$$

for all $z \in U$, respectively. The notions of fuzzy point and fuzzy co-point are dual notions with respect to complementation:

$$(u_\lambda)'(z) = (u_\lambda(z))' = \begin{cases} \lambda', & \text{if } z = u \\ 1, & \text{if } z \neq u \end{cases}$$

that is, $(u_\lambda)' = u^{\lambda'}$ and similarly, we have $(u^\lambda)' = u_{\lambda'}$.

Fuzzy lattices

A lattice $(L, \leq)$ is called *a fuzzy lattice* if it is a complete and completely distributive lattice with an order reversing involution $'$.

Example 1.2.3 (i) *Powerset Fuzzy Lattice* $(\mathcal{P}(U), \subseteq)$
The family $\mathcal{P}(U)$ of all subsets of a domain of discourse U is a complete and completely distributive lattice (see, Theorem 1.2.2).
(ii) *Three valued fuzzy lattice* $L = \{0, \frac{1}{2}, 1\}$.
L is a bounded complete lattice with the order $0 < \frac{1}{2} < 1$ where 0 and 1 are bottom and top elements of L, respectively. Clearly, it is completely distributive and has the order reversing involution defined by $1' = 0, 0' = 1$ and $(\frac{1}{2})' = \frac{1}{2}$. This is equivalent to say that the function $' : L \to L$ defined by $r' = 1 - r$, where $r = 0, \frac{1}{2}, 1$ is a complementation mapping on L.
(iii) *L-Fuzzy lattice* $(\mathcal{F}_L(U), \leq)$
Let us consider the family $\mathcal{F}_L(U) = \{\alpha \mid \alpha : U \to L \text{ is a mapping}\}$ of L-fuzzy sets. Since $(L, \leq)$ is completely distributive, $\mathcal{F}_L(U)$ is also completely distributive. Indeed, if $\alpha, \beta \in \mathcal{F}_L(U)$ and $\alpha \nleq \beta$, then there exists a $u \in U$ such that $\alpha(u) \nleq \beta(u)$ and $\alpha(u), \beta(u) \in L$. Since L is completely distributive, by Raney's theorem, there exist $u_1, u_2 \in L$ such that $\alpha(u) \nleq u_1$ and $u_2 \nleq \beta(u)$, and further, for all $t \in U$, we have $t \leq u_1$ or $u_2 \leq t$. Let us consider the L-fuzzy sets
$\gamma : U \to L, \forall u \in U, \gamma(u) = u_1,$
$\eta : U \to L, \forall u \in U, \eta(u) = u_2.$
Then we clearly have that
$\alpha \nleq \gamma$ and $\eta \nleq \beta$.
Further, for any $\zeta \in \mathcal{F}_L(U)$, if $\zeta \nleq \gamma$, then there exists a $u \in U$ such that
$\zeta(u) \nleq \gamma(u) = u_1$.
Since, L is completely distributive, again by Raney's theorem, for all $u \in U, \eta(u) = u_2 \leq \zeta(u)$. This implies that $\eta \leq \zeta$. That is, $\mathcal{F}_L(U)$ is completely distributive.
(iv) *Unit fuzzy lattice* $([0, 1], \leq)$
One of the primary examples to fuzzy lattices is the unit interval with the ordinary order $\leq$. Let $u, v \in [0, 1]$ and $u \nleq v$. Since $v < u$, we can choose $x, y \in [0, 1]$ such that

$v < y < x < u$.
Then we clearly have that
$u \nleq x$ and $y \nleq v$.

Further, if $z \in [0,1]$ and $z \nleq x$, then $y \leq z$. Hence, x and y satisfy the conditions of Raney's Theorem. Therefore, $([0,1], \leq)$ is a completely distributive lattice.
(v) *Fuzzy lattice $\mathcal{F}(U)$ of fuzzy subsets*
Since $([0,1], \leq)$ is a fuzzy lattice, by (iii),

$$\mathcal{F}(U) = \mathcal{F}_{[0,1]}(U) = \{\alpha \mid \alpha : U \to [0,1] \text{ is a mapping}\}$$

is also fuzzy lattice. However, independently, using Raney's theorem, we can easily show that $\mathcal{F}(U)$ is completely distributive. Let $\alpha, \beta \in \mathcal{F}(U)$ and $\alpha \nleq \beta$. Then $\beta(u) < \alpha(u)$ for some $u \in U$. Let us choose a $\lambda \in [0,1]$ where
$\beta(u) < \lambda < \alpha(u)$.
Then we clearly get that
$\alpha \nleq u^\lambda$ and $u_\lambda \nleq \beta$
where $u_\lambda, u^\lambda \in \mathcal{F}(U)$ are fuzzy point and fuzzy co-point, respectively. Further, let $\gamma \in \mathcal{F}(U)$. If $\gamma \nleq u^\lambda$, then $\lambda < \gamma(u)$ and so we get $u_\lambda \leq \gamma$. This means that $\mathcal{F}(U)$ holds the conditions given in Raney's theorem. The order reversing involution, that is, the complementation $\alpha' = \mathbf{1} - \alpha$ of α is given by $\alpha'(u) = 1 - \alpha(u)$ for all $u \in U$.

A fuzzy point (fuzzy co-point) is also a fuzzy set, and it satisfies similar properties as a crisp point. For instance, every fuzzy subset can be written as a join of fuzzy points or as a meet of fuzzy co-points:

Lemma 1.2.4 *For every fuzzy set $\alpha \in \mathcal{F}(U)$, we have*

$$\alpha = \bigvee_{\lambda < \alpha(u)} u_\lambda = \bigwedge_{\alpha(u) < \lambda} u^\lambda.$$

Proof Let $\lambda < \alpha(u)$. Then $u_\lambda < \alpha$ and so we get $\bigvee_{\lambda<\alpha(u)} u_\lambda \leq \alpha$. Suppose that for some $\beta \in \mathcal{F}(U)$,

$$\bigvee_{\lambda<\alpha(u)} u_\lambda < \beta < \alpha.$$

Then for all $u \in U$ and for all $\lambda \in [0,1]$,
$\lambda < \alpha(u) \implies u_\lambda < \beta$. (*)
Further, $\beta < \alpha$ is equivalent to say that $\beta \leq \alpha$ and $\beta \neq \alpha$. Hence, for some $u' \in U$, we get
$\beta(u') \leq \alpha(u')$ and $\beta(u') \neq \alpha(u')$,
that is, $\beta(u') < \alpha(u')$. Then we can choose a $\lambda' \in [0,1]$ such that
$\beta(u') < \lambda' < \alpha(u')$.
In this case, by (*), $u'_{\lambda'} < \beta$, namely, $\lambda' < \beta(u')$ which is a contradiction. Hence, we conclude that $\alpha = \bigvee_{\lambda<\alpha(u)} u_\lambda$. Using a similar way, it can be shown that

$$\alpha = \bigwedge_{\alpha(u)<\lambda} u^\lambda.$$

□

References

1. G. Birkhoff, *Lattice Theory*, 3rd edn. (American Mathematical Society, Providence, 1967)
2. G.N. Raney, Completely distributive complete lattices. Proc. Am. Math. Soc. **3**(5), 677–680 (1952)
3. G.N. Raney, Tight Galois connections and complete distributivity. Trans. Am. Math. Soc. **97**, 418–426 (1960)
4. V.K. Balachandran, On complete lattices and a problem of Birkhoff and Frink. Proc. Am. Math. Soc. **6**, 548–553 (1955)
5. G. Gierz, K.H. Hofmann, K. Keimel, J.D. Lawson, M.W. Mislove, D.S. Scott, *A Compendium of Continuous Lattices* (Springer, Berlin, 1980)
6. J.A. Goguen, L-fuzzy sets. J. Math. Anal. Appl. **18**, 145–174 (1967)
7. C.K. Wong, Fuzzy points and local properties of fuzzy topology, 1. Math. Anal. Appl. **46**, 316–328 (1974)
8. P. Pao-Ming, L. Ying-Ming, Fuzzy topology II. Product and quotient spaces. J. Math. Anal. Appl. **77**(1), 20–37 (1980)

Chapter 2
Textures

The idea of a texture is motivated from the unit interval [0, 1] with the ordinary ordering of real numbers. This motivation can be extended to the family of fuzzy subsets, namely, to the family of the mappings from a domain of discourse U to [0, 1]. In various areas of general topology or fuzzy topology, the point-set-based setting of textures seriously contributes to the development of the basic results due to the theories of general topology such as compact spaces, closure spaces or uniform spaces with the new descriptions (see, e.g., [1–9]). However, in this chapter, we consider the other side of textures which is also rather effective in rough set theory and fuzzy set theory [7, 10–22]. Here, we do not discuss the covering approach of rough sets with textures [23, 24]. With this aim, in particular, we discuss the main tools of textures at an comprehensible level considering the studies [1, 22, 25–27] in more detail. At the first step, after the motivation of textures, we discuss the basic properties of p-set and q-set and in particular, we give the equivalences of complete distributivity in texture spaces. We consider the important classes called simple and plain textures. It is remarkable to say that the textures corresponding to fuzzy lattices are simple textures. Further, all finite textures are plain. As we shall see in the sequel, no doubt that we also have plain and infinite textures.

2.1 Basic Concepts and Results

Let us consider the fuzzy lattice $([0, 1], \leq)$. Every non-zero element of [0, 1] is a molecule, that is $M_{[0,1]} = (0, 1]$. Using the molecules of [0, 1], we may obtain a complete and completely distributive lattice of sets as follows. For $\lambda \in [0, 1]$, consider the set $\Im(\lambda) = \{\mu \mid \mu \in M_{[0,1]} \text{ and } \mu \leq \lambda\}$. Clearly, $\Im(\lambda) = (0, \lambda]$.
The family $\mathcal{M}_{[0,1]} = \{(0, \lambda] \mid \lambda \in [0, 1]\}$ has the following properties:

M. Diker, *Texture Spaces*, Studies in Fuzziness and Soft Computing 411,
https://doi.org/10.1007/978-3-031-39748-6_2

(i) We have $\lambda \leq \mu \Longleftrightarrow \mathfrak{I}(\lambda) \subseteq \mathfrak{I}(\mu)$, and then the mapping
$\mathfrak{I}: \ [0, 1] \to \mathcal{M}_{[0,1]}, \forall\lambda \in [0, 1], \ \mathfrak{I}(\lambda) = (0, \lambda]$
is a lattice isomorphism where $\mathcal{M}_{[0,1]}$ is ordered by the ordinary set inclusion. Hence, $(\mathcal{M}_{[0,1]}, \ \subseteq)$ is a complete lattice. For $\lambda = 1$ and $\lambda = 0$, we get $M_{[0,1]}, \emptyset \in \mathcal{M}_{\mathcal{F}(U)}$.

(ii) For every family $\{(0, \lambda_j] \mid j \in J\} \subseteq \mathcal{M}_{\mathcal{F}(U)}$, we have
$\bigcap_{j \in J}(0, \lambda_j] = \emptyset$
or
$\bigcap_{j \in J}(0, \lambda_j] = (0, \bigwedge_{j \in J} \lambda_j] = (0, \lambda],$
where $\bigwedge_{j\in J} \lambda_j = \lambda$. Hence, arbitrary meets coincide with intersections. However, for the family $\{(0, 1 - 1/n] \mid n \in N\} \subseteq \mathcal{M}_{[0,1]}$, we have

$$\bigvee_{n\in N} (0, 1 - 1/n] = (0, 1] \text{ and} \bigcup_{n\in N}(0, 1 - 1/n] = (0, 1)$$

and then only finite joins coincide with unions in $\mathcal{M}_{[0,1]}$.

(iii) Since the mapping $\mathfrak{I}: \ [0, 1] \to \mathcal{M}_{[0,1]}$ is a lattice isomorphism and $([0, 1], \leq)$ is completely distributive, $\mathcal{M}_{[0,1]}$ is also completely distributive.

(iv) $\mathcal{M}_{[0,1]}$ separates the points of $[0, 1]$ in the following sense:
If $\lambda, \mu \in [0, 1]$ and $\lambda \neq \mu$, then $\mu \notin (0, \lambda]$ or $\lambda \notin (0, \mu]$.

Basically, the above properties of the complete lattice $\mathcal{M}_{[0,1]}$ lead us to a more general concept related to any family of subsets of a given domain of discourse:

Definition 2.1.1 Let U be a set. Then $\mathcal{U} \subseteq \mathcal{P}(U)$ is called a *texturing* of U, and $(U, \mathcal{U})$ is called a *texture space*, or in brief, a *texture*, if

(T1) $(\mathcal{U}, \ \subseteq)$ is a complete lattice containing U and $\emptyset$,

(T2) $\forall\mathcal{K} \subseteq \mathcal{U}, \ \bigwedge\mathcal{K} = \bigcap\mathcal{K}$ and
$\forall\mathcal{K} \subseteq \mathcal{U}, \ \mathcal{K}$ is finite, $\bigvee\mathcal{K} = \bigcup\mathcal{K}$
where $\bigcap$ is the intersection and $\bigcup$ is the union of sets,

(T3) $\mathcal{U}$ is completely distributive,

(T4) $\mathcal{U}$ separates the points of U. That is, given $u \neq v$ in U there exists $A \in \mathcal{U}$ such that $u \in A, \ v \notin A$, or $u \notin A, \ v \in A$.

A mapping $c : \mathcal{U} \to \mathcal{U}$ is called *a complementation* on $(U, \mathcal{U})$ if
$c(c(A)) = A$, and
$A \subseteq B$ in $\mathcal{U}$ implies $c(B) \subseteq c(A)$
for all $A, B \in \mathcal{U}$. Then the triple $(U, \mathcal{U}, c)$ is called *a complemented texture space*.

Remark 2.1.2 Note that if c is a complementation on a texture $(U, \mathcal{U})$, then by Theorem 1.1.3, c is unique. Therefore, if there is a complementation c on $(U, \mathcal{U})$, in some cases, to prevent the confusion, we use the notation c_U for the complementation.

Remark 2.1.3 Since a texture $(U, \mathcal{U})$ is a complete lattice, note that for every family $\mathcal{K} \subseteq \mathcal{U}$, we already have $\bigvee \mathcal{K} \in \mathcal{U}$. Here, we guarantee the equality

$$\bigvee \mathcal{K} = \bigcup \mathcal{K}$$

only for the finite families.

Example 2.1.4 (i) *Discrete Texture*
A trivial example to a texture is $(\mathcal{P}(U), \subseteq)$ which is called *a discrete texture*. It is a completely distributive lattice as we have shown before. Moreover, a discrete texture is complemented with the ordinary set complementation:
$U \setminus (U \setminus A) = A$, and
$A \subseteq B$ in $\mathcal{U}$ implies $U \setminus B \subseteq U \setminus A$.
Further, $\mathcal{P}(U)$ separates the points, that is, for any $u, v \in U$ with $u \neq v$, we clearly have that
$v \in U \setminus \{u\} \in U$ and $u \notin U \setminus \{u\}$.
(ii) *Fuzzy texture*
As it is observed in the preceding section, the family $\mathcal{M} = \mathcal{M}_{[0,1]} = \{(0, u] \mid u \in [0, 1]\}$ is a complete lattice with respect to ordinary set inclusion satisfying the following properties:
(T1) $\mathcal{M}$ contains the top and bottom elements, that is, $M, \emptyset \in \mathcal{M}$,
(T2) $(\mathcal{M}, \subseteq)$ is a completely distributive lattice,
(T3) Arbitrary meets coincide with intersections, but the only finite joins coincide with unions,
(T4) $\mathcal{M}$ separates the points of $(0, 1]$.
Therefore, the family $\mathcal{M}$ is a texturing on M. Since it is obtained from the fuzzy lattice $([0, 1], \leq)$, we call $(M, \mathcal{M})$ *a fuzzy texture*. The fuzzy texture is complemented. The mapping $c_M : \mathcal{M} \to \mathcal{M}$ defined by

$$c_M((0, \lambda]) = (0, 1 - \lambda], \ \forall \lambda \in [0, 1]$$

is a complementation on $\mathcal{M}$. Indeed, for all $\lambda, \mu \in [0, 1]$,

$$\begin{aligned} (0, \lambda] \subseteq (0, \mu] &\Longrightarrow \lambda \leq \mu \Longrightarrow 1 - \mu \leq 1 - \lambda \\ &\Longrightarrow (0, 1 - \mu] \subseteq (0, 1 - \lambda] \\ &\Longrightarrow c_U((0, \mu]) \subseteq c_U((0, \lambda]) \end{aligned}$$

and
$c_U(c_U((0, \lambda])) = c_U((0, 1 - \lambda]) = (0, 1 - (1 - \lambda]) = (0, \lambda]$
as desired.
(iii) *Unit texture*
Although the motivation of a texture notion arises from a fuzzy lattice, not every texture can be obtained from a fuzzy lattice. One of the important examples is *the unit texture* $(I, \mathcal{I})$ where

$$I = [0, 1] \text{ and } \mathcal{I} = \{[0, r) \mid r \in I\} \cup \{[0, r] \mid r \in I\}.$$

For every family $\mathcal{A} = \{A_j \mid j \in J\} \subseteq \mathcal{I}$, we have $A_j = [0, r_j)$ or $A_j = [0, r_j]$. Therefore,

$$\bigcap \mathcal{A} = \bigcap_{j \in J} A_j = [0, \bigwedge a_j) \text{ or } [0, \bigwedge a_j]$$

where $\bigwedge a_j \in [0, 1]$, that is, $\bigcap \mathcal{A} \in \mathcal{I}$. Accordingly, the family $\mathcal{I}$ is closed under intersection. Further, we also have that $\bigvee A_j \in \mathcal{I}$. Indeed, since $\mathcal{I}$ is closed under intersection, we have

$$A = \bigcap \{C \in \mathcal{I} \mid \bigcup \mathcal{A} \subseteq C\} \in \mathcal{I}.$$

Clearly, $A = \bigvee \mathcal{A}$. As a result $(\mathcal{I}, \subseteq)$ is a complete lattice. For completely distributivity of $\mathcal{I}$, we check the conditions of Raney's Theorem. Let $[0, r), [0, r') \in \mathcal{I}$ and $[0, r) \not\subseteq [0, r')$. Then there exist $s, s' \in [0, 1]$ such that

$$r' < s' < s < r.$$

This follows that

$$[0, r) \not\subseteq [0, s) \text{ and } [0, s') \not\subseteq [0, r').$$

Further, for $[0, z) \in \mathcal{I}$, if $[0, z) \not\subseteq [0, s)$, then we conclude that $[0, s') \subseteq [0, z]$. If

$$[0, r) \not\subseteq [0, r'] \text{ or } [0, r] \not\subseteq [0, r'],$$

then we also see that the conditions of Raney's Theorem also hold. The unit texture $(I, \mathcal{I})$ is also complemented. Because, the mapping $c_I : \mathcal{I} \to \mathcal{I}$ defined by

$$\forall r \in I,\ c_I([0, r]) = [0, 1 - r),\ c_I([0, r)) = [0, 1 - r]$$

is a complementation on $(I, \mathcal{I})$.

(iv) A texture may not be complemented. It is easy to see that the family

$$\mathcal{U} = \{U, \emptyset, \{2\}, \{3\}, \{5\}, \{2, 3\}, \{2, 4\}, \{2, 5\}, \{3, 5\}, \{2, 3, 4\}, \{2, 3, 5\}, \{2, 4, 5\}\}$$

is a texturing on $U = \{2, 3, 4, 5\}$. Note that U consists of molecules which are also divisors of the number 60. However, the texture $(U, \mathcal{U})$ is not complemented. The reason of that the lattice of divisors of 60 is not complemented (see Example 3.1.6 (ii)).

2.2 p-Set and q-Set

The following notions play an important role in the theory of texture spaces:

Definition 2.2.1 Let $(U, \mathcal{U})$ be a texture. For $u \in U$, *the p-set* and *q-set* are defined by

$$P_u = \bigcap \{A \in \mathcal{U} \mid u \in A\} \text{ and } Q_u = \bigvee \{A \in \mathcal{U} \mid u \notin A\},$$

respectively.

The collection of all p-sets is a base for a texture in the following sense:

Theorem 2.2.2 *For every set A in $\mathcal{U}$, $A = \bigcup_{w \in A} P_w = \bigvee_{w \in A} P_w$.*

Proof By definition of a p-set, for $w \in A$, we have $P_w \subseteq A$ and then $\bigcup_{w \in A} P_w \subseteq A$. Further, we clearly have $A \subseteq \bigcup_{w \in A} P_w$. This yields that $A = \bigcup_{w \in A} P_w$. Now, since $\bigcup_{w \in A} P_w \in \mathcal{U}$, and we conclude that $\bigcup_{w \in A} P_w = \bigvee_{w \in A} P_w$.

A q-set has a useful alternative form:

Theorem 2.2.3 *For $u \in U$, $Q_u = \bigvee \{P_v \mid u \notin P_v\}$.*

Proof For some $v \in U$, if $u \notin P_v$, then by definition of Q_u we have $P_v \subseteq Q_u$. Hence, $\bigvee \{P_v \mid u \notin P_v\} \subseteq Q_u$.
For the reverse inclusion, if $u \notin A$ and $A \in \mathcal{U}$, then by Theorem 2.2.2, for all $w \in A$ we have $u \notin P_w$. This implies that for all $w \in A$, we have $P_w \subseteq \bigvee \{P_v \mid u \notin P_v\}$, that is,

$$A = \bigcup_{w \in A} P_w \subseteq \bigvee \{P_v \mid u \notin P_v\}.$$

Then we get $Q_u \subseteq \bigvee \{P_v \mid u \notin P_v\}$.

Example 2.2.4 (i) *Discrete texture*
We can easily determine the p-set and the q-set of a discrete texture $(U, \mathcal{P}(U))$. Note that for every $u \in U$, $\{u\} \in \mathcal{P}(U)$. Hence,
$P_u = \bigcap \{A \in \mathcal{P}(U) \mid u \in A\} = \{u\}$.
Further, since $u \notin U \setminus \{u\} \in \mathcal{P}(U)$,
$Q_u = \bigvee \{A \in \mathcal{P}(U) \mid u \notin A\} = \bigcup \{A \in \mathcal{P}(U) \mid u \notin A\} = U \setminus \{u\}$.
(ii) *Fuzzy texture*
Let us consider the fuzzy texture $(M_{[0,1]}, \mathcal{M}_{[0,1]})$. For all $u \in (0, 1]$, we have
$P_u = \bigcap \{(0, v] \mid u \in (0, v]\} = \bigcap \{(0, v] \mid u \leq v\} = (0, u]$
and
$Q_u = \bigvee \{(0, v] \mid u \notin (0, v]\} = \bigvee \{(0, v] \mid v < u\} = (0, u]$.
Note that $P_u = Q_u$.
(iii) *Unit texture*
Now take the unit texture $(I, \mathcal{I})$. For every $r \in I = [0, 1]$, the p-set and the q-set of $\mathcal{I}$ are

$$P_r = \bigcap \big\{A \in \mathcal{I} \mid (r \in A) \wedge (\exists\, u \in I)(A = [0, u) \text{ or } A = [0, u])\big\} = [0, r]$$

and

$$Q_r = \bigvee (\{A \mid r \notin A, (\exists\, u \in I)(A = [0, u) \text{ or } A = [0, u])\} = [0, r),$$

respectively. Note that every q-set is also a molecule of $\mathcal{I}$.

2.3 Equivalences of Completely Distributivity in Texture Spaces

Core of a Set

Completely distributivity provides us that every set in a texture can be written in terms of the p-set and q-set. An auxiliary notion of *core* of a set $A \in \mathcal{U}$ denoted by $A^\flat$ is a useful tool to construct the equivalences of completely distributivity:
$A^\flat = \bigcap \{ \bigcup_{j\in J} A_j \mid \{A_j\}_{j\in J} \subseteq \mathcal{U},\ A = \bigvee_{j\in J} A_j \}$.
Note that $A^\flat$ may not necessarily be an element of $(U, \mathcal{U})$ as we shall see in the sequel. However, A is the smallest element of $(U, \mathcal{U})$ containing $A^\flat$ for all $A \in \mathcal{U}$.

Theorem 2.3.1 *Let $(\mathcal{U}, \subseteq)$ be completely distributive. Then we have the following propositions:*
(i) $\forall A \in \mathcal{U},\ \forall u \in U,\ u \notin A \implies A \subseteq Q_u \implies u \notin A^\flat$.
(ii) $\forall A \in \mathcal{U},\ A^\flat = \{u \mid A \not\subseteq Q_u\}$.
(iii) *For all* $A, B \in \mathcal{U},\ A \subseteq B \implies A^\flat \subseteq B^\flat$.
(iv) $\{A_j \mid j \in J\} \subseteq \mathcal{U} \implies \left(\bigvee_{j\in J} A_j\right)^\flat = \bigvee_{j\in J} A_j^\flat$.
(v) *For all* $A \in \mathcal{U}$, *A is the smallest element of $(U, \mathcal{U})$ containing $A^\flat$.*

Proof (i) The first implication is immediate from the definition of a q-set. For the second, let $A \subseteq Q_u$. Then by completely distributivity, note that

$$A = A \cap Q_u = A \cap \bigvee\{P_v \mid u \notin P_v\} = \bigvee\{A \cap P_v \mid u \notin P_v\}.$$

Now, by definition of a core, we clearly have $A^\flat \subseteq \bigcup\{A \cap P_v \mid u \notin P_v\}$ and so we conclude that $u \notin A^\flat$.
(ii) By (i), if $u \in A^\flat$, then $A \not\subseteq Q_u$ and thus, $A^\flat \subseteq \{u \mid A \not\subseteq Q_u\}$. For the reverse inclusion, let $A \not\subseteq Q_u$ and $A = \bigvee\{A_j \mid j \in J\}$ where $\{A_j \mid j \in J\} \subseteq \mathcal{U}$. Then for some $j \in J$, $A_j \not\subseteq Q_u$ and so by (i),

$$u \in A_j \subseteq \bigcup\{A_j \mid j \in J\}.$$

Since the family $\{A_j \mid j \in J\}$ is arbitrary, we also obtained that $u \in A^\flat$.
(iii) By definition of a core, the proof is straightforward.

(iv) Note that if $u \in \big(\bigvee_{j\in J} A_j\big)^{\flat}$, then by (i), $\bigvee_{j\in J} A_j \not\subseteq Q_u$ and hence, $A_j \not\subseteq Q_u$ for some $j \in J$. That is, $u \in A_j^{\flat} \subseteq \bigvee_{j\in J} A_j^{\flat}$. This gives that $\big(\bigvee_{j\in J} A_j\big)^{\flat} \subseteq \bigvee_{j\in J} A_j^{\flat}$. For all $j \in J$, $A_j \subseteq \bigvee_{j\in J} A_j$ and by (iii), we get $A_j^{\flat} \subseteq \big(\bigvee_{j\in J} A_j\big)^{\flat}$ whence $\bigvee_{j\in J} A_j^{\flat} \subseteq \big(\bigvee_{j\in J} A_j\big)^{\flat}$.

(v) Without loss of generality, we may take

$$A^{\flat} = \bigcap_{j\in J} \bigcup_{k\in K_j} A_k^j$$

where $\{K_j \mid j \in J\}$ is a family of pairwise disjoint index sets and $A = \bigvee_{k\in K_j} A_k^j$ for all $j \in J$. Now, let $A^{\flat} \subseteq B$ for some $B \in \mathcal{U}$. We show that $A \subseteq B$. Since $\mathcal{P}(U)$ is completely distributive,

$$A^{\flat} = \bigcap_{j\in J} \bigcup_{k\in K_j} A_k^j = \bigcup_{\gamma\in\Pi_j K_j} \bigcap_{j\in J} A_{\gamma(j)}^j \subseteq B.$$

This means that $\bigcap_{j\in J} A_{\gamma(j)}^j \subseteq B$ for all $\gamma \in \Pi_j K_j$. Since $B \in \mathcal{U}$, we also have $\bigvee_{\gamma\in\Pi_j K_j} \bigcap_{j\in J} A_{\gamma(j)}^j \subseteq B$. On the other hand, $\mathcal{U}$ is also completely distributive and so

$$A = \bigcap_{j\in J} \bigvee_{k\in K_j} A_k^j = \bigvee_{\gamma\in\Pi_j K_j} \bigcap_{j\in J} A_{\gamma(j)}^j$$

which implies that $A \subseteq B$ as desired.

Corollary 2.3.2 *Let $(\mathcal{U}, \subseteq)$ be a complete lattice. The following statements are equivalent:*
(i) $(U, \mathcal{U})$ is completely distributive.
(ii) $\forall A, B \in \mathcal{U},\ A \not\subseteq B \Longrightarrow \exists u \in U\ (A \not\subseteq Q_u$ and $P_u \not\subseteq B)$.
(iii) $\forall A \in \mathcal{U},\ A = \bigvee\{P_u | A \not\subseteq Q_u\}$.
(iv) $\forall A \in \mathcal{U},\ A = \bigcap\{Q_u | P_u \not\subseteq A\}$.

Proof First, we prove the equivalence (i)$\Longleftrightarrow$(ii). Let $(U, \mathcal{U})$ be completely distributive and $A \not\subseteq B$. Then by Theorem 2.3.1 (v), we also have that $A^{\flat} \not\subseteq B$. Take a point $u \in U$ such that $u \in A^{\flat}$ and $u \notin B$. Then by Theorem 2.3.1 (i) and definition of p-set, we obtain $A \not\subseteq Q_u$ and $P_u \not\subseteq B$.

For the implication (ii) $\Longrightarrow$ (i), suppose that $A, B \in \mathcal{U}$ and $A \not\subseteq B$. By Theorem 2.3.1 (v), $A^{\flat} \not\subseteq B$ and so there is a $u \in U$ such that $u \in A^{\flat}$ and $u \notin B$. Then by Theorem 2.3.1 (i), $A \not\subseteq Q_u$ and further, we clearly have $P_u \not\subseteq B$. We show that if

$C \in \mathcal{U}$, then $C \subseteq Q_u$ or $P_u \subseteq C$. If $C \not\subseteq Q_u$, then by Theorem 2.3.1 (i), $u \in C$ and hence, $P_u \subseteq C$. This means that the complete lattice $(\mathcal{U}, \subseteq)$ satisfies the conditions given in Raney's theorem, that is, $(\mathcal{U}, \subseteq)$ is completely distributive.
(ii) $\Longrightarrow$ (iii): If $A \not\subseteq Q_u$, by Theorem 2.3.1 (i), $u \in A$, that is $P_u \subseteq A$. Hence we get $\bigvee\{P_u \mid A \not\subseteq Q_u\} \subseteq A$. On the other hand, by the equivalence of (i) and (ii), we may also use the completely distributivity of $(\mathcal{U}, \subseteq)$ and so, in view of Theorem 2.3.1 (ii)

$$A^{\flat} = \{u \mid A \not\subseteq Q_u\} \subseteq \bigvee\{P_u \mid A \not\subseteq Q_u\} \in \mathcal{U}.$$

However, by Theorem 2.3.1 (v), we get $A \subseteq \bigvee\{P_u \mid A \not\subseteq Q_u\}$.
(iii) $\Longrightarrow$ (ii): Suppose that for some $A \in \mathcal{U}$, $A = \bigvee\{P_u | A \not\subseteq Q_u\}$. Let $A \not\subseteq B$. By the assumption, $\bigvee\{P_u | A \not\subseteq Q_u\} \not\subseteq B$. Then we clearly have that $A \not\subseteq Q_u$ and $P_u \not\subseteq B$ for some $u \in U$.
(ii) $\Longrightarrow$ (iv): By Theorem 2.3.1 (i), if $P_u \not\subseteq A$, then $A \subseteq Q_u$ and hence, we automatically have the inclusion $A \subseteq \bigcap\{Q_u \mid P_u \not\subseteq A\}$. For the reverse inclusion, suppose that $\bigcap\{Q_u \mid P_u \not\subseteq A\} \not\subseteq A$. By the assumption, there exists a $u \in U$ such that

$$\bigcap\{Q_u \mid P_u \not\subseteq A\} \not\subseteq Q_u \text{ and } P_u \not\subseteq A$$

and however, $P_u \not\subseteq A$ implies that $\bigcap\{Q_u \mid P_u \not\subseteq A\} \subseteq Q_u$ which is an immediate contradiction.
(iv) $\Longrightarrow$ (ii): Let $A \not\subseteq B$ and $A, B \in \mathcal{U}$. By the assumption

$$A \not\subseteq \bigcap\{Q_u \mid P_u \not\subseteq B\} = B.$$

Then for some $u \in U$, we get $A \not\subseteq Q_u$ and $P_u \not\subseteq B$.

Corollary 2.3.3 *For all $u \in U$, the following implication hold:*
(i) $A \not\subseteq Q_u \implies P_u \subseteq A$.
(ii) $P_u \not\subseteq A \implies A \subseteq Q_u$.

Proof The proof is straightforward from Corollary 2.3.2 (iii) and (iv).

Corollary 2.3.4 *Let $(U, \mathcal{U})$ be a texture space and $A, B \in \mathcal{U}$. Then*
(i) $\bigcup_{u \in U} P_u = \bigvee_{u \in U} P_u = U$.
(ii) $\bigcap_{u \in U} Q_u = \emptyset$.

Proof (i) By definition of a p-set, the proof is trivial.
(ii) Suppose that $\bigcap_{u \in U} Q_u \neq \emptyset$. Let us take a $u' \in \bigcap_{u \in U} Q_u$. Then for all $u \in U$, $u' \in Q_u$ and so we get $P_{u'} \subseteq Q_u$ since $Q_u \in \mathcal{U}$. Hence, by Corollary 2.3.2 (iii), we also have that

$$P_{u'} = \bigvee\{P_u | P_{u'} \not\subseteq Q_u\} = \emptyset,$$

which is a contradiction since $u' \in P_{u'}$.

Corollary 2.3.5 *Let $(U, \mathcal{U})$ be a texture space and $A, B \in \mathcal{U}$. Then the following statements are equivalent.*
(i) $A = B$.
(ii) $\forall u \in U,\ A \not\subseteq Q_u \Longleftrightarrow B \not\subseteq Q_u$.
(iii) $\forall u \in U,\ P_u \not\subseteq A \Longleftrightarrow P_u \not\subseteq B$.

Proof (i) $\Longrightarrow$ (ii) is clear. For (ii) $\Longrightarrow$ (i), suppose that $A \neq B$. Without loss of generality, we may assume that $A \not\subseteq B$. Then by Corollary 2.3.2 (ii), there is a $u \in U$ such that $A \not\subseteq Q_u$ and $P_u \not\subseteq B$. By the assumption we get $B \not\subseteq Q_u$ and so by Theorem 2.3.1(i), we obtain that $P_u \subseteq B$. But this is a contradiction. For (ii) $\Longrightarrow$ (iii), let $P_u \not\subseteq A$. By Corollary 2.3.2 (ii), there exists a $u' \in U$ such that $P_u \not\subseteq Q_{u'}$ and $P_{u'} \not\subseteq A$. By Corollary 2.3.1 (i), $A \subseteq Q_{u'}$ and then by the assumption, we get $B \subseteq Q_{u'}$. However, this gives that $P_u \not\subseteq B$ as desired. If we suppose that $P_u \not\subseteq B$, we also obtain that $P_u \not\subseteq A$. The proof of the implication (iii) $\Longrightarrow$ (ii) is similar.

Theorem 2.3.6 *Let $(U, \mathcal{U}, c)$ be a complemented texture. Then we have the following propositions.*
(i) $\forall A \in \mathcal{U},\ A = \bigvee\{c(Q_u) \mid A \not\subseteq c(P_u)\} = \bigcap\{c(P_u) \mid c(Q_u) \not\subseteq A\}$
(ii) $A \not\subseteq B \Longrightarrow \exists u \in U, A \not\subseteq c(P_u)$ *and* $c(Q_u) \not\subseteq B$.

Proof (i) Since we also have $c(A) \in \mathcal{U}$, by Corollary 2.3.2 (iii), $c(A) = \bigvee\{P_u \mid c(A) \not\subseteq Q_u\}$. If we apply the complementation on both sides, then we get

$$A = c(c(A)) = c(\bigvee\{P_u \mid c(A) \not\subseteq Q_u\}) = \bigcap\{c(P_u) \mid c(Q_u) \not\subseteq A\}$$

as desired. The proof of $A = \bigvee\{c(Q_u) \mid A \not\subseteq c(P_u)\}$ is similar.
(ii) Let $A \not\subseteq B$. By (i), we have $A \not\subseteq \bigcap\{c(P_u) \mid c(Q_u) \not\subseteq B\}$. Hence, for some $u \in U$, $A \not\subseteq c(P_u)$ and $c(Q_u) \not\subseteq B$.

Theorem 2.3.7 *Let $(U, \mathcal{U}, c)$ be a complemented texture. Then we have the following propositions.*
(i) $\forall A \in \mathcal{U},\ c(A) = \bigcap_{u \in c(A)} c(P_u)$.
(ii) $\forall u, v \in U,\ u \in c(P_v)$ *if and only if* $v \in c(P_u)$.

Proof (i) By Theorem 2.2.2, for $A \in \mathcal{U}$, we have $c(A) = \bigcup_{u \in c(A)} P_u$. Applying the complementation both sides, we get $c(c(A)) = A = \bigcap_{u \in c(A)} c(P_u)$.
(ii) Note that $u \in c(P_v)$, that is, $P_u \subseteq c(P_v)$. Applying the complementation on both sides, we obtain $P_v \subseteq c(P_u)$ and so $v \in c(P_u)$. Necessity is similar.

2.4 Simple and Plain Textures

Definition 2.4.1 Let $(U, \mathcal{U})$ be a texture. A non-empty set A is called *a molecule of* $\mathcal{U}$ if

$$\forall B, C \in \mathcal{U}, A \subseteq B \cup C \implies A \subseteq B \text{ or } A \subseteq C.$$

Every p-set is a molecule of a texture space. If $P_u \subseteq A \cup B$, then $u \in A \cup B$ and hence, $P_u \subseteq A$ or $P_u \subseteq B$. This gives an important class of textures:

Definition 2.4.2 Let $\mathcal{M}$ be the family of all molecules of $(U, \mathcal{U})$. If $\mathcal{M} = \{P_u \mid u \in U\}$, then $(U, \mathcal{U})$ is called *a simple texture*.

A texture is closed under arbitrary joins. However, if $\mathcal{A} \subseteq \mathcal{U}$ is an infinite family, then the join and union of elements of $\mathcal{A}$ may not necessarily be equal, in other words, it is not necessary that

$$\bigvee \mathcal{A} = \bigcup \mathcal{A}.$$

The following theorem gives an equivalence for the above equality:

Theorem 2.4.3 *Let $(U, \mathcal{U})$ be a texture space. Then the following statements are equivalent:*
(i) *For all $u \in U$, $P_u \not\subseteq Q_u$.*
(ii) *For all $\mathcal{A} = \{A_j \mid j \in J\} \subseteq \mathcal{U}$, $\bigvee \mathcal{A} = \bigcup \mathcal{A}$.*

Proof (i) $\implies$ (ii): Suppose that $\bigvee \mathcal{A} \neq \bigcup \mathcal{A}$ for some family $\mathcal{A} \subseteq \mathcal{U}$. Clearly, we have $\bigcup \mathcal{A} \subseteq \bigvee \mathcal{A}$ and so $\bigvee \mathcal{A} \not\subseteq \bigcup \mathcal{A}$. Then $u \in \bigvee\limits_{j \in J} A_j$ and $u \notin \bigcup_{j \in J} A_j$ for some $u \in U$. Since $\bigvee \mathcal{A} \in \mathcal{U}$, we also have that $P_u \subseteq \bigvee\limits_{j \in J} A_j$, and for all $j \in J$, $P_u \not\subseteq A_j$. Now, by Corollary 2.3.1 (i), for all $j \in J$, we get $A_j \subseteq Q_u$, that is $\bigcup\limits_{j \in J} A_j \subseteq Q_u$. However, $Q_u \in \mathcal{U}$ and so $\bigvee\limits_{j \in J} A_j \subseteq Q_u$ and this yields that

$$P_u \subseteq \bigvee_{j \in J} A_j \subseteq Q_u$$

which is a contradiction.
(ii) $\implies$ (i): Suppose that $P_u \subseteq Q_u$ for some $u \in U$. By Theorem 2.2.3, we have

$$Q_u = \bigvee \{P_v \mid u \notin P_v\}.$$

By the assumption, we also have that

$$Q_u = \bigcup \{P_v \mid u \notin P_v\},$$

that is, $P_u \subseteq \bigcup \{P_v \mid u \notin P_v\}$. Then we get $u \in P_v$ and $u \notin P_v$ for some $v \in U$, and this is an immediate contradiction.

Definition 2.4.4 A texture space $(U, \mathcal{U})$ is said to be *plain* if it satisfies one of the equivalent conditions given in Theorem 2.4.3.

Corollary 2.4.5 *If* $(U, \mathcal{U})$ *is a plain texture, then for all* $A \in \mathcal{U}$, $A^\flat = A$.

Proof By Theorem 2.3.1 (v), we have $A^\flat \subseteq A$. Suppose that $A \nsubseteq A^\flat$. Take $u \in U$ such that $u \in A$ and $u \notin A^\flat$. Note that $P_u \subseteq A$ and by Theorem 2.3.1 (ii), $A \subseteq Q_u$. Hence, we conclude that $P_u \subseteq Q_u$ which is a contradiction, since $\mathcal{U}$ is plain.

Example 2.4.6 (i) The discrete texture $(U, \mathcal{P}(U))$ is plain. Indeed, for all $u \in U$,

$$P_u = \{u\} \nsubseteq U \setminus \{u\} = Q_u.$$

In other words, we trivially have that $\bigvee \mathcal{A} = \bigcup \mathcal{A}$ for all $\mathcal{A} \subseteq \mathcal{P}(U)$. It is also simple since the singleton sets are the only molecules of $\mathcal{P}(U)$, that is,

$$\mathcal{M} = \{P_u \mid u \in U\} = \{\{u\} \mid u \in U\}.$$

(ii) By Theorem 2.4.3, every finite texture $(U, \mathcal{U})$ is plain since
$\forall \mathcal{K} \subseteq \mathcal{U},\ \mathcal{K}$ is finite, $\bigvee \mathcal{K} = \bigcup \mathcal{K}$.
(iii) The unit texture $(I, \mathcal{I})$ is plain. Because, $P_r = [0, r] \nsubseteq Q_r = [0, r)$. It is easy to see that for every $r \in (0, 1]$, the set $Q_r = [0, r)$ is also a molecule of $\mathcal{I}$. Therefore, unit texture is not simple.
(iv) The fuzzy texture $(M_{[0,1]}, \mathcal{M}_{[0,1]})$ is not plain, since

$$P_r = (0, r] = Q_r \text{ for all } r \in (0, 1].$$

Further, $M = \{P_r = (0, r] \mid r \in (0, 1]\}$ and so it is a simple texture.

Remark 2.4.7 For the discrete texture $(U, \mathcal{P}(U))$, $Q_u = U \setminus \{u\}$ and hence, for all $u \in U$, $u \notin Q_u$. However, for the fuzzy texture $(M_{[0,1]}, \mathcal{M}_{[0,1]})$, $r \in Q_r$ for all $r \in (0, 1]$.

References

1. L.M. Brown, M. Diker, Ditopological texture spaces and intuitionistic sets. Fuzzy Sets Syst. **98**, 217224 (1998)
2. L.M. Brown, M. Diker, Paracompactness and full normality in ditopological texture spaces. J. Math. Anal. Appl. **227**, 144165 (1998)
3. L.M. Brown, R. Ertürk, Ş Dost, Ditopological texture spaces and fuzzy topology. II. Topological considerations. Fuzzy Sets Syst. **147**(2), 201–231 (2004)
4. M. Demirci, Textures and C-spaces. Fuzzy Sets Syst. **158**, 1237–1245 (2007)
5. M. Diker, Connectedness in ditopological texture spaces. Fuzzy Sets Syst. **108**(2), 223–230 (1999)

6. M. Diker, One point compactifications of ditopological texture spaces. Fuzzy Sets Syst. **147**, 233–248 (2004)
7. M. Diker, Ş. Dost, A. Altay Uğur, Interior and closure operators on texture spaces-I: Basic concepts and Céch closure operators. Fuzzy Sets Syst. **161**(7), 935–953 (2010)
8. S. Özçağ, L.M. Brown, A textural view of the distinction between uniformities and quasi-uniformities. Topology Appl. **153**, 3294–3307 (2006)
9. F. Yıldız, L.M. Brown, Extended real dicompactness and an application to Hutton spaces. Fuzzy Sets Syst. **227**, 74–95 (2013)
10. S. Bayhan, Textural dependency and concept lattices. Int. J. Approx. Reason. **136**, 36–65 (2021)
11. M. Diker, A category approach to relation preserving functions in rough set theory. Int. J. Approx. Reason. **56**(1), 71–86 (2015)
12. M. Diker, Categories of rough sets and textures. Theor. Comput. Sci. **488**, 46–65 (2013)
13. M. Diker, Textures and fuzzy unit operations in rough set theory: An approach to fuzzy rough set models. Fuzzy Sets Syst. **336**, 27–53 (2018)
14. M. Diker, Textural approach to rough sets based on relations. Inf. Sci. **180**(8), 1418–1433 (2010)
15. M. Diker, Textures and Fuzzy rough sets. Fundam. Inform. **108**, 305336 (2011)
16. M. Diker, Textures and rough sets, in *Handbook of Neural Computation*. ed. by P. Samui, S.S. Roy, V.E. Balas (Elsevier Science, Oxford/Amsterdam, 2017), pp.17–43
17. M. Diker, Definability and textures. Int. J. Approx. Reason. **53**, 558572 (2012)
18. M. Diker, Categories of direlations and rough set approximation operators, in *Rough Sets and Current Trends in Computing. RSCTC 2010. Lecture Notes in Computer Science*, vol. 6086, eds. by M. Szczuka, M. Kryszkiewicz, S. Ramanna, R. Jensen, Q. Hu (Springer, Berlin, Heidelberg, 2010), pp. 288–298
19. M. Diker, A. Altay Uğur, S. Bayhan, Textures and unit operations in rough set theory. Inf. Sci. **418–419**, 153–168 (2017)
20. M. Diker, A. Altay Uğur, S. Bayhan, Rough sets and textural neigbourhoods, in *Rough Sets and Knowledge Technology. RSKT 2015. Lecture Notes in Computer Science*, vol. 9436, eds. by D. Ciucci, G. Wang, S. Mitra, W.Z. Wu (Springer, 2015), pp. 190–202
21. M. Diker, A. Altay Uğur, S. Bayhan, Textural formal context. Int. J. Approx. Reason. **114**, 182–203 (2019)
22. M. Diker, A. Altay Uğur, Fuzzy rough set models over two universes using textures. Fuzzy Sets Syst. **442**, 155–195 (2022)
23. M. Diker, A. Altay Uğur, Textures and dicovering based rough sets. Inf. Sci. **184**, 44–63 (2012)
24. M. Diker, A. Altay Uğur, Dicovering approximation spaces and definability. Int. J. Approx. Reason. **101**, 255–275 (2018)
25. L.M. Brown, R. Ertürk, Fuzzy sets as texture spaces, I. Representation theorems. Fuzzy Sets Syst. **110**(2), 227–236 (2000)
26. L.M. Brown, R. Ertürk, Ş Dost, Ditopological texture spaces and fuzzy topology. I. Basic concepts. Fuzzy Sets Syst. **147**, 171–199 (2004)
27. A. Altay Uğur, M. Diker, Generalized textural rough Sets: Rough set models over two universes. Inf. Sci. **521**, 398421 (2020)

Chapter 3
Brown's Representation Theorem

In this chapter, we observe that every fuzzy lattice is isomorphic to a simple texture and vice versa. Let U be a domain of discourse. The family of all subsets of U that is the powerset $\mathcal{P}(U)$ is a discrete texture. This provides a route to a link between textural rough sets and Pawlak's rough sets. On the other hand, if $\mathcal{F}(U)$ is the fuzzy power set, that is, the family of all fuzzy subsets of a non-empty domain U, then we need a similar link giving a fruitful connection between textures and fuzzy rough sets. In a more general circumstance, the Representation Theorem of Textures provides this connection with the lattice of L-fuzzy subsets.

3.1 Textural Isomorphisms

Simple textures constitute an important class of textures. There is one-to-one correspondence between simple textures and fuzzy lattices. The following concept will be used in the sequel.

Definition 3.1.1 Let $(U, \mathcal{U})$ and $(V, \mathcal{V})$ be texture spaces. A function $\psi : U \to V$ is called *a textural isomorphism* if

(i) ψ is bijective.

(ii) $\forall A \in \mathcal{U}, \psi(A) \in \mathcal{V}$.

(iii) The mapping $\psi : \mathcal{U} \to \mathcal{V}, A \mapsto \psi(A)$ is bijective.

Any two textures $(U, \mathcal{U})$ and $(V, \mathcal{V})$ are called *isomorphic* if there exists a textural isomorphism between them. If the textures are isomorphic, then we use the notation $(U, \mathcal{U}) \cong (V, \mathcal{V})$. If c_U are c_V are complementations on $(U, \mathcal{U})$ and $(V, \mathcal{V})$, respectively, and ψ satisfies the additional property

M. Diker, *Texture Spaces*, Studies in Fuzziness and Soft Computing 411,
https://doi.org/10.1007/978-3-031-39748-6_3

$$\forall A, \psi(c_U(A)) = c_V(\psi(A)),$$

then ψ is called *a complemented textural isomorphism*. When such an isomorphism exists, we write $(U, \mathcal{U}, c_U) \cong (V, \mathcal{V}, c_V)$.

Proposition 3.1.2 *Let $\psi : U \to V$ be a textural isomorphism and $\{A_j \mid j \in J\} \subseteq \mathcal{U}$. Then*

(i) $\psi(\bigcap_{j\in J} A_j) = \bigcap_{j\in J} \psi(A_j)$.

(ii) $\psi(\bigvee_{j\in J} A_j) = \bigvee_{j\in J} \psi(A_j)$.

Proof (i) The proof is straightforward, since ψ is an injective function.

(ii) Since ψ is a function, we clearly have the inclusion

$$\bigvee_{j\in J} \psi(A_j) \subseteq \psi(\bigvee_{j\in J} A_j).$$

Let us show the reverse inclusion. By Definition 3.1.1 (ii), for all $j \in J$, $\psi(A_j) \in \mathcal{V}$ and hence, $\bigvee_{j\in J} \psi(A_j) \in \mathcal{V}$. Further, since $\psi : \mathcal{U} \to \mathcal{V}$ is bijective, there exists a unique $A \in \mathcal{U}$ such that $\psi(A) = \bigvee_{j\in J} \psi(A_j)$. Then we have $A_j \subseteq A$ for all $j \in J$. Suppose the contrary, that is, let $A_j \nsubseteq A$ for some $j \in J$. Then $A_j \cap A \neq A_j$. Since $\psi : \mathcal{U} \to \mathcal{V}$ is injective, this yields that $\psi(A_j \cap A) \neq \psi(A_j)$ and so by (i) we conclude that $\psi(A_j) \cap \psi(A) \neq \psi(A_j)$ whence $\psi(A_j) \nsubseteq \psi(A)$. However, this implies a contradiction. In this case, $\bigvee_{j\in J} A_j \subseteq A$ and since $\psi : U \to V$ is a function, we obtain that

$$\psi(\bigvee_{j\in J} A_j) \subseteq \psi(A) = \bigvee_{j\in J} \psi(A_j)$$

as desired.

Corollary 3.1.3 *Every textural isomorphism is a lattice isomorphism.*

Proof The proof is immediate by Proposition 3.1.2. □

Now let L be a fuzzy lattice and $M_L = \{m \mid m \text{ is a molecule in } L\}$. Let us consider the family $\mathcal{M}_L = \{\mathfrak{I}(r) \mid r \in L\}$ where

$$\mathfrak{I}(r) = \{m \mid m \in M_L \text{ and } m \leq r\}$$

for every $r \in L$. We may give the following representation theorem:

Theorem 3.1.4 (Brown's Representation Theorem) *The mapping* $\Im : L \to \mathcal{M}_L$ *defined by*

$\forall r \in L,\ r \mapsto \Im(r)$

is a lattice isomorphism and the triple $(M_L, \mathcal{M}_L, c_{M_L})$ *is a complemented simple texture space. Conversely, every complemented simple texture may be obtained in this way from a suitable fuzzy lattice.*

Proof First, we show that

$$r \le s \Longleftrightarrow \Im(r) \subseteq \Im(s)$$

for all $r, s \in L$. The implication $r \le s \Longrightarrow \Im(r) \subseteq \Im(s)$ is trivial. Let $\Im(r) \subseteq \Im(s)$. Since L is completely distributive, every element of L is a join of molecules (see, Theorem 3.15 in [1]) and then we get

$$r = \bigvee_{m \le r} m \text{ and } s = \bigvee_{m \le s} m.$$

By the assumption, $\{m \in M_L \mid m \le r\} \subseteq \{m \in M_L \mid m \le s\}$ and so we conclude that

$$\bigvee_{m \le r} m \le \bigvee_{m \le s} m,$$

which yields $r \le s$ as desired. To show that $\Im$ is a lattice isomorphism, let $r_1, r_2 \in L$ and $m \in \Im(r_1 \vee r_2)$. Then $m \le r_1 \vee r_2$ and since m is a molecule of L, we get $m \le r_1$ or $m \le r_2$. That is, $m \in \Im(r_1)$ or $m \in \Im(r_2)$ yields that $m \in \Im(r_1) \cup \Im(r_2)$. Hence, $\Im(r_1 \vee r_2) \subseteq \Im(r_1) \cup \Im(r_2)$. Further, $r_1 \le r_1 \vee r_2$ and $r_2 \le r_1 \vee r_2$, and so $\Im(r_1) \subseteq \Im(r_1 \vee r_2)$ and $\Im(r_2) \subseteq \Im(r_1 \vee r_2)$. This means that $\Im(r_1) \cup \Im(r_2) \subseteq \Im(r_1 \vee r_2)$. This proves the equality $\Im(r_1 \vee r_2) = \Im(r_1) \cup \Im(r_2)$. Note that

$$\begin{aligned} m \in \Im(r_1 \wedge r_2) &\Longleftrightarrow m \le r_1 \wedge r_2 \Longleftrightarrow m \le r_1 \text{ and } m \le r_2 \\ &\Longleftrightarrow m \in \Im(r_1) \text{ and } m \in \Im(r_2) \Longleftrightarrow m \in \Im(r_1) \cap \Im(r_2). \end{aligned}$$

Hence, we also have $\Im(r_1 \wedge r_2) = \Im(r_1) \cap \Im(r_2)$. The mapping $\Im$ is clearly one-to-one and onto. This yields that $\Im$ is a lattice isomorphism.

Now let us check the conditions (**T1-T4**) to show that $(M_L, \mathcal{M}_L)$ is a texture.

(**T1**) Since L is a complete lattice and $\Im$ is a lattice isomorphism, $(\mathcal{M}_L, \subseteq)$ is also a complete. Further, $\Im(0) = \emptyset$ and $\Im(1) = M_L$ and so $\emptyset, M_L \in \mathcal{M}_L$.

(**T2**) Take $\{r_j \mid j = 1, 2, ..., n\} \subseteq L$. For all $j \in \{1, 2, ..., n\}$, $r_j \le \bigvee_{j=1}^{n} r_j$ and then $\Im(r_j) \subseteq \Im\left(\bigvee_{j=1}^{n} r_j\right)$, that is,

$$\bigcup_{j=1}^{n} \mathfrak{I}(r_j) \subseteq \mathfrak{I}(\bigvee_{j=1}^{n} r_j).$$

Now let $m \in \mathfrak{I}(\bigvee_{j=1}^{n} r_j)$. Then $m \leq \bigvee_{j=1}^{n} r_j$ and so $m \leq r_j$ for some $j \in \{1, 2, ..., n\}$.

Then $m \in \mathfrak{I}(r_j)$ implies that $m \in \bigcup_{j=1}^{n} \mathfrak{I}(r_j)$. This gives that $\bigcup_{j=1}^{n} \mathfrak{I}(r_j) = \mathfrak{I}(\bigvee_{j=1}^{n} r_j)$. Since $\bigcup_{j=1}^{n} \mathfrak{I}(r_j) \in \mathcal{M}_L$, we get

$$\bigvee_{j=1}^{n} \mathfrak{I}(r_j) = \bigcup_{j=1}^{n} \mathfrak{I}(r_j)$$

for every finite index set $J = \{1, 2, ..., n\}$. On the other hand, for any $\{r_j \mid j \in J\} \subseteq L$,

$$\begin{aligned} m \in \bigcap_{j \in J} \mathfrak{I}(r_j) &\Longleftrightarrow \forall j \in J,\ m \in \mathfrak{I}(r_j) \\ &\Longleftrightarrow \forall j \in J,\ m \leq r_j \Longleftrightarrow m \leq \bigwedge_{j \in J} r_j \\ &\Longleftrightarrow m \in \mathfrak{I}(\bigwedge_{j \in J} r_j), \end{aligned}$$

that is, $\bigcap_{j \in J} \mathfrak{I}(r_j) = \mathfrak{I}(\bigwedge_{j \in J} r_j)$. As a result, for any index set J, we obtain that

$$\bigcap_{j \in J} \mathfrak{I}(r_j) = \bigwedge_{j \in J} \mathfrak{I}(r_j).$$

(T3) Since L is complete and completely distributive, $(\mathcal{M}_L, \subseteq)$ is also completely distributive.

(T4) Let $m_1, m_2 \in M_L$ and $m_1 \neq m_2$. This implies that $m_1 \nleq m_2$ or $m_2 \nleq m_1$. This gives that

$m_1 \notin \mathfrak{I}(m_2)$ or $m_2 \notin \mathfrak{I}(m_1)$.

As a result, $\mathcal{M}_L$ separates the points.

$(M_L, \mathcal{M}_L)$ is complemented. Indeed, the mapping $c_{\mathcal{M}_L} : \mathcal{M}_L \to \mathcal{M}_L$ defined by

$$c_{\mathcal{M}_L}(\mathfrak{I}(r)) = \mathfrak{I}(r')$$

for all $r \in L$ where $'$ is the order reversing involution of L is a complementation on $\mathcal{M}_L$. Let $\mathfrak{I}(r_1) \subseteq \mathfrak{I}(r_2)$, Then $r_1 \leq r_2$ and so $r_2' \leq r_1'$. Hence, $\mathfrak{I}(r_2') \subseteq \mathfrak{I}(r_1')$ and the definition of the mapping c_{M_L} yields that $c_{M_L}(\Im(r_2) \subseteq c_{M_L}(\Im(r_1)$. Note that we also have

$$c_{M_L}(c_{M_L}(\Im(r)) = (c_{M_L}(\Im(r')) = \Im((r')') = \Im(r).$$

Let m be a molecule in L and let $\Im(m) \subseteq \Im(r_1) \cup \Im r_2)$ for some $r_1, r_2 \in L$. Note that

$$\Im(r_1) \cup \Im(r_2) = \Im(r_1 \vee r_2).$$

Then we have $m \leq r_1 \vee r_2$ and so by the assumption $m \leq r_1$ or $m \leq r_2$. This implies that $\Im(m)$ is also a molecule since

$$\Im(m) \subseteq \Im(r_1) \text{ or } \Im(m) \subseteq \Im(r_2).$$

On the other hand, for every $m \in M_L$, $m \in \Im(m)$ and hence, $P_m \subseteq \Im(m)$. If $m' \in \Im(m)$, then $m' \leq m$ and this also gives the inclusion $\Im(m') \subseteq \Im(m)$. Now we have

$$m' \in P_{m'} \subseteq \Im(m') \subseteq \Im(m).$$

This means that $P_m = \Im(m)$. Further, if $\Im(r)$ is a molecule of $\mathcal{M}_L$, then r is clearly a molecule of L and hence, it should be that $\Im(r) = P_r$. Therefore, the family of all molecules of $\mathcal{M}_L$ is $\{P_m \mid m \in M_L\}$ and so the texture $(M_L, \mathcal{M}_L)$ is simple.

Now, let $(U, \mathcal{U}, c_U)$ be a simple complemented texture. Note that $(\mathcal{U}, \subseteq)$ is complete and completely distributive, and hence it is a fuzzy lattice with the order reversing involution $A' = c_U(A)$ where $A \in \mathcal{U}$. Let $(U^*, \mathcal{U}^*, c_{U^*})$ be the complemented simple texture corresponding to $(U, \mathcal{U}, c_U)$ so that

$$\mathcal{U}^* = \{\Im(A) \mid A \in \mathcal{U}\},\ U^* = \{P_u \mid u \in U\}$$

where

$$\Im(A) = \{P_u \mid P_u \subseteq A\} = \{P_u \mid u \in A\},\ c_{U^*}(\Im(A)) = \Im(c_U(A))$$

for each $A \in \mathcal{U}$. Then the mapping $\psi : U \to U^*$ defined by

$$\forall u \in U,\ \psi(u) = P_u$$

is clearly a bijection. Further, for all $A \in \mathcal{U}$, $\psi(A) = \{P_u \mid u \in A\} \in \mathcal{U}^*$. The mapping $\psi : \mathcal{U} \to \mathcal{U}^*$ defined by $\psi(A) = \{P_u \mid u \in A\}$ for all $A \in \mathcal{U}$, is also a bijection. As a result, $\psi : U \to U^*$ is a textural isomorphism. Then by Proposition 3.1.2, $\psi : \mathcal{U} \to \mathcal{U}^*$ is a lattice isomorphism. This completes the proof of the theorem.

Definition 3.1.5 The texture $(M_L, \mathcal{M}_L, c_{M_L})$ given in Brown's Representation Theorem is called *the fuzzy texture of the fuzzy lattice L*.

Example 3.1.6 *Finite Textures.*

(i) *Divisors of* 30

Let us consider the lattice $L = \{1, 2, 3, 5, 6, 10, 15, 30\}$ of divisors of 30.

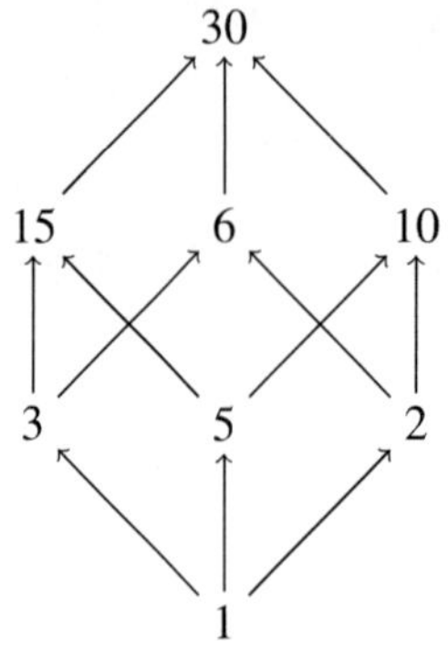

We know that L is complete, distributive (trivially, it is also completely distributive) and complemented. Hence, L is a fuzzy lattice. By Brown's Representation Theorem, there exists a simple texture corresponding to L. We may determine this texture. The set of all molecules of L is $M_L = \{2, 3, 5\}$. The mapping $\mathfrak{I} : L \to \mathcal{M}_L = \{\mathfrak{I}(u) \mid u \in L\}$ defined by

$$u \mapsto \mathfrak{I}(u) = \{m \mid m \leq u, m \in M_L\}$$

for all $u \in L$ is a lattice isomorphism. We clearly have that

$$\mathfrak{I}(1) = \emptyset,\ \mathfrak{I}(2) = \{2\},$$

$$\mathfrak{I}(3) = \{3\},\ \mathfrak{I}(5) = \{5\},$$

$$\mathfrak{I}(6) = \{2, 3\},\ \mathfrak{I}(10) = \{2, 5\},$$

$$\mathfrak{I}(15) = \{3, 5\},\ \mathfrak{I}(30) = \{2, 3, 5\}.$$

Then the family

$$\mathcal{M}_L = \{M_L, \emptyset,\ \{2\}, \{3\}, \{5\}, \{2, 3\}, \{2, 5\}, \{3, 5\}\} = \mathcal{P}(M_L)$$

is a discrete texturing on $M_L = \{2, 3, 5\}$. Clearly, $(M_L, \mathcal{M}_L)$ is simple and complemented with the ordinary set complementation.

(ii) *Divisors of* 60

Let us consider the lattice $L = \{1, 2, 3, 4, 5, 6, 12, 15, 20, 30, 60\}$ of divisors of 60 (Note that L is complete and distributive). By Brown's Representation Theorem, L corresponds to a simple texture. Let us determine this texture.

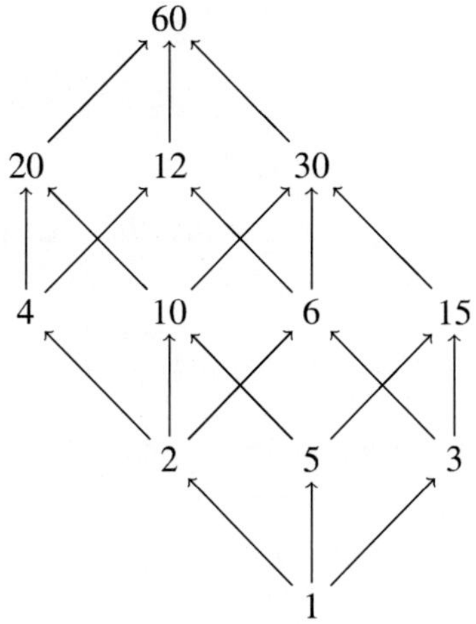

It is easy to see that the set of all molecules of L is $M_L = \{2, 3, 4, 5\}$. Then the mapping $\mathfrak{I} : L \to \{\mathfrak{I}(r) \mid r \in L\}$ defined by

$$\mathfrak{I}(1) = \emptyset,\ \mathfrak{I}(2) = \{2\},$$
$$\mathfrak{I}(3) = \{3\},\ \mathfrak{I}(4) = \{2, 4\},$$
$$\mathfrak{I}(5) = \{5\},\ \mathfrak{I}(6) = \{2, 3\},$$
$$\mathfrak{I}(10) = \{2, 5\},\ \mathfrak{I}(12) = \{2, 3, 4\},$$
$$\mathfrak{I}(15) = \{3, 5\},\ \mathfrak{I}(20) = \{2, 4, 5\},$$
$$\mathfrak{I}(30) = \{2, 3, 5\},\ \mathfrak{I}(60) = \{2, 3, 4, 5\}$$

is a lattice isomorphism where

$$\mathcal{U} = \{M_L, \emptyset,\ \{2\}, \{3\}, \{5\}, \{2, 3\}, \{2, 4\}, \{2, 5\}, \{3, 5\}, \{2, 3, 4\}, \{2, 3, 5\}, \{2, 4, 5\}\}$$

is a texturing on M_L. For the finite texture space $(M_L, \mathcal{M}_L)$, we have

$$P_2 = \{2\}, P_3 = \{3\}, P_4 = \{2, 4\}, P_5 = \{5\}$$

and

$$Q_2 = \{3, 5\}, Q_3 = \{2, 4, 5\}, Q_4 = \{2, 3, 5\}, Q_5 = \{2, 3, 4\}.$$

Note that $\mathcal{M} = \{P_2, P_3, P_4, P_5\}$ is the set of all molecules of $\mathcal{M}_L$. Therefore, $(M_L, \mathcal{M}_L)$ is a simple texture. However, the lattice L is not complemented, and hence, $(M_L, \mathcal{M}_L)$ is not a complemented texture.

3.2 Product of Textures

Ordinary product of textures may not be a texture. For instance, let us consider the textures $(U_1, \mathcal{U}_1)$ and $(U_2, \mathcal{U}_2)$ where

$$U_1 = \{a, b, c\},\ \mathcal{U}_1 = \{U_1, \emptyset, \{a, c\}, \{b\}, \{c\}, \{b, c\}\}$$

and

$$U_2 = \{d, e, f, g\},\ \mathcal{U}_2 = \{U, \emptyset, \{d\}, \{d, f\}, \{d, e\}, \{d, f, e\}\},$$

respectively. Considering the ordinary cross product of the families $\mathcal{U}_1$ and $\mathcal{U}_2$, we see that $\{b\} \times \{d\}, \{c\} \times \{d, f\} \in \mathcal{U}_1 \times \mathcal{U}_2$, and however,

$$(\{b\} \times \{d\}) \cup (\{c\} \times \{d, f\}) = \{(b, d), (c, d), (c, f)\} \notin \mathcal{U}_1 \times \mathcal{U}_2.$$

This means that $\mathcal{U}_1 \times \mathcal{U}_2$ is not closed under union and hence, it cannot be a texture on $U_1 \times U_2$.

Now, for $A \subseteq U_j$, $j = 1, 2$, let $E(k, A) = B_1 \times B_2,\ k = 1, 2$ where

$$B_k = \begin{cases} A, & k = j \\ U_k, & k \neq j. \end{cases}$$

Let us take the family

$$\mathcal{B} = \{\bigcup_{k \in K} E(k, A_k) \mid K \subseteq \{1, 2\}, A_k \in \mathcal{U}_k\}.$$

Note that every element of $\mathcal{B}$ has one of the following forms:

$$(U_1 \times B) \cup (A \times U_2),\ U_1 \times B,\ A \times U_2$$

where $A \in \mathcal{U}_1$ and $B \in \mathcal{U}_2$.

The following lemma will be used in the sequel.

Lemma 3.2.1 *For any $A \subseteq U_1$ and $B \subseteq U_2$, we have*

$$(A \times B) \subseteq (C \times U_2) \cup (U_1 \times D) \Longrightarrow A \times B \subseteq C \times U_2 \ \text{or}\ A \times B \subseteq U_1 \times D.$$

Proof Immediate. □

We have the following theorem:

Proposition 3.2.2 *Let $(U_1, \mathcal{U}_1)$ and $(U_2, \mathcal{U}_2)$ be any two texture spaces and let $\{E_j \mid j \in J\} \subseteq \mathcal{B}$ where*

$$A_j \in \mathcal{U}_1,\ B_j \in \mathcal{U}_2 \text{ and } E_j = (A_j \times U_2) \cup (U_1 \times B_j)$$

for all $j \in J$. Then

$$\bigvee_{j \in J} E_j = ((\bigvee_{j \in J} A_j) \times U_2) \bigcup (U_1 \times (\bigvee_{j \in J} B_j)).$$

Proof Let us take

$$Q = \big((\bigvee_{j\in J} A_j) \times U_2\big) \bigcup \big(U_1 \times (\bigvee_{j\in J} B_j)\big).$$

Since $\mathcal{U}_1$ and $\mathcal{U}_2$ are textures, we have

$$\bigvee_{j\in J} A_j \in \mathcal{U}_1 \text{ and } \bigvee_{j\in J} B_j \in \mathcal{U}_2$$

and hence, by definition of $\mathcal{B}$, $Q \in \mathcal{B}$. Let $p \in \bigcup_{j\in J} E_j$. Then for some $j \in J$, $p \in E_j$, and so we get

$$p \in A_j \times U_2 \text{ or } p \in U_1 \times B_j.$$

Therefore,

$$p \in \bigvee_{j\in J} A_j \times U_2 \text{ or } p \in U_1 \times \bigvee_{j\in J} B_j,$$

that is,

$$p \in \big((\bigvee_{j\in J} A_j) \times U_2\big) \bigcup \big(U_1 \times (\bigvee_{j\in J} B_j)\big) = Q.$$

This yields that $\bigcup_{j\in J} E_j \subseteq Q$. Therefore, for all $j \in J$, $E_j \subseteq Q$, and so we obtain $\bigvee_{j\in J} E_j \subseteq Q$.

Let $(T_1 \times U_2) \cup (U_1 \times T_2) = P$ where $\bigcup_{j\in J}(A_j \times U_2) \cup (U_1 \times B_j) \subseteq P$. Note that $P \in \mathcal{B}$. For $j \in J$, $(A_j \times U_2) \cup (U \times B_j) \subseteq P$ and hence, $(A_j \times U_2) \subseteq P$ and $(U_1 \times B_j) \subseteq P$. Then by Lemma 3.2.1, for all $j \in J$, we conclude that $A_j \subseteq T_1$ and $B_j \subseteq T_2$ (here we take $A_j \neq U_1$ and $B_j \neq U_2$). This yields that $\bigcup_{j\in J} A_j \subseteq T_1$ and $\bigcup_{j\in J} B_j \subseteq T_2$ and so we get $\bigvee_{j\in J} A_j \subseteq T_1$ and $\bigvee_{j\in J} B_j \subseteq T_2$, since $T_1 \in \mathcal{U}_1$ and $T_2 \in \mathcal{U}_2$. As a result, we get

$$Q = \big((\bigvee_{j\in J} A_j) \times U_2\big) \bigcup \big(U_1 \times (\bigvee_{j\in J} B_j)\big) \subseteq (T_1 \times U_2) \cup (U_1 \times T_2) = P.$$

Since

$$\bigvee_{j\in J} E_j = \bigcap \{P \mid \bigcup_{j\in J} E_j \subseteq P \in \mathcal{B}\},$$

we obtain that $Q \subseteq \bigvee_{j\in J} E_j$. □

Now, we may give the following immediate result:

Corollary 3.2.3

(i) *The family $\mathcal{B}$ is closed under arbitrary joins, that is, for all $\{E_j \mid j \in J\} \subseteq \mathcal{B}$, $\bigvee_{j\in J} E_j \in \mathcal{B}$.*

(ii) *If $\{E_j \mid j \in J\}$ is a finite family, then $\bigvee_{j\in J} E_j = \bigcup_{j\in J} E_j$.*

Proof (ii) By Proposition 3.2.2,

$$\begin{aligned}\bigvee_{j\in J} E_j &= ((\bigvee_{j\in J} A_j)\times U_2)\bigcup(U_1\times(\bigvee_{j\in J} B_j))\\ &= ((\bigcup_{j\in J} A_j)\times U_2)\bigcup(U_1\times(\bigcup_{j\in J} B_j))\\ &= (\bigcup_{j\in J}(A_j\times U_2))\bigcup(\bigcup_{j\in J}(U_1\times B_j))\\ &= \bigcup_{j\in J}((A_j\times U_2)\cup(U_1\times B_j))\\ &= \bigcup_{j\in J} E_j.\end{aligned}$$

Theorem 3.2.4 *The family*

$$\mathcal{U}_1\otimes\mathcal{U}_2=\{\bigcap\mathcal{D}\mid\forall\mathcal{D},\ \mathcal{D}\subseteq\mathcal{B}\}$$

is a texture on $U\times V$ *which is called a product of* $(U,\mathcal{U})$ *and* $(V,\mathcal{V})$.

Proof **(T1)** The family $\mathcal{U}_1\otimes\mathcal{U}_2$ is clearly closed under intersections. Therefore, by Theorem1.1.1, $(\mathcal{U}\otimes\mathcal{V},\subseteq)$ is a complete lattice. Clearly, $U_1\times U_2,\emptyset\in\mathcal{U}_1\otimes\mathcal{U}_2$.
(T2) By the same reason, for every family $\{D_j\mid j\in J\}\subseteq\mathcal{U}_1\otimes\mathcal{U}_2$ we have

$$\bigwedge_{j\in J} D_j=\bigcap_{j\in J} D_j\in\mathcal{U}_1\otimes\mathcal{U}_2.$$

Further,

$$\begin{aligned}\bigvee_{j\in J} D_j &= \bigcap\{Q\mid\bigcup_{j\in J} D_j\subseteq Q\in\mathcal{U}_1\otimes\mathcal{U}_2\}\\ &= \bigcap\{Q\mid\bigcup_{j\in J} D_j\subseteq Q\in\mathcal{B}\}\in\mathcal{U}_1\otimes\mathcal{U}_2.\end{aligned}$$

(T3) Let $D_1,D_2\in\mathcal{U}_1\otimes\mathcal{U}_2$ and $D_1\not\subseteq D_2$. Without loss of generality, we may take

$$D_1=\bigcap_{j\in J}(A_j\times U_2)\cup(U_1\times B_j)\text{ and }D_2=\bigcap_{k\in K}(A_k\times U_2)\cup(U_1\times B_k)$$

for some index sets J and K. For all $j\in J$, $(A_j\times U_2)\cup(U_1\times B_j)\not\subseteq D_2$ and so for all j and for some $k\in K$, we get

$$(A_j\times U_2)\cup(U_1\times B_j)\not\subseteq(A_k\times U_2)\cup(U_1\times B_k).$$

This yields that

$$(A_j \times U_2) \not\subseteq (A_k \times U_2) \cup (U_1 \times B_k)$$

or

$$(U_1 \times B_j) \not\subseteq (A_k \times U_2) \cup (U_1 \times B_k),$$

that is,

$$A_j \times U_2 \not\subseteq A_k \times U_2$$

or

$$U_1 \times B_j \not\subseteq U_1 \times B_k.$$

Finally, we conclude that $A_j \not\subseteq A_k$ or $B_j \not\subseteq B_k$. Since $(U_1, \mathcal{U}_1)$ and $(U_2, \mathcal{U}_2)$ are completely distributive, by Corollary 2.3.2, for some $u_1 \in U_1$ and $u_2 \in U_2$, we get

$$A_j \not\subseteq Q_{u_1} \text{ and } P_{u_1} \not\subseteq A_k \text{ or } B_j \not\subseteq Q_{u_2} \text{ and } P_{u_2} \not\subseteq B_k.$$

Then

$$A_j \times U_2 \not\subseteq Q_{u_1} \times U_2 \text{ or } U_1 \times B_j \not\subseteq U_1 \times Q_{u_2},$$

and hence, by Lemma 3.2.1,

$$(A_j \times U_2) \cup (U_1 \times B_j) \not\subseteq (Q_{u_1} \times U_2) \cup (U_1 \times Q_{u_2}),$$

and

$$P_{u_1} \times P_{u_2} \not\subseteq A_k \times U_2 \text{ or } P_{u_1} \times P_{u_2} \not\subseteq U_1 \times B_k.$$

Again by Lemma 3.2.1, we get

$$P_{u_1} \times P_{u_2} \not\subseteq (A_k \times U_2) \cup (U_1 \times B_k).$$

Therefore, by Theorem 3.2.6, we obtain that

$$D_1 = (A_j \times U_2) \cup (U_1 \times B_j) \not\subseteq Q_{(u_1,u_2)} \text{ and } P_{(u_1,u_2)} \not\subseteq (A_k \times U_2) \cup (U_1 \times B_k) = D_2.$$

Finally, by Corollary 2.3.2, $\mathcal{U}_1 \otimes \mathcal{U}_2$ is completely distributive.

(T4) Let $(u_1, u_2), (t_1, t_2) \in U_1 \times U_2$ and $(u_1, u_2) \neq (t_1, t_2)$. Then $u_1 \neq t_1$ or $u_2 \neq t_2$. Without loss of generality, if $u_1 \neq t_1$, there is a set $A \in \mathcal{U}_1$ with $u_1 \in A$ and $t_1 \notin A$. This yields that $(u_1, u_2) \in A \times U_2$ and $(t_1, t_2) \notin A \times U_2$ where $A \times U_2 \in \mathcal{U}_1 \otimes \mathcal{U}_2$. □

Theorem 3.2.5 *For all $A \in \mathcal{U}_1$ and for all $B \in \mathcal{U}_2$, $A \times B \in \mathcal{U} \otimes \mathcal{V}$.*

Proof Note that $A \times U_2, U_1 \times B \in \mathcal{B}$ and so by definition of product texture,

$$(A \times U_2) \cap (U_1 \times B) = A \times B \in \mathcal{U}_1 \otimes \mathcal{U}_2.$$

Now, we may determine the p-set and the q-set of a product texture:

Theorem 3.2.6 *For the product texture* $(U_1 \times U_2, \mathcal{U}_1 \otimes \mathcal{U}_2)$, *we have the following statements:*

(i) $P_{(u_1,u_2)} = P_{u_1} \times P_{u_2}$.

(ii) $Q_{(u_1,u_2)} = (U_1 \times Q_{u_2}) \cup (Q_{u_1} \times U_2)$.

Proof (i) By definition of the p-set, we have

$$P_{(u_1,u_2)} = \bigcap \{A \times B \in \mathcal{U}_1 \otimes \mathcal{U}_2 \mid (u_1, u_2) \in A \times B\}.$$

Since $u_1 \in P_{u_1}$ and $u_2 \in P_{u_1}$, $(u_1, u_2) \in P_{u_1} \times P_{u_2}$. Further, $P_{u_1} \times U_2, U_1 \times P_{u_2} \in \mathcal{U}_1 \otimes \mathcal{U}_2$ and so

$$(P_{u_1} \times U_2) \cap (U_1 \times P_{u_2}) = P_{u_1} \times P_{u_2} \in \mathcal{U}_1 \otimes \mathcal{U}_2.$$

This implies that $P_{(u_1,u_2)} \subseteq P_{u_1} \times P_{u_2}$. Suppose that $P_{u_1} \times P_{u_2} \not\subseteq P_{(u_1,u_2)}$. Then for some $A \in \mathcal{U}_1$ and $B \in \mathcal{U}_2$ with $u_1 \in A$ and $u_2 \in B$, we have $P_{u_1} \times P_{u_2} \not\subseteq A \times B$. However, this gives that $P_{u_1} \not\subseteq A$ or $P_{u_2} \not\subseteq B$, that is, we obtain the contradiction $u_1 \notin A$ or $u_2 \notin B$.

(ii) Suppose that $Q_{(u_1,u_2)} \not\subseteq (U_1 \times Q_{u_2}) \cup (Q_{u_1} \times U_2)$. Then by Theorem 2.2.3, for some $w_1 \in U_1$ and $w_2 \in U_2$, we have $P_{(u_1,u_2)} \not\subseteq P_{(w_1,w_2)}$ and $P_{(w_1,w_2)} \not\subseteq (U_1 \times Q_{u_2}) \cup (Q_{u_1} \times U_2)$. By (i), we have $P_{u_1} \not\subseteq P_{w_1}$ or $P_{u_2} \not\subseteq P_{w_2}$, that is, $u_1 \notin P_{w_1}$ or $u_2 \notin P_{w_2}$. By definition of a q-set, $P_{w_1} \subseteq Q_{u_1}$ or $P_{w_2} \subseteq Q_{u_2}$. However, we know that $P_{(w_1,w_2)} \not\subseteq (U_1 \times Q_{u_2}) \cup (Q_{u_1} \times U_2)$ and we obtain the contradiction $P_{w_2} \not\subseteq Q_{u_2}$ and $P_{w_1} \not\subseteq Q_{u_1}$.

Now let us show that $(U_1 \times Q_{u_2}) \cup (Q_{u_1} \times U_2) \subseteq Q_{(u_1,u_2)}$. Suppose the contrary, that is, let $(U_1 \times Q_{u_2}) \cup (Q_{u_1} \times U_2) \not\subseteq Q_{(u_1,u_2)}$. By Corollary 2.3.2 (ii), there exists some $u' \in U_1$ and $u'' \in U_2$ such that

$$(U_1 \times Q_{u_2}) \cup (Q_{u_1} \times U_2) \not\subseteq Q_{u',u''}) \text{ and } P_{(u',u'')} \not\subseteq Q_{(u_1,u_2)}.$$

By the first statement above, we have two cases:

$$U_1 \times Q_{u_2} \not\subseteq Q_{(u',u'')} = (Q_{u'} \times U_2) \cup (U_1 \times Q_{u''}) \quad (1)$$

or

$$Q_{u_1} \times U_2 \not\subseteq Q_{(u',u'')} = (Q_{u'} \times U_2) \cup (U_1 \times Q_{u''}). \quad (2)$$

By the second, we get

$$P_{u''} \not\subseteq Q_{u_2} \text{ and } P_{u'} \not\subseteq Q_{u_1}.$$

For the case (1), we clearly have that $Q_2 \not\subseteq Q_{u''}$, that is $P_{u''} \subseteq Q_{u_2}$ and this is an immediate contradiction. If the case (2) holds, since $Q_{u_1} \not\subseteq Q_{u'}$, we obtain a similar contradiction.

Example 3.2.7 Let us consider the textures $(U_1, \mathcal{U}_1)$ and $(U_2, \mathcal{U}_2)$ where

$$U_1 = \{a, b, c\}, \ \mathcal{U}_1 = \{U_1, \emptyset, \{a\}, \{b\}, \{b, c\}, \{a, b\}\}$$

and

$$U_2 = \{d, e\}, \ \mathcal{U}_2 = \{U_2, \emptyset, \{d\}\},$$

respectively. Clearly,

$$\begin{aligned}\mathcal{B} &= \{\textstyle\bigcup_{k \in K} E(k, A_k) \mid K \subseteq \{1, 2\}, A_k \in \mathcal{U}_k\} \\ &= \{U_1 \times U_2, \ \emptyset, \ \{a\} \times U_2, \ \{b\} \times U_2, \ \{a, b\} \times U_2, \ \{b, c\} \times U_2, \ U_1 \times \{d\}\}.\end{aligned}$$

Finally, we obtain the product texture as

$$\begin{aligned}\mathcal{U}_1 \otimes \mathcal{U}_2 = \{&\textstyle\bigcap \mathcal{D} \mid \forall \mathcal{D}, \ \mathcal{D} \subseteq \mathcal{B}\} \\ = \{&U_1 \times U_2, \ \emptyset, \ \{a\} \times U_2, \ \{b\} \times U_2, \ \{a, b\} \times U_2, \ \{b, c\} \times U_2, U_1 \times \{d\}, \\ &\{a\} \times \{d\}, \{b\} \times \{d\}, \{a, b\} \times \{d\}, \ \{b, c\} \times \{d\}\}.\end{aligned}$$

Note that for the texture $\mathcal{U}_1$,

$$P_a = \{a\}, P_b = \{b\}, P_c = \{b, c\},$$

$$Q_a = \{b, c\}, Q_b = \{a\}, Q_c = \{a, b\},$$

and for the texture $\mathcal{U}_2$,

$$P_d = \{d\}, P_e = U_2 \text{ and}$$

$$Q_d = \emptyset, Q_e = \{d\}.$$

Hence, the p-sets and q-sets of the product $\mathcal{U}_1 \otimes \mathcal{U}_2$ are

$$P_{(a,d)} = P_a \times P_d = \{(a, d)\}, P_{(a,e)} = P_a \times P_e = \{a\} \times U_2,$$

$$P_{(b,d)} = P_b \times P_d = \{(b, d)\}, P_{(b,e)} = P_b \times P_e = \{b\} \times U_2,$$

$$P_{(c,d)} = P_c \times P_d = \{(b, d), (c, d)\}, P_{(c,e)} = P_c \times P_e = \{b, c\} \times U_2.$$

$$Q_{(a,d)} = (Q_a \times U_2) \cup (U_1 \times Q_d) = \{b, c\} \times U_2,$$

$$Q_{(a,e)} = (Q_a \times U_2) \cup (U_1 \times Q_e) = (\{b, c\} \times U_2) \cup (U_1 \times \{d\}),$$

$$Q_{(b,d)} = (Q_b \times U_2) \cup (U_1 \times Q_d) = (\{a\} \times U_2) \cup (U_1 \times \emptyset) = \{a\} \times U_2$$

$$Q_{(b,e)} = (Q_b \times U_2) \cup (U_1 \times Q_e) = (\{a\} \times U_2) \cup (U_1 \times \{d\}),$$

$$Q_{(c,d)} = (Q_c \times U_2) \cup (U_1 \times Q_d) = (\{a, b\} \times U_2) \cup (U_1 \times \emptyset) = \{a, b\} \times U_2,$$

$$Q_{(c,e)} = (Q_c \times U_2) \cup (U_1 \times Q_e) = (\{a, b\} \times U_2) \cup (U_1 \times \{d\}). \qquad \square$$

Theorem 3.2.8 *Let $(U_1 \times U_2, \mathcal{U}_1 \otimes \mathcal{U}_2)$ be the product of the complemented textures $(U_1, \mathcal{U}_1, c_1)$ and $(U_2, \mathcal{U}_2, c_2)$. Then the mapping $c : \mathcal{U}_1 \otimes \mathcal{U}_2 \to \mathcal{U}_1 \otimes \mathcal{U}_2$ defined by*

$$\forall D \in \mathcal{U}_1 \otimes \mathcal{U}_2,\ c(D) = \bigcap_{(u_1,u_2)\in D} ((c_1(P_{u_1}) \times U_2) \cup (U_1 \times c_2(P_{u_2})))$$

is a complementation on $\mathcal{U}_1 \otimes \mathcal{U}_2$.

Proof If $D \subseteq E$, then it is clear that $c(E) \subseteq c(D)$. Let us show that $c(c(D)) = D$ for all $D \in \mathcal{U}_1 \otimes \mathcal{U}_2$. Take $(u_1, u_2) \in D$. By definition of $c(D)$, there exists a $(t_1, t_2) \in c(D)$ such that $(t_1, t_2) \in (c_1(P_{u_1}) \times U_2) \cup (U_1 \times c_2(P_{u_2}))$. This implies that $(t_1, t_2) \in (c_1(P_{u_1}) \times U_2)$ or $(t_1, t_2) \in (U_1 \times c_2(P_{u_2}))$ and then $t_1 \in c_1(P_{u_1})$ or $t_2 \in c_2(P_{u_2})$. Now by Theorem 2.3.7 (ii), $u_1 \in c_1(P_{t_1})$ or $u_2 \in c_2(P_{t_2})$ whence

$$(u_1, u_2) \in (c_1(P_{t_1}) \times U_2) \text{ or } (u_1, u_2) \in (U_1 \times c_2(P_{t_2})).$$

Therefore,

$$\begin{aligned}(u_1, u_2) \in &(c_1(P_{t_1}) \times U_2) \cup (U_1 \times c_2(P_{t_2}))\\ &\subseteq \bigcap_{(t_1,t_2)\in c(D)} ((c_1(P_{t_1}) \times U_2) \cup (U_1 \times c_2(P_{t_2}))) = c(c(D)).\end{aligned}$$

We have proved that $D \subseteq c(c(D))$. For the reverse inclusion, let $r = (r_1, r_2) \in c(c(D))$ and $r \notin D$. Note that

$$\begin{aligned}D = \bigvee_{(d_1,d_2)\in D} P_{(d_1,d_2)} &= \bigvee_{(d_1,d_2)\in D} (P_{d_1} \times P_{d_2})\\ &= \bigvee_{(d_1,d_2)\in D} ((P_{d_1} \times U_2) \cap (U_1 \times P_{d_2}))\\ &= \Big(\bigvee_{(d_1,d_2)\in D} (P_{d_1} \times U_2)\Big) \bigcap \Big(\bigvee_{(d_1,d_2)\in D} (U_1 \times P_{d_2})\Big)\\ &= \Big(\bigvee_{(d_1,d_2)\in C} (P_{d_1} \times U_2)\Big) \bigcap \Big((U_1 \times \bigvee_{(d_1,d_2)\in D} P_{d_2})\Big).\end{aligned}$$

By the assumption, we have

$$r \notin \Big(\bigvee_{(d_1,d_2)\in D} P_{d_1}\Big) \times U_2 \text{ or } r \notin U_1 \times \Big(\bigvee_{(d_1,d_2)\in D} P_{d_2}\Big).$$

Then

$$r_1 \notin \bigvee_{(d_1,d_2)\in D} P_{d_1} \text{ or } r_2 \notin \bigvee_{(d_1,d_2)\in D} P_{d_2}.$$

Without loss of generality, we may suppose that $r_1 \notin \bigvee_{(d_1,d_2)\in D} P_{d_1}$. Let $P = \bigvee_{(d_1,d_2)\in D} P_{d_1}$. Note that by Theorem 2.3.7 (i), we also have

$$P = \bigvee_{(d_1,d_2)\in D} P_{d_1} = \bigcap_{t_1 \in c_1(P)} c_1(P_{t_1}).$$

Hence, for some $t_1 \in c_1(P)$, $r_1 \notin c_1(P_{t_1})$. Further, by definition of $c(D)$ and completely distributivity of $\mathcal{U}_1 \otimes \mathcal{U}_2$, we get

$$c(D) = (\bigcap_{(d_1,d_2)\in D} c_1(P_{d_1})) \times U_2) \cup (\bigcap_{(d_1,d_2)\in D} U_1 \times c_2(P_{d_2})).$$

Therefore, for every $t_2 \in U_2$, we clearly have that $t = (t_1, t_2) \in c(D)$. On the other hand,

$$r = (r_1, r_2) \in c(c(D)) = \bigcap_{(u_1,u_2)\in c(D)} (c_1(P_{u_1}) \times U_2) \cup (U_1 \times c_2(P_{u_2})),$$

that is, for all $(u_1, u_2) \in c(D)$

$$r = (r_1, r_2) \in (c_1(P_{u_1}) \times U_2) \cup (U_1 \times c_2(P_{u_2})).$$

However, $(t_1, t_2) \in c(D)$ and hence, we also have

$$r = (r_1, r_2) \in (c_1(P_{t_1}) \times U_2) \cup (U_1 \times c_2(P_{t_2})).$$

Finally, we obtain that $r_1 \in c_1(P_{t_1})$ and this is an immediate contradiction.

Lemma 3.2.9

$\forall A \in \mathcal{U}_1, \bigcap_{u_2 \in U_2} (A \times c_2(P_{u_2})) = \emptyset,$

$\forall B \in \mathcal{U}_2, \bigcap_{u_1 \in U_1} (c_1(P_{u_1}) \times B)) = \emptyset.$

Proof (i) We easily observe that

$$\begin{aligned}\bigcap_{u_2 \in U_2} (A \times c_2(P_{u_2})) &= A \times \bigcap_{u_2 \in U_2} c_2(P_{u_2}) \\ &= A \times c_2(\bigcup_{u_2 \in U_2} P_{u_2}) \\ &= A \times c_2(U_2) \\ &= A \times \emptyset = \emptyset.\end{aligned}$$

The proof of (ii) is similar.

Corollary 3.2.10 *Let $(U_1 \times U_2, \mathcal{U}_1 \otimes \mathcal{U}_2, c)$ be the product of the complemented textures $(U_1, \mathcal{U}_1, c_1)$ and $(U_2, \mathcal{U}_2, c_2)$. Then we have the following propositions.*

(i) *If $A \in \mathcal{U}_1$, then $c(A \times U_2) = c_1(A) \times U_2$.*

(ii) *If $B \in \mathcal{U}_2$, then $c(U_1 \times B) = U_2 \times c_2(B)$.*

Proof (i) By Theorem 1.1.2, we observe that

$$\begin{aligned} c(A \times U_2) &= \bigcap_{(u_1,u_2)\in A\times U_2} ((c_1(P_{u_1}) \times U_2) \cup (U_1 \times c_2(P_{u_2})) \\ &= (\bigcap_{(u_1,u_2)\in A\times U_2} c_1(P_{u_1}) \times U_2) \cup (\bigcap_{(u_1,u_2)\in A\times U_2} U_1 \times c_2(P_{u_2})) \\ &= c_1(\bigvee_{u_1\in A\times U_2} P_{u_1}) \times U_2 \\ &= c_1(\bigvee_{u_1\in A} P_{u_1}) \times U_2 = c_1(A) \times U_2. \end{aligned}$$

(ii) The proof is left to the interested reader.

Theorem 3.2.11 *Let $(U_1 \times U_2, \mathcal{U}_1 \otimes \mathcal{U}_2)$ be the product of the textures $(U_1, \mathcal{U}_1)$ and $(U_2, \mathcal{U}_2)$. Then*

(i) $(U_1 \times U_2)^\flat = U_1^\flat \times U_2^\flat$.

(ii) *For all $u = (u_1, u_2)$, $w = (w_1, w_2) \in U_1 \times U_2$*

$$P_u \not\subseteq Q_w \iff P_{u_1} \not\subseteq Q_{w_1} \text{ and } P_{u_2} \not\subseteq Q_{w_2}.$$

Proof (i) Note that if $(u_1, u_2) \in (U_1 \times U_2)^\flat$, then by definition of core of a set, $U_1 \times U_2 \not\subseteq Q_{(u_1,u_2)}$. Then recalling

$$\overline{Q}_{(u_1,u_2)} = (Q_{u_1} \times U_2) \cup (U_1 \times Q_2),$$

we obtain that

$$U_1 \times U_2 \not\subseteq Q_{u_1} \times U_2 \text{ and } U_1 \times U_2 \not\subseteq U_2 \times Q_{u_2}.$$

This gives $U_1 \not\subseteq Q_{u_1}$ and $U_2 \not\subseteq Q_{u_2}$, that is $u_1 \in U_1^\flat$ and $u_2 \in U_2^\flat$ whence $(u_1, u_2) \in U_1^\flat \times U_2^\flat$. We have showed that $(U_1 \times U_2)^\flat \subseteq U_1^\flat \times U_2^\flat$. The reverse inclusion is similar.

(ii) For the necessity, let $P_u \not\subseteq Q_w$. Then $P_{u_1} \times P_{u_2} \not\subseteq (Q_{w_1} \times U_2) \cup (U_1 \times Q_{w_2})$ and so clearly, $P_{u_1} \not\subseteq Q_{w_1}$ and $P_{u_2} \not\subseteq Q_{w_2}$. The proof of the sufficiency is similar. □

3.3 L-Fuzzy Set Texturing $\mathcal{P}(U) \otimes \mathcal{M}_L$

Let $(L, \leq)$ be a fuzzy lattice. If U is a domain of discourse, then the family of all L-fuzzy subset of U, that is, $(\mathcal{F}_L(U), \leq)$ is a fuzzy lattice. Every fuzzy point u_r of $\mathcal{F}_L(U)$ is a molecule and vice versa. That is, the family of all molecules of $\mathcal{F}(U)$ is

$$M_{\mathcal{F}_L(U)} = \{u_r \mid u_r \text{ is a fuzzy point in } \mathcal{F}_L(U)\ \}.$$

Let us take the sets

$$\mathfrak{I}(\alpha) = \{u_r \mid u_r \leq \alpha\},$$

$$\mathcal{M}_{\mathcal{F}_L(U)} = \{\mathfrak{I}(\alpha) \mid \alpha \in \mathcal{F}_L(U)\}$$

where $\alpha \in \mathcal{F}_L(U)$. By Brown's Representation Theorem, $(M_{\mathcal{F}_L(U)}, \mathcal{M}_{\mathcal{F}_L(U)}, c_{\mathcal{F}_L(U)})$ is a complemented simple texture where the complementation mapping $c_{\mathcal{F}_L(U)} : \mathcal{M}_{\mathcal{F}_L(U)} \to \mathcal{M}_{\mathcal{F}_L(U)}$ defined by

$$\forall \mathfrak{I}(\alpha),\ c_{\mathcal{F}_L(U)}(\mathfrak{I}(\alpha)) = \mathfrak{I}(\alpha').$$

Essentially, the texture $(M_{\mathcal{F}_L(U)}, \mathcal{M}_{\mathcal{F}_L(U)})$ may not be adequate for further developments and applications due to operations between the fuzzy subsets and textures. However, as we will prove in the following, $\mathcal{M}_{\mathcal{F}_L(U)}$ is isomorphic to the product of the discrete texturing $\mathcal{P}(U)$ and fuzzy texturing $\mathcal{M}_L$ on M_L.

Theorem 3.3.1 *The textures $(M_{\mathcal{F}_L(U)}, \mathcal{M}_{\mathcal{F}_L(U)})$ and $\mathcal{P}(U) \otimes \mathcal{M}_L$ are isomorphic.*

Proof By definition of the product of textures, every element of $\mathcal{P}(U) \otimes \mathcal{M}_L$ is an arbitrary intersection of the sets

$$(A \times M_L) \cup (U \times \mathfrak{I}(r))$$

where $A \subseteq U$ and $r \in M_L$ (see, Theorem 3.2.4). We show that $(A \times M_L) \cup (U \times \mathfrak{I}(r)) \in \mathcal{M}_{\mathcal{F}(U)}$. Consider the fuzzy set $\mu : U \to L$ defined by

$$\mu(u) = \begin{cases} 1, & \text{if } u \in A \\ r, & \text{if } u \in U \setminus A. \end{cases}$$

Note that the fuzzy points of $\mathcal{F}_L(U)$ are the only molecules of $\mathcal{F}_L(U)$. Clearly, for all $u \in A$ and $r \in M_L$, we have $u_r \leq \mu$. Further, we also have $v_s \leq \mu$ for all $v \in U$ and $s \leq r$. Then we obtain

$$\begin{aligned} \mathfrak{I}(\mu) &= \{v_s \mid v_s \leq \mu\} \\ &= \{(v, s) \mid v_s \leq \mu\}) \\ &= (A \times M_L) \cup (U \times \mathfrak{I}(r)) \in \mathcal{M}_{\mathcal{F}_L(U)}. \end{aligned}$$

This means that $\mathcal{M}_{\mathcal{F}_L(U)}$ is coarser than $\mathcal{P}(U) \otimes \mathcal{M}_L$. Now we show that for $\alpha \in \mathcal{F}_L(U)$,

$$\Im(\alpha) = \bigcap_{u \in U} (U \setminus \{u\} \times M_L) \cup (U \times \Im(\alpha(u)).$$

Suppose that for some fuzzy point v_r, we have $v_r \in \Im(\alpha)$, that is,

$$\Im(\alpha) = \bigcap_{u \in U} (U \setminus \{u\} \times M_L) \cup (U \times \Im(\alpha(u))).$$

Then we have $v_r \notin (U \setminus \{u\} \times M_L) \cup (U \times \Im(\alpha(u))$ for some $u \in U$. Namely,

$$v_r \notin ((U \setminus \{u\}) \times M_L)) \text{ and } v_r \notin (U \times \Im(\alpha(u)).$$

This clearly implies that $v = u$ and $\alpha(u) < r$. However, this is a contradiction since $v_r \leq \alpha$. The reverse inclusion is similar. In this case, from definition of product texture, $\Im(\alpha) \in \mathcal{P}(U) \otimes \mathcal{M}_L$. Since $\Im$ is a lattice isomorphism, the above arguments prove the existence of a textural isomorphism between the textures $\mathcal{M}_{\mathcal{F}_L(U)}$ and $\mathcal{P}(U) \otimes \mathcal{M}_L$. □

Note that the mapping $f : M_{\mathcal{F}(U)} \to U \times M_L$ defined by $v_r \mapsto (v, r)$, for all $v_r \in M_{\mathcal{F}(U)}$ is a set isomorphism, that is, f is an injective and surjective function. Therefore, $\mathcal{M}'_{\mathcal{F}(U)}$ is also a texturing on $U \times M_L$ where $\mathcal{M}'_{\mathcal{F}(U)} = \{\Im'(\alpha) \mid \alpha \in \mathcal{F}(U)\}$ with $\Im'(\alpha) = \{(v, s) \mid v_s \leq \mu\}$ for all $\alpha \in \mathcal{F}(U)$. Therefore, the textures $(M_{\mathcal{F}_L(U)}, \mathcal{M}_{\mathcal{F}(U)})$ and $(M'_{\mathcal{F}_L(U)}, \mathcal{M}'_{\mathcal{F}_L(U)})$ are clearly textural isomorphic. Hence, in the proof of the above theorem, every fuzzy point u_r can be regarded as an ordered pair (u, r) where $u \in U$ and $r \in M_L$.

Fuzzy Subset Texturing $\mathcal{P}(U) \otimes \mathcal{M}$

We denote the p-sets and q-sets of the texture $\mathcal{P}(U) \otimes \mathcal{M}$ by $\overline{P}_{(u,\lambda)}$ and $\overline{Q}_{(u,\lambda)}$. It is easy to see that

$$\overline{P}_{(u,\lambda)} = \{u\} \times (0, \lambda] \text{ and } \overline{Q}_{(u,\lambda)} = ((U \setminus \{u\}) \times U) \cup (U \times (0, \lambda]).$$

By the lattice isomorphism $\Im : \mathcal{F}(U) \to \mathcal{P}(U) \otimes \mathcal{M}$, we have the following result:

Lemma 3.3.2

$$\Im(u_\lambda) = \overline{P}_{(u,\lambda)} \text{ and } \Im(u^\lambda) = \overline{Q}_{(u,\lambda)}.$$

Proof For the first equality, we easily observe that

$$\begin{aligned} P_{(u,\lambda)} &= \bigcap \{\Im(\alpha) \mid u_\lambda \in \Im(\alpha)\} = \bigcap \{\Im(\alpha) \mid u_\lambda < \alpha\} \\ &= \bigcap \{\Im(\alpha) \mid \lambda < \alpha(u)\} = \Im(\bigwedge \{\alpha \mid \lambda < \alpha(u)\}) \\ &= \Im(u_\lambda). \end{aligned}$$

$$
\begin{aligned}
Q_{(u,\lambda)} &= \bigvee\{\mathfrak{I}(\alpha) \mid u_\lambda \notin \mathfrak{I}(\alpha)\} = \bigvee\{\mathfrak{I}(\alpha) \mid \alpha < u_\lambda\} \\
&= \bigvee\{\mathfrak{I}(\alpha) \mid \alpha(u) < \lambda\} = \mathfrak{I}(\bigvee\{\alpha \mid \lambda < \alpha(u)\}) \\
&= \mathfrak{I}(u^\lambda).
\end{aligned}
$$

Note that if $A \in \mathcal{W}_U$, then for some $\alpha \in \mathcal{F}(U)$, we have $A = \mathfrak{I}(\alpha)$. As a result, we have the following equalities.

Corollary 3.3.3

$$
\mathfrak{I}(\alpha) = \bigvee_{\mathfrak{I}(\alpha) \not\subseteq \overline{Q}_{(u,\lambda)}} \overline{P}_{(u,\lambda)} = \bigcap_{\overline{P}_{(u,\lambda)} \not\subseteq \mathfrak{I}(\alpha)} \overline{Q}_{(u,\lambda)}.
$$

Remark 3.3.4 By Theorem 3.2.8, note that for all $D \in \mathcal{P}(U) \otimes \mathcal{M}$,

$$
\begin{aligned}
c_{U\times(0,1]}(D) &= \bigcap_{(u,\lambda)\in D} ((c_1(P_u) \times U_2) \cup (U_1 \times c_2(P_\lambda))) \\
&= \bigcap_{(u,\lambda)\in D} ((c_1(\{u\})) \times U_2) \cup (U_1 \times c_2((0,\lambda])) \\
&= \bigcap_{(u,\lambda)\in D} ((U \setminus \{u\})) \times U_2) \cup (U_1 \times (0, 1-\lambda]).
\end{aligned}
$$

3.4 Orthopair (Intuitionistic) Textures

Fuzzy lattice of orthopairs

The notion of orthopair is used for the representation of uncertainty by D. Ciucci (See, e.g, [2–4].) It is a classical version of intuitionistic fuzzy sets produced by Atanassov [5] which is also called a double set by Çoker in [6]. Let U be a domain of discourse and $A, B \subseteq U$. Then the pair (A, B) is called *an orthopair* if $A \cap B = \emptyset$. Let us consider the collection

$$
O(U) = \{(A, B) \mid A, B \subseteq U, A \cap B = \emptyset\}
$$

of all orthopairs. Orthopair texture was first presented in [7]. A revised version was given in [8]. Every rough set uniquely determines an orthopair since the pair of lower approximation and exterior of a set is an orthopair, and vice versa (see Sect. 5.1). This makes the orthopair texture interesting since this argument may give different directions in the framework of textures and the rough set theory. Here, we observe that the family $O(U)$ is a fuzzy lattice. Therefore, by Brown's Representation Theorem, there exists a complemented fuzzy texture denoted by

$$
(M_{O(U)}, \mathcal{M}_{O(U)}, c_{M_{O(U)}})
$$

corresponding to $O(U)$. It is remarkable to say that this texture is the same as the product texture $\mathcal{P}(U) \otimes \mathcal{M}_L$ where $L = \{0, \frac{1}{2}, 1\}$. Moreover, since the three valued fuzzy lattice $\mathcal{F}_{\frac{1}{2}}(U)$ and $O(U)$ are isomorphic fuzzy lattices, the corresponding texture to $\mathcal{F}_{\frac{1}{2}}(U)$ is an orthopair texture as we will prove in the sequel.

Now, we may give the following result.

Theorem 3.4.1 *The family $O(U)$ is a fuzzy lattice with the order defined by*

$$(A, B) \subseteq (C, D) \Longleftrightarrow A \subseteq C \text{ and } D \subseteq B.$$

Proof The top and bottom elements of $O(U)$ are $(U, \emptyset)$ and $(\emptyset, U)$. Further, for any subfamily $\mathcal{A} = \{(A_j, B_j) \mid j \in J\} \subseteq O(U)$, it is easy to see that the least upper bound and greatest lower bound of $\mathcal{A}$ are

$$\bigvee_{j\in J}(A_j, B_j) = (\bigcup_{j\in J} A_j, \bigcap_{j\in J} B_j)$$

and

$$\bigwedge_{j\in J}(A_j, B_j) = (\bigcap_{j\in J} A_j, \bigcup_{j\in J} B_j),$$

respectively. Therefore, $(O(U), \subseteq)$ is a complete lattice. Further, trivially, the mapping $' : O(U) \to O(U)$ defined by

$$\forall (A, B) \in O(U),\ (A, B)' = (B, A)$$

is a complementation on $O(U)$. Now, using Raney's theorem, we show that the lattice $(O(U), \subseteq, ')$ is completely distributive. Let $(A, B), (C, D) \in O(U)$ and $(A, B) \not\subseteq (C, D)$. Then we have $A \not\subseteq C$ or $D \not\subseteq B$. Without loss of generality, we may suppose that $A \not\subseteq C$. Then there exists a $u \in U$ such that $u \in A$ and $u \notin C$. We clearly have

$$(A, B) \not\subseteq (U \setminus \{u\}, \emptyset\}) \text{ and } (\{u\}, U \setminus \{u\}) \not\subseteq (C, D).$$

That is the first condition Raney's theorem holds. For the second condition, note that the implication

$$(K, L) \in O(U) \text{ and } (K, L) \not\subseteq (U \setminus \{u\}, \emptyset) \implies (\{u\}, U \setminus \{u\}) \subseteq (K, L)$$

also holds since $K \cap L = \emptyset$. As a result, the family $O(U)$ of orthopairs with the given order above is a fuzzy lattice. □

Orthopair Texture

The fuzzy lattice $O(U)$ of orthopairs has two types of molecules. Therefore, the collection of all molecules of $O(U)$ is

$$M_{O(U)} = \{(U \setminus \{u\}, \{u\}) \mid u \in U\} \bigcup \{(\emptyset, U \setminus \{u\}) \mid u \in U\}.$$

Then by Brown's Representation Theorem, the family

$$\mathcal{M}_{O_U} = \{\Im(A, B) \mid (A, B) \in O_U\}$$

is a simple complemented texture on M_{O_U} corresponding to the fuzzy lattice $O(U)$ where the lattice isomorphism $\Im : O_U \to \mathcal{M}_{O_U}$ is defined by

$$\Im(A, B) = \{(C, D) \mid (C, D) \subseteq (A, B) \text{ and } (C, D) \in M_{O_U}\}$$

for all $(A, B) \in O(U)$. It is easy to see that

$$\Im(A, B) = \{(\{u\}, U \setminus \{u\}) \mid u \in A\} \cup \{(\emptyset, U \setminus \{u\}) \mid u \in U \setminus B\}.$$

The mapping $c_{M_{O(U)}} : \mathcal{M}_{O(U)} \to \mathcal{M}_{O(U)}$ defined by

$$\forall (A, B) \in \mathcal{M}_{O(U)},\ (\Im(A, B))' = \Im(B, A)$$

is a complementation on $\mathcal{M}_{O(U)}$ where $(A, B) \in O(U)$.

As we will see in the sequel, the texture given in the following theorem is same as the orthopair texture $(M_{O(U)}, \mathcal{M}_{O(U)}, c_{M_{O(U)}})$ up to textural isomorphism.

Theorem 3.4.2 *The triple $(D_U, \mathcal{D}_U, c_{D(U)})$ is a simple complemented texture where*

$$D_U = (U \times \{1\}) \cup (U \times \{\tfrac{1}{2}\}) = U \times \{1, \tfrac{1}{2}\},$$

$$\mathcal{D}_U = \{(A \times \{1\}) \cup ((U \setminus B) \times \{\tfrac{1}{2}\}) \mid (A, B) \in O(U)\} \text{ and}$$

$$c_{D_U}((A \times \{1\}) \cup ((U \setminus B) \times \{\tfrac{1}{2}\}) = (B \times \{1\}) \cup ((U \setminus A) \times \{\tfrac{1}{2}\})$$

for all $(A, B) \in O(U)$.

Proof Let $\{(A_j \times \{1\}\{\}) \cup ((U \setminus B_j) \times \{\tfrac{1}{2}\}) \mid j \in J\} \subseteq \mathcal{D}_U$. Then

$$\bigcup_{j \in J}((A_j \times \{1\}) \cup ((U \setminus B_j) \times \{\tfrac{1}{2}\}))$$
$$= ((\bigcup_{j \in J} A_j) \times \{1\}) \cup ((U \setminus \bigcap_{j \in J} B_j) \times \{\tfrac{1}{2}\}) \in \mathcal{D}_U$$

since we have $(\bigcup_{j \in J} A_j, \bigcap_{j \in J} B_j) \in O_U$. Using a similar way, we also conclude that

$$\bigcap_{j \in J}((A_j \times \{1\}\}) \cup (U \setminus B_j) \times \{\tfrac{1}{2}\})) \in \mathcal{D}_U.$$

Then by Theorem 1.1.1, $(\mathcal{D}_U, \subseteq)$ is a complete lattice. Note that $\mathcal{D}_U$ is closed under arbitrary unions and intersections. Therefore, $(\mathcal{D}_U, \subseteq)$ is also completely distributive,. since $(\mathcal{P}(D_U), \subseteq)$ is completely distributive and $\mathcal{D}_U \subseteq \mathcal{P}(D_U)$ (see, Theorem 1.2.2).

On the other hand, $\mathcal{D}_U$ separates the points of $U \times \{1, \frac{1}{2}\}$. Take $(u_1, 1), (u_2, 1) \in D_U$ with $u_1 \neq u_2$. From definition, $(U \setminus \{u_1\}) \times \{1\} \in \mathcal{D}_U$ and $(u_2, 1) \in (U \setminus \{u_1\}) \times \{1\}$ with $(u_1, 1) \notin (U \setminus \{u_1\}) \times \{1\}$. Further, if $(u_1, \frac{1}{2}), (u_2, \frac{1}{2}) \in D_U$ with $u_1 \neq u_2$, then from definition of $\mathcal{D}_U$,

$$(\{u_1\} \times \{\tfrac{1}{2}\}) \cup (\{u_1\} \times \{1\}) = \{(u_1, \tfrac{1}{2}), (u_1, 1)\} \in \mathcal{D}_U.$$

Finally, $(u_2, 1) \notin \{(u_1, \frac{1}{2}), (u_1, 1)\}$, and $(u_1, \frac{1}{2}) \in \{(u_1, \frac{1}{2}), (u_1, 1)\}$, that is, $\mathcal{D}_U$ separates the points of D_U. As a result, $\mathcal{D}_U$ is a texturing on D_U.

Now let us determine the p-sets and q-sets of the texturing $\mathcal{D}_U$. Take a point $(u, 1) \in D_U$. Since $(\{ u\}, U \setminus \{u\}) \in O(U)$,

$$(\{u\} \times \{1\}) \cup (\{u\} \times \{\tfrac{1}{2}\}) = \{(u, 1), (u, \tfrac{1}{2})\}$$

and hence, $P_{(u,1)} = \{(u, 1), (u, \frac{1}{2})\}$. It is easy to see that for a point $(u, \frac{1}{2}) \in D_U$, the p-set is $P_{(u,\frac{1}{2})} = \{(u, \frac{1}{2})\}$. Let $X = A \times \{\frac{1}{2}\} \cup ((U \setminus B) \times \{1\}) \in \mathcal{D}_U$ where $(A, B) \in O(U)$. and $A \neq \emptyset$. Clearly, there exists a $u \in U$ such that $(u, 1) \in X$. Take $v \in U$ with $v \neq u$. Let us consider the sets

$$Y = ((A \setminus \{u\}) \times \{\tfrac{1}{2}\}) \cup (U \setminus (B \cup \{u\})) \times \{1\}),$$

$$Z = ((A \setminus \{v\}) \times \{\tfrac{1}{2}\}) \cup (U \setminus (B \cup \{v\})) \times \{1\}).$$

Note that $X \subseteq Y \cup Z$ and $X \not\subseteq Y$. If $v \in A$, then $(v, \frac{1}{2}) \in X$ and $(v, \frac{1}{2}) \notin Z$ and so, $X \not\subseteq Z$. Similarly, if $v \notin B$, then $(v, 1) \in X$ and hence, we also have $X \not\subseteq Z$. This means that $X = \{(u, 1)\} = P_{(u,1)}$ or $X = \{(u, \frac{1}{2}), (u, 1)\} = P_{(u,\frac{1}{2})}$. Consequently, $(D_U, \mathcal{D}_U)$ is a simple texture.

Let us consider the mapping $c_{D_U} : \mathcal{D}_U \to \mathcal{D}_U$ defined by

$$c_{D_U}((A \times \{\tfrac{1}{2}\}) \cup ((U \setminus B) \times \{1\}) = (B \times \{\tfrac{1}{2}\}) \cup ((U \setminus A) \times \{1\})$$

for all $(A, B) \in O_U$. Clearly, for all $X \in \mathcal{D}_U$, $c_{D_U}(c_{D_U}(X)) = X$. If

$$X = (A \times \{\tfrac{1}{2}\}) \cup ((U \setminus B) \times \{1\}) \subseteq (C \times \{\tfrac{1}{2}\}) \cup ((U \setminus D) \times \{1\}) = Y,$$

then $A \subseteq C$ and $U \setminus B \subseteq U \setminus D$, that is, $U \setminus C \subseteq U \setminus A$ and $D \subseteq B$. Therefore,

$$(D \times \{\tfrac{1}{2}\}) \cup ((U \setminus C) \times \{1\}) \subseteq (B \times \{\tfrac{1}{2}\}) \cup ((U \setminus A) \times \{1\})$$

implies that $c_{D_U}(Y) \subseteq c_{D(U)}(X)$. We have proved that $(D_U, \mathcal{D}_U, c_{D(U)})$ is complemented and this completes the proof.

Theorem 3.4.3 *The textures $(M_{O(U)}, \mathcal{M}_{O(U)}, c_{\mathcal{M}_{O(U)}})$ and $(D_U, \mathcal{D}_U, c_{D(U)})$ are isomorphic.*

Proof Let us consider the mapping $\psi : M_{O(U)} \to D_U$ defined by

$$\psi(X) = \begin{cases} (u, 1), & \text{if } X = (\{u\}, U \setminus \{u\}) \\ (u, \frac{1}{2}), & \text{if } X = (\emptyset, U \setminus \{u\}) \end{cases}$$

for all $X \in M_{O(U)}$. We show that ψ is a complemented textural isomorphism. Clearly, $\psi : M_{O(U)} \to D_U$ is one-to-one and onto. Further, for the mapping ψ, we also have

$$\begin{aligned} \psi(\mathfrak{I}(A, B)) &= \psi(\{(\{u\}, U \setminus \{u\}) \mid u \in A\} \cup \{(\emptyset, U \setminus \{u\}) \mid u \in U \setminus B\}) \\ &= A \times \{1\} \cup ((U \setminus B) \times \{\tfrac{1}{2}\}) \in \mathcal{D}_U \end{aligned}$$

for all $\mathfrak{I}(A, B) \in \mathcal{M}_{O(U)}$. We can also easily see that $\psi : \mathcal{M}_{O(U)} \to \mathcal{D}_U$ is one-to-one and onto. Consequently, ψ is a textural isomorphism.

Now, let us consider the fuzzy lattice $L = \{0, \frac{1}{2}, 1\}$ and the three valued fuzzy lattice

$$\mathcal{F}_L(U) = \mathcal{F}_{\frac{1}{2}}(U) = \{f \mid f : U \to \{0, \tfrac{1}{2}, 1\} \text{ is a mapping}\}.$$

Theorem 3.4.4 *The fuzzy lattices $\mathcal{F}_{\frac{1}{2}}(U)$ and $O(U)$ are isomorphic.*

Proof Let us consider the mapping $\varphi : \mathcal{F}_{\frac{1}{2}}(U) \to O(U)$ defined by

$$\forall \alpha \in M_{\mathcal{F}_{\frac{1}{2}}(U)}, \quad \varphi(\alpha) = (A_\alpha, B_\alpha)$$

where

$$A_\alpha = \{u \in U \mid \alpha(u) = 1\} \text{ and } B_\alpha = \{u \in U \mid \alpha(u) = 0\}.$$

Note that $A_\alpha \cap B_\alpha = \emptyset$ and hence, $(A_\alpha, B_\alpha) \in O(U)$. Clearly, φ is injective. For surjectivity, let $(A, B) \in O(U)$. Let us take the mapping $\alpha : U \to \{0, \frac{1}{2}, 1\}$ defined by

$$\alpha(u) = \begin{cases} 1, & \text{if } u \in A, \\ \frac{1}{2}, & \text{if } u \in U \setminus (A \cup B), \\ 0, & \text{if } u \in B. \end{cases}$$

Then $A = A_\alpha$ and $B = B_\alpha$. That is, $\varphi(\alpha) = (A, B)$. Further, note that for some subfamily $\{\alpha_j \mid j \in J\} \subseteq \mathcal{F}_{\frac{1}{2}}(U)$,

$$A_{\bigvee_{j \in J} \alpha_j} = \bigcup_{j \in J} A_{\alpha_j} \text{ and } A_{\bigwedge_{j \in J} \alpha_j} = \bigcap_{j \in J} A_{\alpha_j}.$$

Indeed, if $u \in A_{\bigvee_{j \in J} \alpha_j}$, then $\bigvee_{j \in J} \alpha_j(u) = 1$, and since $L = \{0, \frac{1}{2}, 1\}$ is finite, for some $j_0 \in J$, we have $\alpha_{j_0}(u) = 1$. Hence, $u \in A_{\alpha_{j_0}}$ and this implies that $u \in A_{\alpha_{j_0}} \subseteq \bigcup_{j \in J} A_{\alpha_j}$. The proof of the reverse inclusion and the second equality is similar. Hence, ϕ is a lattice isomorphism. □

Theorem 3.4.5 *The fuzzy texture of* $\mathcal{F}_{\frac{1}{2}}(U)$ *is the orthopair texture* $(D_U, \mathcal{D}_U, c_{D(U)})$.

Proof We know that there exists a fuzzy texture $(M_{\mathcal{F}_{\frac{1}{2}}(U)}, \mathcal{M}_{\mathcal{F}_{\frac{1}{2}}(U)}, c_{M_{\mathcal{F}_{\frac{1}{2}}(U)}})$ of $\mathcal{F}_{\frac{1}{2}}(U)$ obtained via Brown's Representation Theorem. We show that this texture is the same as $(D_U, \mathcal{D}_U, c_{D(U)})$. Let $\alpha \in M_{\mathcal{F}_{\frac{1}{2}}(U)}$. Then $\mathfrak{I}(\alpha) = \{(u, r) \mid r \leq \alpha(u)\}$. Define the sets A_α, B_α where

$$A_\alpha = \{u \in U \mid \alpha(u) = 1\} \text{ and } B_\alpha = \{u \in U \mid \alpha(u) = 0\}.$$

Let us take a $(u, r) \in \mathfrak{I}(\alpha)$. Note that we have two cases: $\alpha(u) = \frac{1}{2}$, $\alpha(u) = 1$. If $\alpha(u) = \frac{1}{2}$, then we also have $r = \frac{1}{2}$. This implies that $(u, r) \in (U \setminus B_\alpha) \times \{\frac{1}{2}\}$. Let $\alpha(u) = 1$. Then $r = 1$ and $r = \frac{1}{2}$. In this case, $(u, 1) \in A_\alpha \times \{1\}$ and $(u, \frac{1}{2}) \in (U \setminus B_\alpha) \times \{\frac{1}{2}\}$. This gives the inclusion

$$\mathfrak{I}(\alpha) \subseteq (A_\alpha \times \{1\}) \cup ((U \setminus B_\alpha) \times \{\tfrac{1}{2}\}).$$

For the reverse inclusion, let $(u, r) \in (A_\alpha \times \{1\}) \cup ((U \setminus B_\alpha) \times \{\frac{1}{2}\})$. Then

$$u \in A_\alpha,\ r = 1 \ \text{ or } \ u \in U \setminus B_\alpha,\ r = \frac{1}{2}.$$

Both of the cases above, gives us $r \leq \alpha(u)$, namely, $(u, r) \in \mathfrak{I}(\alpha)$. This means that

$$\mathfrak{I}(\alpha) = (A_\alpha \times \{1\}) \cup ((U \setminus B_\alpha) \times \{\tfrac{1}{2}\}) \in \mathcal{D}_U.$$

Therefore, we get $\mathcal{M}_{\mathcal{F}_{\frac{1}{2}}(U)} = \{\mathfrak{I}(\alpha) \mid \alpha \in M_{\mathcal{F}_{\frac{1}{2}}(U)}\} \subseteq \mathcal{D}_U$. Now, let $(A, B) \in O(U)$. Take $(A \times \{1\}) \cup ((U \setminus B) \times \{\frac{1}{2}\}) \in \mathcal{D}_U$. Define the mapping $\alpha : U \to \{0, \frac{1}{2}, 1\}$ as

$$\alpha(u) = \begin{cases} 1, & \text{if } u \in A, \\ \frac{1}{2}, & \text{if } u \in U \setminus (A \cup B), \\ 0, & \text{if } u \in B. \end{cases}$$

Then $A = A_\alpha$ and $B = B_\alpha$. It is easy to see that

$$(A_\alpha \times \{1\}) \cup ((U \setminus B_\alpha) \times \{\tfrac{1}{2}\}) = \mathfrak{I}(\alpha) \in \mathcal{M}_{\mathcal{F}_{\frac{1}{2}}(U)}.$$

Consequently, we conclude that $\mathcal{M}_{\mathcal{F}_{\frac{1}{2}}(U)} = \mathcal{D}_U$. For the complementation, note that

$$A_{\alpha'} = B_\alpha \text{ and } B_{\alpha'} = A_\alpha.$$

For the first equality, it is enough to observe that

$$u \in A_{\alpha'} \Longleftrightarrow \alpha'(u) = 1 \Longleftrightarrow 1 - \alpha(u) = 1 \Longleftrightarrow \alpha(u) = 0 \Longleftrightarrow u \in B_\alpha.$$

The proof of the second equality is similar. Therefore,

$$\begin{aligned}\mathfrak{I}(\alpha') &= (A_{\alpha'} \times \{1\}) \cup ((U \setminus B_{\alpha'}) \times \{\tfrac{1}{2}\}) \\ &= (B_{\alpha} \times \{1\}) \cup ((U \setminus A_{\alpha}) \times \{\tfrac{1}{2}\}) \\ &= c_{D_U}((A_{\alpha} \times \{1\}) \cup ((U \setminus B_{\alpha}) \times \{\tfrac{1}{2}\})) \\ &= c_{D_U}(\mathfrak{I}(\alpha)).\end{aligned}$$

This completes the proof.

Corollary 3.4.6 *The textures* $(M_{\mathcal{F}_{\frac{1}{2}}(U)}, \mathcal{M}_{\mathcal{F}_{\frac{1}{2}}(U)})$ *and* $(U \times M_L, \mathcal{P}(U) \otimes \mathcal{M}_L)$ *are isomorphic where* $L = \{0, \frac{1}{2}, 1\}$.

Proof From Theorem 3.3.1, the proof is immediate.

References

1. G. Gierz, K.H. Hofmann, K. Keimel, J.D. Lawson, M.W. Mislove, D.S. Scott, *A Compendium of Continuous Lattices* (Springer, Berlin, 1980)
2. D. Ciucci, Orthopairs: a simple and widely used way to model uncertainty. Fundam. Inform. **108**(34), 287304 (2011)
3. D. Ciucci, Orthopairs in the 1960s: historical remarks and new ideas, in *Proc RSCTC, LNCS*, vol. 8536, eds. by C. Cornelis, M. Kryszkiewicz, D. Slezak, E.M. Ruiz, R. Bello, L. Shang (Cham, 2014)
4. D. Ciucci, Orthopairs and granular computing. Granul. Comput. **1**, 159–170 (2016)
5. K.T. Atanassov, Intuitionistic fuzzy sets. Fuzzy Sets Syst. **20**(1), 87–96 (1986)
6. D. Çoker, A note on intuitionistic sets and intuitionistic points. Turk J. Math. **20**, 343351 (1996)
7. L.M. Brown, M. Diker, Ditopological texture spaces and intuitionistic sets. Fuzzy Sets Syst. **98**, 217224 (1998)
8. Ş. Dost, L.M. Brown, Intuitionistic textures revisited. Hacettepe J. Math. Stat. **34**(S), 115–130 (2005)

Chapter 4
Direlations

4.1 Basic Concepts

Let us consider the product texture $(U \times V, \mathcal{P}(U) \otimes \mathcal{V})$ where $(U, \mathcal{P}(U))$ is discrete and $(V, \mathcal{V})$ is a general texture. The texturing $\mathcal{P}(U) \otimes \mathcal{V}$ also plays an important role for the morphisms between textures. An ordinary *binary relation* r from a set U to a set V is a subset of $U \times V$. In other words, $r \in \mathcal{P}(U \times V) = \{A \mid A \subseteq U \times V\}$. Trivially, we have $\mathcal{P}(U \times V) = \mathcal{P}(U) \times \mathcal{P}(V)$. If we take the discrete textures $(U, \mathcal{P}(U))$ and $(V, \mathcal{P}(V))$, then we also have that $r \in \mathcal{P}(U) \otimes \mathcal{P}(V)$ since

$$\mathcal{P}(U) \times \mathcal{P}(V) = \mathcal{P}(U) \otimes \mathcal{P}(V)$$

where $\times$ is the ordinary cartesian product and $\otimes$ is the textural product. In this case, we may probably expect that $r \in \mathcal{U} \otimes \mathcal{V}$ where $(U, \mathcal{U})$ and $(V, \mathcal{V})$ are any two textures. However, this choice leads us to some inappropriate restrictions. For instance, for the fuzzy texture

$$\mathcal{M}_{[0,1]} = \{(0, u] \mid u \in [0, 1]\}$$

on $M_{[0,1]} = (0, 1]$, the textural product $\mathcal{M}_{[0,1]} \otimes \mathcal{M}_{[0,1]}$ is not sufficiently large since for all $D \in \mathcal{M}_{[0,1]} \otimes \mathcal{M}_{[0,1]}$,

$$\Delta = \{(u, u) \mid u \in (0, 1]\} \not\subseteq D$$

where $D \neq (0, 1] \times (0, 1]$. Therefore, the imposition of the product $\mathcal{U} \otimes \mathcal{V}$ may not be convenient for a useful theory. As a result, in place of $\mathcal{U} \otimes \mathcal{V}$, we take the product $\mathcal{P}(U) \otimes \mathcal{V}$ where $\mathcal{P}(U)$ is the discrete texture on U. However, in this case, we have to establish the connection between the textures $\mathcal{U}$ and $\mathcal{V}$ using additional conditions in terms of dual notions of p-set and q-set. For this establishment, let us first determine the p-set and q-set of the product texture $\mathcal{P}(U) \otimes \mathcal{V}$ on $U \times V$.

M. Diker, *Texture Spaces*, Studies in Fuzziness and Soft Computing 411,
https://doi.org/10.1007/978-3-031-39748-6_4

For any $(u, v) \in U \times V$, we denote the p-set and q-set of $\mathcal{P}(U) \otimes \mathcal{V}$ by $\overline{P}_{(u,v)}$ and $\overline{Q}_{(u,v)}$, respectively.

Corollary 4.1.1 *For the product texture $(U \times V, \mathcal{P}(U) \otimes \mathcal{V})$, we have the following equalities.*

$$\forall (u, v) \in U \times V,\ \overline{P}_{(u,v)} = \{u\} \times P_v \text{ and } \overline{Q}_{(u,v)} = ((U \setminus \{u\}) \times V) \cup (U \times Q_v).$$

Proof For the discrete texture $\mathcal{P}(U)$, we have $P_u = \{u\}$ and $Q_u = U \setminus \{u\}$ (see Example 2.2.4 (i)). Then by Theorem 3.2.6, we get
$\overline{P}_{(u,v)} = P_u \times P_v = \{u\} \times P_v$
and
$\overline{Q}_{(u,v)} = (U \times Q_v) \cup (Q_u \times V) = (U \times Q_v) \cup (U \setminus \{u\} \times V)$.
For the p-set and q-set of the product $\mathcal{P}(U) \otimes \mathcal{V}$, the following equivalence will be useful in the sequel.

Lemma 4.1.2 *Let $u, u' \in U$ and $v, v' \in V$. Then*

$$\overline{P}_{(u,v)} \not\subseteq \overline{Q}_{(u',v')} \text{ if and only if } u = u' \text{ and } P_v \not\subseteq Q_{v'}.$$

Proof Let $\overline{P}_{(u,v)} \not\subseteq \overline{Q}_{(u',v')}$. By Corollary 4.1.1,

$$\{u\} \times P_v \not\subseteq ((U \setminus \{u'\}) \times V) \cup (U \times Q_{v'})),$$

that is, $\{u\} \times P_v \not\subseteq U \setminus \{u'\} \times V$ and $\{u\} \times P_v \not\subseteq U \times Q_{v'}$. Then we clearly get $u = u'$ and $P_v \not\subseteq Q_{v'}$. For the sufficiency, let $P_v \not\subseteq Q_{v'}$. Then

$$\{u\} \times P_v \not\subseteq U \setminus \{u'\} \times Q_{v'},$$

where $u = u'$. That is, $\overline{P}_{(u,v)} \not\subseteq \overline{Q}_{(u',v')}$.

Example 4.1.3 Let us consider the product texture $\mathcal{P}(U) \otimes \mathcal{M}_{[0,1]}$ on $U \times (0, 1]$ where $\mathcal{P}(U)$ is the discrete texturing on U and $\mathcal{M}_{[0,1]}$ is the fuzzy texturing on $(0, 1]$. Take $u, u' \in U$ and $\lambda, \lambda' \in (0, 1]$. Then we have

$$\overline{P}_{(u,\lambda)} \not\subseteq \overline{Q}_{(u',\lambda')} \text{ if and only if } u = u' \text{ and } \lambda' < \lambda.$$

Indeed, if $\overline{P}_{(u,\lambda)} \not\subseteq \overline{Q}_{(u',\lambda')}$, then

$$\{u\} \times (0, \lambda] \not\subseteq ((U \setminus \{u'\}) \times (0, 1]) \cup (U \times (0, \lambda']).$$

Hence, we get

$$\{u\} \times (0, \lambda] \not\subseteq (U \setminus \{u'\}) \times (0, 1]) \text{ and } \{u\} \times (0, \lambda] \not\subseteq U \times (0, \lambda'].$$

This clearly gives that $u = u'$ and $\lambda' < \lambda$. The proof of the sufficiency part is similar.

Now, we are ready to give the following definition.

Definition 4.1.4

$r \in \mathcal{P}(U) \otimes \mathcal{V}$ is called a *relation* from $(U, \mathcal{U})$ to $(V, \mathcal{V})$ if

R1 $r \not\subseteq \overline{Q}_{(u,v)}$, $P_{u'} \not\subseteq Q_u \implies r \not\subseteq \overline{Q}_{(u',v)}$, and

R2 $r \not\subseteq \overline{Q}_{(u,v)} \implies \exists u' \in U$ such that $P_u \not\subseteq Q_{u'}$ and $r \not\subseteq \overline{Q}_{(u',v)}$.

Note that in the above definition, for $u, u' \in U$, $\overline{Q}_{(u',v)}$ and $\overline{Q}_{(u,v)}$ are the q-sets of the texture $\mathcal{P}(U) \otimes \mathcal{V}$. However, P_u and $P_{u'}$ are the p-sets of the texture $(U, \mathcal{U})$ and this provides a plausible connection between the textures $(U, \mathcal{U})$ and $(V, \mathcal{V})$.

As we know, the notions of p-set and q-set are dual concepts and hence, naturally we may obtain dual correspondence of the relation r as follows:

Definition 4.1.5

$R \in \mathcal{P}(U) \otimes \mathcal{V}$ is called a *corelation* from $(U, \mathcal{U})$ to $(V, \mathcal{V})$ if

CR1 $\overline{P}_{(u,v)} \not\subseteq R$, $P_u \not\subseteq Q_{u'} \implies \overline{P}_{(u',v)} \not\subseteq R$, and

CR2 $\overline{P}_{(u,v)} \not\subseteq R \implies \exists u' \in U$ such that $P_{u'} \not\subseteq Q_u$ and $\overline{P}_{(u',v)} \not\subseteq R$.

Considering the Definitions 4.1.4 and 4.1.5, we obtain the desired morphisms between textures:

Definition 4.1.6 A pair (r, R) where r is a relation and R a corelation from $(U, \mathcal{U})$ to $(V, \mathcal{V})$ is called *a direlation* from $(U, \mathcal{U})$ to $(V, \mathcal{V})$. If (r, R) is a direlation from $(U, \mathcal{U})$ to $(U, \mathcal{U})$, then we say that (r, R) is a direlation on $(U, \mathcal{U})$.

Example 4.1.7 Let us take the texture $(U, \mathcal{U})$ where

$$U = \{a, b, c\} \text{ and } \mathcal{U} = \{\{a\}, \{a, b\}, \emptyset, U\}$$

and the texture $(V, \mathcal{V})$ where

$$V = \{d, e\} \text{ and } \mathcal{V} = \{\{d\}, \emptyset, V\}.$$

For the texturing $\mathcal{V}$, p-sets and q-sets are

$$P_e = V,\ P_d = \{d\},\ Q_e = \{d\} \text{ and } Q_d = \emptyset.$$

Now, it is easy to see that p-sets and q-sets of the product texturing $\mathcal{P}(U) \otimes \mathcal{V}$ are as follows:

$\overline{P}_{(a,d)} = \{a\} \times P_d = \{(a, d)\},\ \overline{P}_{(a,e)} = \{a\} \times V = \{(a, d), (a, e)\},$

$\overline{P}_{(b,d)} = \{a, b\} \times P_d = \{(a, d), (b, d)\}$, $\overline{P}_{(b,e)} = \{a, b\} \times V = \{(a, b), (b, d), (a, e), (b, e)\}$,

$\overline{P}_{(c,d)} = U \times P_d = \{(c, d)\}$, $\overline{P}_{(c,e)} = U \times V$,

and

$$\begin{aligned}\overline{Q}_{(a,d)} &= ((U \setminus \{a\}) \times V) \cup (U \times Q_d), & \overline{Q}_{(a,e)} &= ((U \setminus \{a\}) \times V) \cup (U \times Q_e)\\ &= (\{(b, c)\} \times V) \cup (U \times \emptyset) & &= (\{(b, c)\} \times V) \cup (U \times \{d\})\\ &= \{(b, d), (b, e), (c, d), (c, e)\}, & &= \{(b, d), (b, e), (c, d), (c, e), (a, d)\},\end{aligned}$$

$$\begin{aligned}\overline{Q}_{(b,d)} &= ((U \setminus \{b\}) \times V) \cup (U \times Q_d) & \overline{Q}_{(b,e)} &= (U \setminus \{b\} \times V) \cup (U \times Q_e)\\ &= (\{(a, c)\} \times V) \cup (U \times \emptyset) & &= (\{(a, c)\} \times V) \cup (U \times \{d\})\\ &= \{(a, d), (a, e), (c, d), (c, e)\}, & &= \{(a, d), (a, e), (c, d), (c, e), (b, d)\},\end{aligned}$$

$$\begin{aligned}\overline{Q}_{(c,d)} &= ((U \setminus \{c\}) \times V) \cup (U \times Q_d), & \overline{Q}_{(c,e)} &= ((U \setminus \{c\}) \times V) \cup (U \times Q_e)\\ &= (\{(a, b)\} \times V) \cup (U \times \emptyset) & &= (\{(a, b)\} \times V) \cup (U \times \{d\}),\\ &= \{(a, d), (a, e), (b, d), (b, e)\}, & &= \{(a, d), (a, e), (b, d), (b, e), (c, d)\}.\end{aligned}$$

Now let us show that the pair (r, R) where

$$r = \{(a, d), (b, e), (c, d), (c, e)\} \text{ and } R = \{(a, d), (b, d), (c, d)\}$$

is a direlation from $(U, \mathcal{U})$ to $(V, \mathcal{V})$. Since $P_a \not\subseteq Q_a$, $P_b \not\subseteq Q_b$ and $P_c \not\subseteq Q_c$, conditions R2 and CR2 are clear.

Let us check condition R1 for r. Consider the following cases:

$$r \not\subseteq \overline{Q}_{(a,d)},\ r \not\subseteq \overline{Q}_{(b,d)},\ r \not\subseteq \overline{Q}_{(b,e)},\ r \not\subseteq \overline{Q}_{(c,d)} \text{ and } r \not\subseteq \overline{Q}_{(c,e)}.$$

For $r \not\subseteq \overline{Q}_{(a,d)}$, we have $P_a \not\subseteq Q_a$, $P_b \not\subseteq Q_a$, $P_c \not\subseteq Q_a$. But, we already have $r \not\subseteq \overline{Q}_{(a,d)}$, $r \not\subseteq \overline{Q}_{(b,d)}$, $r \not\subseteq \overline{Q}_{(c,d)}$. For $r \not\subseteq \overline{Q}_{(b,d)}$, we have $P_b \not\subseteq Q_b$, $P_c \not\subseteq Q_b$ and $r \not\subseteq \overline{Q}_{(b,d)}$, $r \not\subseteq \overline{Q}_{(c,d)}$. Since we have $P_c \not\subseteq Q_c$, for $r \not\subseteq \overline{Q}_{(c,d)}$ and $r \not\subseteq \overline{Q}_{(c,e)}$, the condition R1 is clear. For $r \not\subseteq \overline{Q}_{(b,e)}$, we have $P_b \not\subseteq Q_b$, $P_c \not\subseteq Q_b$ and $r \not\subseteq \overline{Q}_{(b,e)}$, $r \not\subseteq \overline{Q}_{(c,e)}$.

For condition CR1 we have to check the following cases:

$$\overline{P}_{(a,e)} \not\subseteq R,\ \overline{P}_{(b,e)} \not\subseteq R \text{ and } \overline{P}_{(c,e)} \not\subseteq R.$$

For $\overline{P}_{(a,e)} \not\subseteq R$, we only have $P_a \not\subseteq Q_a$. For $\overline{P}_{(b,e)} \not\subseteq R$ we have $P_b \not\subseteq Q_a$. Then we already have $\overline{P}_{(a,e)} \not\subseteq R$. For $\overline{P}_{(c,e)} \not\subseteq R$, we have $P_c \not\subseteq Q_a$, $P_c \not\subseteq Q_b$, $\overline{P}_{(a,e)} \not\subseteq R$ and $\overline{P}_{(b,e)} \not\subseteq R$. This means that condition CR1 holds. Then we conclude that (r, R) is a direlation from $(U, \mathcal{U})$ to $(V, \mathcal{V})$.

Example 4.1.8 Consider the fuzzy texture $(M, \mathcal{M})$ and the unit texture $(I, \mathcal{I})$ where

$M = (0, 1],\ \mathcal{M} = \{(0, u] \mid r \in [0, 1]\},$

and

$I = [0, 1],\ \mathcal{I} = \{[0, u] \mid u \in [0, 1]\} \cup \{[0, u) \mid u \in [0, 1]\}.$

The pair (r, R) is a direlation from $(M, \mathcal{M})$ to $(I, \mathcal{I})$ where

$$r = \{(u, v) \mid v < u\} \text{ and } R = (0, 1] \times \{0\}.$$

Indeed, for $R1$, let $r \not\subseteq \overline{Q}_{(a,b)}$ and $P_{u'} \not\subseteq Q_a$. Then

$$\{(u, v) \mid v < u\} \not\subseteq ((0, 1] \setminus \{a\} \times [0, 1]) \cup ((0, 1] \times [0, b))$$

and $(0, u'] \not\subseteq (0, a]$. This gives that $b \leq v < u = a < u'$ for some $u \in (0, 1]$ and $v \in [0, 1]$. Note that $(u', v) \in r$ and $b \leq v$ implies that

$$r \not\subseteq \overline{Q}_{(u',b)} = ((0, 1] \setminus \{u'\} \times [0, 1]) \cup ((0, 1] \times [0, b))$$

as desired. For R2, let $r \not\subseteq \overline{Q}_{(a,b)}$. Then we have $b < a$ for some $v \in V$. Choose a $a' \in (0, 1]$ such that $0 < a' < a$. Since $a' < b$, $(a', b) \in r$. Note that we also have $P_a \not\subseteq Q_{a'}$. Therefore,

$$r \not\subseteq \overline{Q}_{(a',b)} = ((0, 1] \setminus \{a'\} \times [0, 1]) \cup ((0, 1] \times [0, b)),$$

that is, $R2$ holds. The rest of the proof is similar.

If (r, R) is a direlation from $(U, \mathcal{U})$ to $(V, \mathcal{V})$, then by the inclusion

$\mathcal{P}(U) \otimes \mathcal{V} \subseteq \mathcal{P}(U) \otimes \mathcal{P}(V) = \mathcal{P}(U \times V),$

we have $r, R \in \mathcal{P}(U \times V)$ and so r and R are also ordinary relations from U to V.

Conversely, every relation from U to V can be also regarded as a relation or corelation in the textural sense:

Theorem 4.1.9 *If $r, R \subseteq U \times V$, then (r, R) is a direlation from $(U, \mathcal{P}(U))$ to $(V, \mathcal{P}(V))$.*

Proof Clearly, $r, R \in \mathcal{P}(U) \otimes \mathcal{P}(V)$. For R1, let $r \not\subseteq \overline{Q}_{(u,v)}$ and $P_{u'} \not\subseteq Q_u$. Since the textures are discrete, we note that

$$r \not\subseteq (U \setminus \{u\} \times V) \cup (U \times (V \setminus \{v\})), \text{ and } \{u'\} \not\subseteq U \setminus \{u\}.$$

This implies that $(u, v) \in r$ and $u = u'$, that is, $(u', v) \in r$ and so we get

$$r \not\subseteq (U \setminus \{u'\} \times V) \cup (U \times (V \setminus \{v\})) = \overline{Q}_{(u',v)}$$

as desired. For R2, let $r \not\subseteq \overline{Q}_{(u,v)}$. Note that $\{u\} \not\subseteq U \setminus \{u\}$ and hence, $P_u \not\subseteq Q_u$. If we take $u = u'$, then we find $P_u \not\subseteq Q_{u'}$ and $r \not\subseteq \overline{Q}_{(u',v)}$. Using a similar way, the conditions CR1 and CR2 can also be proved. □

The Identity Direlation

Theorem 4.1.10 *Let $(U, \mathcal{U})$ be a texture. The pair (i, I) defined by*

$$i = \bigvee\{\overline{P}_{(u,u)} \mid u \in U\} \text{ and } I = \bigcap\{\overline{Q}_{(u,u)} \mid u \in U^\flat\}$$

is a direlation from $(U, \mathcal{U})$ to $(U, \mathcal{U})$.

Proof For R1, let $i \not\subseteq \overline{Q}_{(u,v)}$ and $P_{u'} \not\subseteq Q_u$ where $u, u', v \in U$. Then for some $u'' \in U$, we have $\overline{P}_{(u'',u'')} \not\subseteq \overline{Q}_{(u,v)}$. By Lemma 4.1.2, we get $u = u''$ and $P_{u''} \not\subseteq Q_v$, that is $P_u \not\subseteq Q_v$. However, by Corollary 2.2.4 (i), $P_{u'} \not\subseteq Q_u$ implies that $P_u \subseteq P_{u'}$ and hence, $P_{u'} \not\subseteq Q_v$. Again by Lemma 4.1.2, we obtain that $\overline{P}_{(u',u')} \not\subseteq \overline{Q}_{(u',v)}$. Since $\overline{P}_{(u',u')} \subseteq i$, we conclude that $i \not\subseteq \overline{Q}_{(u',v)}$. That is, the condition R1 holds. Leaving the proofs of R2 and CR2, we only prove the condition CR1. Let $\overline{P}_{(u,v)} \not\subseteq I$ and $P_u \not\subseteq Q_{u'}$. By definition of I, there is some $u'' \in U^\flat$ such that $\overline{P}_{(u,v)} \not\subseteq \overline{Q}_{(u'',u'')}$. By Lemma 4.1.2, $u = u''$ and $P_v \not\subseteq Q_{u''}$. Thus we get $P_v \not\subseteq Q_u$ and so $P_v \not\subseteq Q_{u'}$. As a result, we conclude that $\overline{P}_{(u',v)} \not\subseteq \overline{Q}_{(u',u')}$. Moreover, since $U \not\subseteq Q_{u'}$, by Theorem 2.3.1 (i), we have $u' \in U^\flat$. Then $\overline{P}_{(u',v)} \not\subseteq I$ and so CR1 holds as desired.

Corollary 4.1.11 *Let (i, I) be the identity direlation on $(U, \mathcal{U})$. Then*

(1) $i \not\subseteq \overline{Q}_{(u,v)}$ *if and only if* $P_u \not\subseteq Q_v$.

(2) $\overline{P}_{(u,v)} \not\subseteq I$ *if and only if* $P_v \not\subseteq Q_u$.

Proof Immediate. □

Corollary 4.1.12 *For a discrete texture $(U, \mathcal{P}(U))$, we have*

$$i = \Delta \text{ and } I = (U \times U) \setminus \Delta$$

where $\Delta = \{(u, u) \mid u \in U\}$ is the diagonal.

Proof

$$\begin{aligned} i &= \bigvee\{\overline{P}_{(u,u)} \mid u \in U\} = \bigcup\{\{u\} \times P_u \mid u \in U\} \\ &= \bigcup\{\{u\} \times \{u\} \mid u \in U\} = \bigcup\{\{(u, u)\} \mid u \in U\} \\ &= \{(u, u) \mid u \in U\} = \Delta \end{aligned}$$

and

$$\begin{aligned} I &= \bigcap\{\overline{Q}_{(u,u)} \mid u \in U^\flat\} = \bigcap\{(U \times Q_u) \cup (Q_u \times U) \mid u \in U\} \\ &= \bigcap\{(U \times (U \setminus \{u\})) \cup ((U \setminus \{u\}) \times U) \mid u \in U\} \\ &= (U \times U) \setminus \{(u, u) \mid u \in U\} = (U \times U) \setminus \Delta. \end{aligned}$$

Reflexivity
Using identity direlations, we may consider the reflexivity in a textural sense:

Definition 4.1.13 Let (r, R) be a direlation on $(U, \mathcal{U})$. Then

(i) r is called *reflexive* if $i \subseteq r$,

(ii) R is called *reflexive* if $R \subseteq I$ and

(ii) (r, R) is called *reflexive* if r and R are reflexive.

Theorem 4.1.14 *A direlation (r, R) on a discrete texture $(U, \mathcal{P}(U))$ is reflexive if and only if r is an ordinary reflexive relation on U where $R \subseteq (U \times U) \setminus \Delta$.*

Proof By Corollary 4.1.12, for discrete textures, we have $i = \{(u, u) \mid u \in U\} = \Delta$ and $I = (U \times U) \setminus i$ and so the proof is immediate.

Example 4.1.15 For the texture $(U, \mathcal{U})$ where $U = \{a, b, c\}$ and $\mathcal{U} = \{\{a\}, \{a, b\}, \emptyset, U\}$, the identity direlation (i, I) is as follows:

$$i = \overline{P}_{(a,a)} \cup \overline{P}_{(b,b)} \cup \overline{P}_{(c,c)}$$
$$= (\{a\} \times \{a\}) \cup (\{b\} \times \{a, b\}) \cup (\{c\} \times U)$$
$$=\{(a, a), (b, a), (b, b), (c, a), (c, b), (c, c)\}$$

and

$$I = \overline{Q}_{(a,a)} \cap \overline{Q}_{(b,b)} \cap \overline{Q}_{(c,c)}$$
$$= ((U \setminus \{a\} \times U) \cup (U \times \emptyset)) \cap ((U \setminus \{b\} \times U) \cup (U \times Q_b)) \cap ((U \setminus \{c\} \times U) \cup (U \times Q_c))$$
$$= \{(b, a), (c, a), (c, b)\}.$$

Note that $i \neq \Delta$ and $I \neq (U \times U) \setminus \Delta$. □

Inverse Direlation

If r is an ordinary relation from a set U to a set V, then the inverse of r is given by $r^{-1} = \{(v, u) \mid (u, v) \in r\}$. As we know, a (textural) relation r from a texture $(U, \mathcal{U})$ to a texture $(V, \mathcal{V})$ is defined in terms of p-set and q-set. Therefore, the duality between the concepts p-set and q-set leads us to a different inverse of a relation r. In general, a set $h \in \mathcal{P}(U) \otimes \mathcal{V}$ may not be a relation or a corelation in a textural sense if it does not hold one of the conditions of R1-R2 or CR1-CR2, respectively. However, it is interesting that we may obtain a relation or a corelation using a set $h \in \mathcal{P}(U) \otimes \mathcal{V}$.

Lemma 4.1.16 *Let $h \in \mathcal{P}(U) \otimes \mathcal{V}$. Then*

(i) *The set $\bigvee\{\overline{P}_{(v,u)} \mid \overline{P}_{(u,v)} \not\subseteq h\}$ is a relation from $(V, \mathcal{V})$ to $(U, \mathcal{V})$.*

(ii) *The set $\bigcap\{\overline{Q}_{(v,u)} \mid h \not\subseteq \overline{Q}_{(u,v)}\}$ is a corelation from $(V, \mathcal{V})$ to $(U, \mathcal{V})$.*

Proof Let $r = \bigvee\{\overline{P}_{(v,u)} \mid \overline{P}_{(u,v)} \nsubseteq h\}$. For the condition R1, let $r \nsubseteq \overline{Q}_{(v,u)}$ and $P_{v'} \nsubseteq Q_v$. By definition of r, there exist $u' \in U$ and $v'' \in V$ such that

$$\overline{P}_{(v'',u')} \nsubseteq \overline{Q}_{(v,u)} \text{ and } \overline{P}_{(u',v'')} \nsubseteq h.$$

By Lemma 4.1.2, $v'' = v$ and $P_{u'} \nsubseteq Q_u$. Further, by Theorem 2.3.1 (i), $P_{v'} \nsubseteq Q_v$ implies that $P_v \subseteq P_{v'}$ and hence,

$$\overline{P}_{(u',v'')} = \overline{P}_{(u',v)} = \{u'\} \times P_v \subseteq \{u'\} \times P_{v'} = \overline{P}_{(u',v')} \nsubseteq h.$$

Now, by definition of r, we get $\overline{P}_{(v',u')} \subseteq r$. However, since $P_{u'} \nsubseteq Q_u$, by Lemma 4.1.2, we have $\overline{P}_{(v',u')} \nsubseteq \overline{Q}_{(v',u)}$. Then we conclude that $r \nsubseteq \overline{Q}_{(v',u)}$ as desired. That is, r holds the condition R1.

For R2, let $r \nsubseteq \overline{Q}_{(v,u)}$. By definition of r, for some $u' \in U$ and $v' \in V$,

$$\overline{P}_{(v',u')} \nsubseteq \overline{Q}_{(v,u)} \text{ and } \overline{P}_{(u',v')} \nsubseteq h.$$

By Lemma 4.1.2, $v = v'$ and $P_{u'} \nsubseteq Q_u$, and so we conclude that $\overline{P}_{(u',v)} \nsubseteq h$. Now, by Corollary 2.3.2 (ii), there exist $u'' \in U$ and $v'' \in V$ such that

$$\overline{P}_{(u',v)} \nsubseteq \overline{Q}_{(u'',v'')} \text{ and } \overline{P}_{(u'',v'')} \nsubseteq h.$$

Again by Lemma 4.1.2, we have $u' = u''$ and $P_v \nsubseteq Q_{v''}$. Then $\overline{P}_{(u',v'')} \nsubseteq h$ and so by definition of r, $\overline{P}_{(v'',u')} \subseteq r$. By Lemma 4.1.2, this means that $r \nsubseteq \overline{Q}_{(v'',u)}$. The proofs of CR1 and CR2 are left to the reader. □

In view of Lemma 4.1.16, we may give the following definition.

Definition 4.1.17 Let (r, R) be a direlation from $(U, \mathcal{U})$ to $(V, \mathcal{V})$. Then the inverses of r and R are defined by

$$r^{\leftarrow} = \bigcap\{\overline{Q}_{(v,u)} \mid r \nsubseteq \overline{Q}_{(u,v)}\}$$

and

$$R^{\leftarrow} = \bigvee\{\overline{P}_{(v,u)} \mid \overline{P}_{(u,v)} \nsubseteq R\}.$$

Corollary 4.1.18 *The pair* $(R^{\leftarrow}, r^{\leftarrow})$ *is also a direlation from* $(V, \mathcal{V})$ *to* $(U, \mathcal{V})$.

Proof By Lemma 4.1.16, the proof is immediate. □

Example 4.1.19 Consider the direlation (r, R) given in Example 4.1.7.

(i) We may determine the inverse $(R^{\leftarrow}, r^{\leftarrow})$ of (r, R) as follows:

$$\begin{aligned} r^{\leftarrow} &= \bigcap\{\overline{Q}_{(u,v)} \mid r \nsubseteq \overline{Q}_{(v,u)}\} \\ &= \overline{Q}_{(d,a)} \cap \overline{Q}_{(d,b)} \cap \overline{Q}_{(d,c)} \cap \overline{Q}_{(e,b)} \cap \overline{Q}_{(e,c)} \end{aligned}$$

$= ((\{e\} \times U) \cup (V \times \emptyset)) \cap ((\{e\} \times U) \cup (V \times \{a\})) \cap ((\{e\} \times U) \cup (V \times \{a, b\}))$

$\cap((\{d\} \times U) \cup (V \times \{a\})) \cap ((\{d\} \times U) \cup (V \times \{a, b\})) = \{(e, a)\}$

and

$$\begin{aligned} R^{\leftarrow} &= \bigvee\{\overline{P}_{(v,u)} \mid \overline{P}_{(u,v)} \not\subseteq R\} \\ &= \overline{P}_{(e,a)} \cup \overline{P}_{(e,b)} \cup \overline{P}_{(e,c)} \\ &= \{(e, a)\} \cup \{(e, a), (e, b)\} \cup \{(e, a), (e, b), (e, c)\} \\ &= \{(e, a), (e, b), (e, c)\}. \end{aligned}$$

For discrete textures, textural inverse of a relation or a corelation is the complement of the inverse of the crisp relation.

Theorem 4.1.20 *If* (r, R) *is a direlation from* $(U, \mathcal{P}(U))$ *to* $(V, \mathcal{P}(V))$, *then*

$$r^{\leftarrow} = (V \times U) \setminus r^{-1} \text{ and } R^{\leftarrow} = (V \times U) \setminus R^{-1}$$

where $r^{-1} = \{(v, u) \mid (u, v) \in r\}$ *and* $R^{-1} = \{(v, u) \mid (u, v) \in R\}$.

Proof Note that

$$\begin{aligned} r^{\leftarrow} &= \bigcap\{\overline{Q}_{(v,u)} \mid r \not\subseteq \overline{Q}_{(u,v)}\} \\ &= \bigcap\{((V \setminus \{v\}) \times U) \cup (V \times (U \setminus \{u\})) \mid (u, v) \in r\} \\ &= (V \times U) \setminus (\bigcup\{(\{v\} \times U) \cap (V \times \{u\}) \mid (u, v) \in r\}) \\ &= (V \times U) \setminus (\bigcup\{\{(v, u)\} \mid (u, v) \in r\}) \\ &= (V \times U) \setminus \{(v, u) \mid (u, v) \in r\} = (V \times U) \setminus r^{-1} \end{aligned}$$

and

$$\begin{aligned} R^{\leftarrow} &= \bigvee\{\overline{P}_{(v,u)} \mid \overline{P}_{(u,v)} \not\subseteq R\} \\ &- \bigcup\{\{v\} \times P_u \mid \{u\} \times P_v \not\subseteq R\} \\ &= \bigcup\{\{v\} \times \{u\} \mid \{u\} \times \{v\} \not\subseteq R\} \\ &= \{(v, u) \mid (u, v) \notin R\} \\ &= \{(v, u) \mid (u, v) \in (V \times U) \setminus R\} \\ &= ((V \times U) \setminus R)^{-1} = (V \times U) \setminus R^{-1}. \end{aligned}$$

Remark 4.1.21 Note that if the textures are discrete and (r, R) is a complemented direlation, then the textural inverse of a corelation is nothing but the inverse of an ordinary relation, that is, $R^{\leftarrow} = r^{-1}$ (See Corollary 4.2.26). □

The above observation leads us to the counterparts of symmetric relations in textures.

Definition 4.1.22 A direlation (r, R) from $(U, \mathcal{U})$ to $(V, \mathcal{V})$ is called *symmetric* if $r = R^{\leftarrow}$ and $R = r^{\leftarrow}$.

Lemma 4.1.23 *Let (r, R) be a direlation from $(U, \mathcal{U})$ to $(V, \mathcal{V})$. Then*

(i) $r \not\subseteq \overline{Q}_{(u,v)} \iff \overline{P}_{(v,u)} \not\subseteq r^{\leftarrow}$.

(ii) $\overline{P}_{(u,v)} \not\subseteq R \iff R^{\leftarrow} \not\subseteq \overline{Q}_{(v,u)}$.

(iii) $(r^{\leftarrow})^{\leftarrow} = r$ *and* $(R^{\leftarrow})^{\leftarrow} = R$.

(iv) *Let (h, H) be a direlation from $(U, \mathcal{U})$ to $(V, \mathcal{V})$.*

(a) *If $r \subseteq h$, then $h^{\leftarrow} \subseteq r^{\leftarrow}$,*

(b) *If $R \subseteq H$, then $H^{\leftarrow} \subseteq h^{\leftarrow}$.*

Proof (i) Let $r \not\subseteq \overline{Q}_{(u,v)}$. By R2, there exists a $u' \in U$ such that $r \not\subseteq \overline{Q}_{(u',v)}$ and $P_u \not\subseteq Q_{u'}$. Now, by definition of $r^{\leftarrow}$, we have $r^{\leftarrow} \subseteq \overline{Q}_{(v,u')}$. On the other hand, $\overline{P}_{(v,u)} \not\subseteq \overline{Q}_{(v,u')}$. This gives that $\overline{P}_{(v,u)} \not\subseteq r^{\leftarrow}$. For the sufficiency, let $\overline{P}_{(v,u)} \not\subseteq r^{\leftarrow}$. Again, from definition of $r^{\leftarrow}$, there exists a $u' \in U$ such that $P_u \not\subseteq Q_{u'}$ and $r \not\subseteq \overline{Q}_{(u',v)}$. As a result, by R1, we get that $r \not\subseteq Q_{(u,v)}$.

(ii) The proof is left to the reader.

(iii) Note that $r^{\leftarrow}$ is a corelation and so for the proof of the first equality, it is enough to observe that

$$(r^{\leftarrow})^{\leftarrow} \not\subseteq \overline{Q}_{(u,v)} \iff \overline{P}_{(v,u)} \not\subseteq r^{\leftarrow} \iff r \not\subseteq \overline{Q}_{(u,v)}.$$

Then by Corollary 2.3.5 (i), we conclude that $(r^{\leftarrow})^{\leftarrow} = r$. The proof of $(R^{\leftarrow})^{\leftarrow} = R$ is similar.

(iv) Let us prove the first implication, leaving the second one to the interested reader. Suppose that $r \subseteq h$ and $h^{\leftarrow} \not\subseteq r^{\leftarrow}$. Take a $u \in U$ and $v \in V$ such that $h^{\leftarrow} \not\subseteq \overline{Q}_{(v,u)}$ and $\overline{P}_{(v,u)} \not\subseteq r^{\leftarrow}$. Then by definition of $r^{\leftarrow}$, the inclusion $h \subseteq \overline{Q}_{(u,v)}$ holds. Further by (i), we have $r \not\subseteq \overline{Q}_{(u,v)}$ and so by the assumption we obtain that $h \not\subseteq \overline{Q}_{(u,v)}$ which is a contradiction.

Note that if $r = R^{\leftarrow}$, then by Lemma 4.1.23 (iii), we also have $r^{\leftarrow} = R$ and vice versa. Indeed, $r = R^{\leftarrow}$ implies that $r^{\leftarrow} = (R^{\leftarrow})^{\leftarrow} = R$.

Corollary 4.1.24 *Let (r, R) be a direlation from $(U, \mathcal{U})$ to $(V, \mathcal{V})$. Then*

$$r^{\leftarrow} = R \text{ if and only if } r = R^{\leftarrow}.$$

Theorem 4.1.25 *A direlation (r, R) on a discrete texture $(U, \mathcal{P}(U))$ is symmetric if and only if r is an ordinary symmetric relation on U.*

Proof It is enough to observe that $r^{\leftarrow} = R \iff (U \times U) \setminus r^{-1} = (U \times U) \setminus r \iff r = r^{-1}$.

Theorem 4.1.26 *The identity direlation (i, I) on $(U, \mathcal{U})$ is symmetric.*

Proof Assume that $i^{\leftarrow} \not\subseteq I$. Then for some $u, u' \in U$, we have $i^{\leftarrow} \not\subseteq \overline{Q}_{(u,u')}$ and $\overline{P}_{(u,u')} \not\subseteq I$. Hence, by definition of $r^{\leftarrow}$, $i \subseteq \overline{Q}_{(u',u)}$. Further, $\overline{P}_{(u,u')} \not\subseteq I$ implies that $P_{u'} \not\subseteq Q_u$. However, $\overline{P}_{(u',u')} \not\subseteq \overline{Q}_{(u',u)}$ implies that $i \not\subseteq \overline{Q}_{(u',u)}$. This gives a contradiction and we obtain that $i^{\leftarrow} \subseteq I$. By Lemma 4.1.23 (iii) and (iv)(a), we immediately conclude the reverse inclusion as $I^{\leftarrow} \subseteq (i^{\leftarrow})^{\leftarrow} = i$. □

Lemma 4.1.27 *Let (r, R) be a direlation on $(U, \mathcal{U})$ Then the following conditions are equivalent.*

(i) (r, R) *is reflexive.*

(ii) $(R^{\leftarrow}, r^{\leftarrow})$ *is reflexive.*

Proof (i) $\Longrightarrow$ (ii) : Since (r, R) is reflexive, then $i \subseteq r$ and $R \subseteq I$. By Lemma 4.1.23 (iv), we may write that $r^{\leftarrow} \subseteq i^{\leftarrow}$ and $I^{\leftarrow} \subseteq R^{\leftarrow}$. By Theorem 4.1.26, $i^{\leftarrow} = I$ and $I^{\rightarrow} = i$, that is (i, I) is symmetric. Then we obtain that $r^{\leftarrow} \subseteq I$ and $i \subseteq$ and $i^{\leftarrow}$. Therefore $(R^{\leftarrow}, r^{\leftarrow})$ is also reflexive.

(ii) $\Longrightarrow$ (i): The proof is similar. □

4.2 Textural Image of a Set: Sections

Let r be a relation from a set U to a set V. If we consider the ordinary image of a subset $A \subseteq U$, then we may observe that

$$\begin{aligned} r(A) &= \{v \in V \mid \exists u \in A, (u, v) \in r\} \\ &= V \setminus \{v \mid \forall u, (u, v) \in r \implies u \notin A\} \\ &= \bigcap \{V \setminus \{v\} \mid \forall u, (u, v) \in r \implies u \notin A\}. \end{aligned}$$

Recall that for discrete textures, we have the following statements

$V \setminus \{v\} = Q_v$,
$(u, v) \in r \iff r \not\subseteq \overline{Q}_{(u,v)}$ and
$u \notin A \iff A \subseteq Q_u$.

Then for any discrete texture $(U, \mathcal{P}(U))$, the image set $r(A)$ can be stated as

$$r(A) = \bigcap\{Q_v \mid \forall u \in U, r \not\subseteq \overline{Q}_{(u,v)} \implies A \subseteq Q_u\}.$$

In view of the above argument, a suitable textural image of a set can be defined in terms of the p-set and q-set of a texture as follows.

Definition 4.2.1 Let (r, R) be a direlation from $(U, \mathcal{U})$ to $(V, \mathcal{V})$ and $A \in \mathcal{U}$. Then the sets

$$r^{\to} A = \bigcap\{Q_v \mid \forall u, r \not\subseteq \overline{Q}_{(u,v)} \Rightarrow A \subseteq Q_u\}, \text{ and}$$

$$R^{\to} A = \bigvee\{P_v \mid \forall u, \overline{P}_{(u,v)} \not\subseteq R \Rightarrow P_u \subseteq A\}$$

are called *the* A*-section* with respect to relation r and *the* A*-section* with respect to corelation R, respectively.

Note that in Definition 4.2.1, the A-section with respect to corelation R is exactly considered as dual correspondence of the A-section with respect to relation r.

Example 4.2.2 Let us consider the direlation (r, R) in Example 4.1.7. The sections with respect to (r, R) are as follows:

$$r^{\to}\{a\} = \bigcap\{Q_v \mid \forall u,\ r \not\subseteq \overline{Q}_{(u,v)} \Rightarrow\ \{a\} \subseteq Q_u\} = Q_e = \{d\},$$

$$R^{\to}\{a\} = \bigvee\{P_v \mid \forall u,\ \overline{P}_{(u,v)} \not\subseteq R \Rightarrow\ P_u \subseteq \{a\}\} = P_d = \{d\},$$

$$r^{\to}\{a, b\} = \bigcap\{Q_v \mid \forall u,\ r \not\subseteq \overline{Q}_{(u,v)} \Rightarrow\ \{a, b\} \subseteq Q_u\} = \bigcap_{v\in\emptyset} Q_v = V = \{d, e\},$$

$$R^{\to}\{a, b\} = \bigvee\{P_v \mid \forall u,\ \overline{P}_{(u,v)} \not\subseteq R \Rightarrow\ P_u \subseteq \{a, b\}\} = P_d = \{d\}.$$ □

For the discrete textures, we have the following remarkable equalities:

Theorem 4.2.3 *Let* (r, R) *be a direlation from* $(U, \mathcal{P}(U))$ *to* $(V, \mathcal{P}(V))$ *where* $(U, \mathcal{P}(U))$ *and* $(V, \mathcal{P}(V))$ *are any two discrete textures. Then for all* $A \subseteq \mathcal{U}$*, we have the following equalities.*

(i) $r^{\to} A = r(A)$.

(ii) $R^{\to} A = V \setminus ((U \times V) \setminus R)(U \setminus A)$.

Proof (i) By the above argument, we observe that

$$\begin{aligned} r^{\to} A &= \bigcap\{Q_v \mid \forall u, r \not\subseteq \overline{Q}_{(u,v)} \Rightarrow A \subseteq Q_u\} \\ &= \bigcap\{U \setminus \{v\} \mid \forall u, (u, v) \in r \implies u \notin A\} \\ &= U \setminus \{v \mid \forall u, (u, v) \in r \implies u \notin A\} \\ &= \{v \in V \mid \exists u \in A, (u, v) \in r\} \\ &= r(A). \end{aligned}$$

(ii) Using a similar argument, we also see that

$$\begin{aligned}
R^{\to}A &= \bigvee\{P_v \mid \forall u, \overline{P}_{(u,v)} \not\subseteq R \Rightarrow P_u \subseteq A\} \\
&= \bigcup\{\{v\} \mid \forall u \in U,\ (u,v) \notin R \implies u \in A\} \\
&= \{v \mid \forall u,\ (u,v) \notin R \implies u \in A\} \\
&= V \setminus \{v \mid \exists u \in U, (u,v) \notin R \text{ and } u \in U \setminus A\} \\
&= V \setminus ((U \times V) \setminus R)(U \setminus A).
\end{aligned}$$

Remark 4.2.4 If (r, R) given in Theorem 4.2.3 is complemented, we clearly have that $R^{\to}A = V \setminus r(U \setminus A)$. □

Theorem 4.2.5 *Let (i, I) be the identity direlation on $(U, \mathcal{U})$. Then for all $A \in \mathcal{U}$,*

$$i^{\to}A = A \textit{ and } I^{\to}A = A.$$

Proof For the first equality, suppose that $i^{\to}A \not\subseteq A$. Let us choose a $u \in U$ such that $i^{\to}A \not\subseteq Q_u$ and $P_u \not\subseteq A$. Then by definition of a section, we have some $u' \in U$ such that $i \not\subseteq \overline{Q}_{(u',u)}$ and $A \not\subseteq Q_{u'}$. Further, by definition of identity relation, the first statement implies that $P_{u'} \not\subseteq Q_u$. Then by Theorem 2.3.1 (i), $Q_u \subseteq Q_{u'}$ whence $A \not\subseteq Q_u$. Again by Theorem 2.3.1 (i), we obtain $P_u \subseteq A$ which is an immediate contradiction. For the reverse inclusion, let $A \not\subseteq i^{\to}A$. Then there exists a $u \in U$ with $A \not\subseteq Q_u$ and $P_u \not\subseteq i^{\to}A$. Then by definition of a section, there exists a $u' \in U$ with $P_u \not\subseteq Q_{u'}$ such that

$$\forall u'' \in U,\ i \not\subseteq \overline{Q}_{(u'',u')} \implies A \subseteq Q_{u''}. \quad (1)$$

Now, by the implication (1), we obtain $i \subseteq \overline{Q}_{(u,u')}$. However, this is a contradiction, since that $\overline{P}_{(u,u)} \subseteq i$ and $\overline{P}_{(u,u)} \not\subseteq \overline{Q}_{(u,u')}$. □

The following lemmas will be useful in the sequel.

Lemma 4.2.6 *Let (r, R) be a direlation from $(U, \mathcal{U})$ and $(V, \mathcal{V})$. Then for all $u \in U$ and $v \in V$, we have the following equivalences:*

(i) $r \not\subseteq \overline{Q}_{(u,v)} \iff r^{\to}P_u \not\subseteq Q_v$,

(ii) $\overline{P}_{(u,v)} \not\subseteq R \iff P_v \not\subseteq R^{\to}Q_u$.

Proof We prove (ii) leaving the proof of (i). For the necessity, let $\overline{P}_{(u,v)} \not\subseteq R$. By CR2, there exists a $u' \in U$ such that $P_{u'} \not\subseteq Q_u$ and $\overline{P}_{(u',v)} \not\subseteq R$. Further, by Lemma 4.1.2, we may choose a $v' \in V$ such that $P_v \not\subseteq Q_{v'}$ and $\overline{P}_{(u',v')} \not\subseteq R$. If $P_v \subseteq R^{\to}Q_u$, then $R^{\to}Q_u \not\subseteq Q_{v'}$. By definition of Q_u-section of R, there exists a $v'' \in V$ such that $P_{v''} \not\subseteq Q_{v'}$ and

$$\forall u'' \in U,\ \overline{P}_{(u'',v'')} \not\subseteq R \implies P_{u''} \subseteq Q_u. \quad (*)$$

Since $P_{v''} \not\subseteq Q_{v'}$, by Theorem 2.3.1 (i), $P_{v'} \subseteq P_{v''}$ and hence, we get

$$\overline{P}_{(u',v')} \subseteq \overline{P}_{(u',v'')} \not\subseteq R.$$

Then by (*), we obtain $P_{u'} \subseteq Q_u$ which is an immediate contradiction. For the sufficiency if $P_v \not\subseteq R^{\rightarrow} Q_u$, then for some $u' \in U$, we have

$$\overline{P}_{(u',v)} \not\subseteq R \text{ and } P_{u'} \not\subseteq Q_u.$$

Now, by CR1, we obtain $\overline{P}_{(u,v)} \not\subseteq R$ as desired.

□

Lemma 4.2.7 *Let* (r_1, R_1) *and* (r_2, R_2) *be any two direlations from* $(U, \mathcal{U})$ *and* $(V, \mathcal{V})$. *Then we have*

(i) $r_1 \subseteq r_2 \iff \forall A \in \mathcal{U},\ r_1^{\rightarrow} A \subseteq r_2^{\rightarrow} A,$

(ii) $R_1 \subseteq R_2 \iff \forall A \in \mathcal{U},\ R_1^{\rightarrow} A \subseteq R_2^{\rightarrow} A.$

Proof (i) Let $r_1 \subseteq r_2$ and suppose that $r_1^{\rightarrow} A \not\subseteq r_2^{\rightarrow} A$ for some $A \in \mathcal{U}$. By Corollary 2.3.2 (ii), there is a $v \in V$ such that $r_1^{\rightarrow} A \not\subseteq Q_v$ and $P_v \not\subseteq r_2^{\rightarrow} A$. By definition of sections, for some $u \in U$,

$$r_1 \not\subseteq \overline{Q}_{(u,v)} \text{ and } A \not\subseteq Q_u,$$

and for some $v' \in V$,

$$P_v \not\subseteq Q_{v'}, \text{ and } r_2 \not\subseteq \overline{Q}_{(u,v')} \implies A \subseteq Q_u.$$

Now, we clearly have $r_1 \not\subseteq \overline{Q}_{(u,v')}$ and so we obtain the contradiction of $A \subseteq Q_u$. For the sufficiency, for all $A \in \mathcal{U}$, let $r_1^{\rightarrow} A \subseteq r_2^{\rightarrow} A$ and $r_1 \not\subseteq r_2$. By Corollary 2.3.2 (ii), there exists a $(u, v) \in U \times V$ such that $r_1 \not\subseteq \overline{Q}_{(u,v)}$ and $\overline{P}_{(u,v)} \not\subseteq r_2$. Then by Lemma 4.2.3 (i), we have $r_1^{\rightarrow} P_u \not\subseteq Q_v$ and $r_2^{\rightarrow} P_u \subseteq Q_v$. However, by the assumption, $r_1^{\rightarrow} P_u \not\subseteq r_2^{\rightarrow} P_u$ is a contradiction.

(ii) The proof is dual to (i). □

Textural Inverse Image of a Set: Presections

Recall that if r is an ordinary relation from a set U to a set V, then the inverse image of a subset $B \subseteq V$ is defined by

$$r^{-1}(B) = \{u \mid \forall v \in V, (u, v) \in r \implies v \in B\}.$$

For any two discrete textures $(U, \mathcal{P}(U))$ and $(V, \mathcal{P}(V))$, the inverse image of B can be stated as

$$\begin{aligned} r^{-1}(B) &= \bigcup\{\{u\} \mid \forall v \in V, (u, v) \in r \implies v \in B\} \\ &= \bigvee\{P_u \mid \forall v \in V, r \notin \overline{Q}_{(u,v)} \implies P_v \subseteq B\}. \end{aligned}$$

This gives a hint for a convenient definition due to a textural inverse image:

Definition 4.2.8 Let (r, R) be a direlation from $(U, \mathcal{U})$ to $(V, \mathcal{V})$ and $B \in \mathcal{U}$. Then the sets

$r^{\leftarrow}B = \bigvee\{P_u \mid \forall v, r \not\subseteq \overline{Q}_{(u,v)} \Rightarrow P_v \subseteq B\}$, and

$R^{\leftarrow}B = \bigcap\{Q_u \mid \forall v, \overline{P}_{(u,v)} \not\subseteq R \Rightarrow B \subseteq Q_v\}$

are called *the B-presection* with respect to relation r and *the B-section* with respect to corelation R, respectively.

Example 4.2.9 Let us take the direlation (r, R) in Example 4.1.7. Then

$r^{\leftarrow}\{d\} = \bigvee\{P_u \mid \forall v,\ r \not\subseteq \overline{Q}_{(u,v)} \Rightarrow P_v \subseteq \{d\}\} = P_a = \{a\}$,

$R^{\leftarrow}\{d\} = \bigcap\{Q_u \mid \forall v,\ \overline{P}_{(u,v)} \not\subseteq R \Rightarrow \{d\} \subseteq Q_v\} = Q_a = \emptyset$. □

Theorem 4.2.10 *Let (r, R) be a direlation from $(U, \mathcal{P}(U))$ to $(V, \mathcal{P}(V))$ where $(U, \mathcal{P}(U))$ and $(V, \mathcal{P}(V))$ are any two discrete textures and $R = (U \times V) \setminus r$. Then for all $A \subseteq \mathcal{U}$ and $B \subseteq V$, we have the following equalities.*

(i) $r^{\leftarrow}B = r^{-1}(B)$.

(ii) $R^{\leftarrow}B = U \setminus r^{-1}(U \setminus B)$.

Theorem 4.2.11

$$R^{\leftarrow}B = (R^{\leftarrow})^{\rightarrow}B \ \text{ and } \ r^{\leftarrow}B = (r^{\leftarrow})^{\rightarrow}B.$$

Proof Note that by Lemma 4.1.16, $r^{\leftarrow}$ is a corelation and so by definition of a section of corelation, we observe that

$$\begin{aligned} (r^{\leftarrow})^{\rightarrow}B &= \bigvee\{P_u \mid \forall v \in V, \overline{P}_{(v,u)} \not\subseteq r^{\leftarrow} \implies P_v \subseteq B\} \\ &= \bigvee\{P_u \mid \forall v \in V, (r^{\leftarrow})^{\leftarrow} \not\subseteq \overline{Q}_{(u,v)} \implies P_v \subseteq B\} \\ &= \bigvee\{P_u \mid \forall v \in V, r \not\subseteq \overline{Q}_{(u,v)} \implies P_v \subseteq B\} \\ &= r^{\leftarrow}B. \end{aligned}$$

The proof of the second equality is similar. □

Theorem 4.2.12 *Let (i, I) be the identity direlation on a texture $(U, \mathcal{U})$. Then for all $A \in \mathcal{U}$,*

$$i^{\leftarrow}A = I^{\leftarrow}A = A.$$

Proof Note that as we have proved in Theorem 4.1.26, (i, I) is symmetric and so $i^{\leftarrow} = I$. Therefore, by Theorem 4.2.11 and Theorem 4.2.5, we obtain that $i^{\leftarrow}A = (i^{\leftarrow})^{\rightarrow}A = I^{\rightarrow}A = A$. The proof of $I^{\leftarrow}A = A$ is similar. □

Presections can be written in terms of sections and conversely, sections can be written in terms of presections.

Theorem 4.2.13 *Let $(U, \mathcal{U})$ and $(V, \mathcal{V})$ be any two textures and (r, R) be a direlation from $(U, \mathcal{U})$ to $(V, \mathcal{V})$. For all $A \in \mathcal{U}$, $B \in \mathcal{V}$, we have the following equalities.*

(i) $r^{\leftarrow}B = \bigvee\{P_u \mid r^{\rightarrow}P_u \subseteq B\}$ *and*

$R^{\leftarrow}B = \bigcap\{Q_u \mid B \subseteq R^{\rightarrow}Q_u\}$.

(ii) $R^{\rightarrow}A = \bigvee\{P_v \mid R^{\leftarrow}P_v \subseteq A\}$ *and*

$r^{\rightarrow}A = \bigcap\{Q_v \mid A \subseteq r^{\leftarrow}Q_v\}$.

Proof (i) Since $r^{\leftarrow}B = \bigvee\{P_u \mid \forall v, r \not\subseteq \overline{Q}_{(u,v)} \Rightarrow P_v \subseteq B\}$, for the proof of the first equality, it is enough to prove the equivalence

$$(\forall v, r \not\subseteq \overline{Q}_{(u,v)} \Rightarrow P_v \subseteq B) \Longleftrightarrow (r^{\rightarrow}P_u \subseteq B).$$

Suppose that $r^{\rightarrow}P_u \not\subseteq B$. Take a $v \in V$ with $r^{\rightarrow}P_u \not\subseteq Q_v$ and $P_v \not\subseteq B$. By Lemma 4.2.3(1), we get $r \not\subseteq \overline{Q}_{(u,v)}$. By the assumption, we obtain the contradiction $P_v \subseteq B$.

Now let $r^{\rightarrow}P_u \subseteq B$. Assume that $r \not\subseteq \overline{Q}_{(u,v)}$ and $P_v \not\subseteq B$ for some $v \in V$. Again by Lemma 4.2.3(i), we have $r^{\rightarrow}P_u \not\subseteq Q_v$ and so $B \not\subseteq Q_v$ which is a contradiction. The proof of the second equality is similar.

(ii) Since $R^{\rightarrow}A = \bigvee\{P_v \mid \forall u, \overline{P}_{(u,v)} \not\subseteq R \Rightarrow P_u \subseteq A\}$, we only show that

$$(\forall u, \overline{P}_{(u,v)} \not\subseteq R \Rightarrow P_u \subseteq A) \Longleftrightarrow (R^{\leftarrow}P_v \subseteq A).$$

Assume that $R^{\leftarrow}P_v \not\subseteq A$. Then there exists a $u \in U$ such that $R^{\leftarrow}P_v \not\subseteq Q_u$ and $P_u \not\subseteq A$. Since $R^{\leftarrow}P_v$ is the P_v-section of the relation $R^{\leftarrow}$, by Lemma 4.2.3(i), we get $R^{\leftarrow} \not\subseteq \overline{Q}_{(v,u)}$. By Lemma 4.1.23 (i), we have $\overline{P}_{(u,v)} \not\subseteq (R^{\leftarrow})^{\leftarrow} = R$. However, by the assumption, we obtain the contradiction $P_u \subseteq A$. For the sufficiency, let $R^{\leftarrow}P_v \subseteq A$. Suppose that $\overline{P}_{(u,v)} \not\subseteq R$ and $P_u \not\subseteq A$ for some $u \in U$. By Lemma 4.1.23 (ii), we have $R^{\leftarrow} \not\subseteq \overline{Q}_{(v,u)}$. By Lemma, we get $R^{\leftarrow}P_v \not\subseteq Q_u$. Since $R^{\leftarrow}P_v \subseteq A$, we get $A \not\subseteq Q_u$ which is a contradiction.

Lemma 4.2.14 *Let (r, R) be a direlation from $(U, \mathcal{U})$ to $(V, \mathcal{V})$. Then for all $A, B \in \mathcal{U}$ and $C, D \in \mathcal{V}$, the following statements hold:*

(i) $A \subseteq B \implies r^{\rightarrow}A \subseteq r^{\rightarrow}B$.

(ii) $C \subseteq D \implies r^{\leftarrow}C \subseteq r^{\leftarrow}D$.

(iii) $A \subseteq B \implies R^{\rightarrow}A \subseteq R^{\rightarrow}B$.

(iv) $C \subseteq D \implies R^{\leftarrow}C \subseteq R^{\leftarrow}D$.

Proof (i) Let$A \subseteq B$ and $r^{\to} A \nsubseteq r^{\to} B$. Take $v \in V$ such that $r^{\to} A \nsubseteq Q_v$ and $P_v \nsubseteq r^{\to} B$. Then by Theorem 4.2.13 (ii), there exists a $v' \in V$ such that $P_v \nsubseteq Q_{v'}$ and $B \subseteq r^{\leftarrow} Q_{v'}$. Not that we have $Q_{v'} \subseteq Q_v$ and hence, $r^{\to} A \nsubseteq Q_{v'}$. Again by Theorem 4.2.13 (ii), this implies that $A \nsubseteq r^{\leftarrow} Q_{v'}$. However, it is a contradiction since $A \subseteq B$. The proofs of the rest of items are similar.

Lemma 4.2.15 *Let* (r, R) *be a direlation from* $(U, \mathcal{U})$ *to* $(V, \mathcal{V})$*. Then the following statements hold:*

(1) (i) $r^{\to} \emptyset = \emptyset$.

(ii) $\forall A \in \mathcal{U},\ A \subseteq r^{\leftarrow}(r^{\to} A)$.

(iii) $\forall B \in \mathcal{V},\ r^{\to}(r^{\leftarrow} B) \subseteq B$.

(2) (i) $R^{\to} U = V$.

(ii) $\forall B \in \mathcal{V},\ B \subseteq R^{\to}(R^{\leftarrow} B)$.

(iii) $\forall A \in \mathcal{U},\ R^{\leftarrow}(R^{\to} A) \subseteq A$.

Proof (1) (i) Note that for all $v \in V$, the inclusion $\emptyset \subseteq r^{\to} Q_v$ holds. Then by Corollary 2.2.4 (ii) and Theorem 4.2.13 (i), we observe that

$$r^{\to} \emptyset = \bigcap \{Q_v \mid \emptyset \subseteq r^{\leftarrow} Q_v\} = \bigcap \{Q_v \mid v \in V\} = \emptyset.$$

(ii) By Theorem 2.2.2, Lemma 4.2.14 (i) and Theorem 4.2.13 (i), we have

$$A = \bigvee \{P_u \mid P_u \subseteq A\} \subseteq \bigvee \{P_u \mid r^{\to} P_u \subseteq r^{\to} A\} = r^{\leftarrow}(r^{\to} A).$$

(iii) Note that by Corollary 2.3.2 (iv), we have

$$B = \bigcap_{P_v \nsubseteq B} Q_v \supseteq \bigcap_{B \subseteq Q_v} Q_v.$$

Then by Theorem 4.2.13 (ii), we obtain

$$\begin{aligned} r^{\to}(r^{\leftarrow} B) &= \bigcap \{Q_v \mid r^{\leftarrow} B \subseteq r^{\leftarrow} Q_v\} \\ &\subseteq \bigcap \{Q_v \mid B \subseteq Q_v\} \\ &\subseteq \bigcap \{Q_v \mid P_v \nsubseteq B\} = B. \end{aligned}$$

(2) (i) Note that by Theorem 4.2.13 (ii), we obtain

$$R^{\to} U = \bigvee \{P_v \mid R^{\leftarrow} P_v \subseteq U\} = \bigvee \{P_v \mid v \in V\} = V.$$

The rest of the proof is left to the reader.

Theorem 4.2.16 *Let (r, R) be a direlation on $(U, \mathcal{U})$.*

(i) *The following conditions are equivalent.*

(1) *r is reflexive.*

(2) $\forall A \in \mathcal{U}$, $A \subseteq r^{\rightarrow} A$.

(3) $\forall A \in \mathcal{U}$, $r^{\leftarrow} A \subseteq A$.

(ii) *The following conditions are equivalent.*

(1) *R is reflexive.*

(2) $\forall A \in \mathcal{U}$, $R^{\rightarrow} A \subseteq A$.

(3) $\forall A \in \mathcal{U}$, $A \subseteq R^{\leftarrow} A$.

Proof (i) (1)$\Longleftrightarrow$ (2): Let r be reflexive. Then $i \subseteq r$ and by Lemma 4.2.7 and Theorem 4.2.5, for all $A \in \mathcal{U}$, we get $A = i^{\rightarrow} A \subseteq r^{\rightarrow} A$. Conversely, if for all $A \in \mathcal{U}$, $A \subseteq r^{\rightarrow} A$, then $i^{\rightarrow} A = A \subseteq r^{\rightarrow} A$, hence, $i \subseteq r$, that is r is reflexive.

(1)$\Longleftrightarrow$ (3): For the necessity, if r reflexive, then $i \subseteq r$ and then by Lemma 4.1.23 (iv)(a), $r^{\leftarrow} \subseteq i^{\leftarrow}$. Hence, by Lemma 4.2.7 (ii) and Theorem 4.2.11, for all $A \in \mathcal{U}$, we conclude that

$$r^{\leftarrow} A \subseteq i^{\leftarrow} A = (i^{\leftarrow})^{\rightarrow} A = I^{\rightarrow} A = A.$$

For the sufficiency, note that if $r^{\leftarrow} A \subseteq A$, then $r^{\leftarrow} A \subseteq i^{\leftarrow} A$, since $i^{\leftarrow} A = A$ and so $r^{\leftarrow} \subseteq i^{\leftarrow}$ whence $i \subseteq r$. That is, r is reflexive.

Theorem 4.2.17 *Let (r, R) be a direlation from $(U, \mathcal{U})$ to $(V, \mathcal{V})$. Then we have the following equalities:*

(i) $r^{\rightarrow} \bigvee_{j \in J} A_j = \bigvee_{j \in J} r^{\rightarrow} A_j$ *and* $R^{\leftarrow} \bigvee_{j \in J} B_j = \bigvee_{j \in J} R^{\leftarrow} B_j$.

(ii) $r^{\leftarrow} \bigcap_{j \in J} B_j = \bigcap_{j \in J} r^{\leftarrow} B_j$ *and* $R^{\rightarrow} \bigcap_{j \in J} A_j = \bigcap_{j \in J} R^{\rightarrow} A_j$.

Proof (i) Note that for all $j \in J$, $A_j \subseteq \bigvee_{j \in J} A_j$. Then by Lemma 4.2.14 (i), for all $j \in J$, we get $r^{\rightarrow} A_j \subseteq r^{\rightarrow} \bigvee_{j \in J} A_j$ whence $\bigvee_{j \in J} r^{\rightarrow} A_j \subseteq r^{\rightarrow} \bigvee_{j \in J} A_j$. For the reverse inclusion, suppose that $r^{\rightarrow} \bigvee_{j \in J} A_j \not\subseteq \bigvee_{j \in J} r^{\rightarrow} A_j$. Take a $v \in V$ such that $r^{\rightarrow} \bigvee_{j \in J} A_j \not\subseteq Q_v$ and $P_v \not\subseteq \bigvee_{j \in J} r^{\rightarrow} A_j$. By Theorem 4.2.13 (ii), $\bigvee_{j \in J} A_j \not\subseteq r^{\leftarrow} Q_v$. Then for some $j \in J$ we also have that $A_j \not\subseteq r^{\leftarrow} Q_v$. By the second statement, for all $j \in J$, $P_v \not\subseteq r^{\rightarrow} A_j$, that is $r^{\leftarrow} A_j \subseteq Q_v$ and by Lemma 4.2.14 (ii), we get $r^{\leftarrow} r^{\rightarrow} A_j \subseteq r^{\leftarrow} Q_v$. Finally, by Lemma 4.2.15 (1)(ii), we find that $A_j \subseteq r^{\leftarrow} Q_v$ which is an immediate contradiction. The proof of the second equality is similar.

(ii) We prove the second equality leaving the first to the interested reader. Suppose that $\bigcap_{j\in J} R^{\to} A_j \not\subseteq R^{\to} \bigcap_{j\in J} A_j$. Take a $v \in V$ such that $\bigcap_{j\in J} R^{\to} A_j \not\subseteq Q_v$ and $P_v \not\subseteq R^{\to} \bigcap_{j\in J} A_j$. By the first statement, for all $j \in J$, we get $R^{\to} A_j \not\subseteq Q_v$ and so $P_v \subseteq R^{\to} A_j$. By Lemma 4.2.14 (2)(iv), for all $j \in J$, we obtain $R^{\leftarrow} P_v \subseteq R^{\leftarrow} R^{\to} A_j \subseteq A_j$. and by Theorem 4.2.13 (i) $R^{\leftarrow} P_v \not\subseteq \bigcap_{j\in J} A_j$. This means that for some $j \in J$, we have $R^{\leftarrow} P_v \not\subseteq A_j$. However, this is an immediate contradiction.

Theorem 4.2.18 *Let (r, R) be a direlation from $(U, \mathcal{U})$ to $(V, \mathcal{V})$. Then the following inclusions hold:*

(i) $\bigvee_{j\in J} r^{\leftarrow} B_j \subseteq r^{\leftarrow} \bigvee_{j\in J} B_j$ *and* $R^{\leftarrow} \bigcap_{j\in J} B_j \subseteq \bigcap_{j\in J} R^{\leftarrow} B_j$.

(ii) $\bigvee_{j\in J} R^{\to} A_j \subseteq R^{\to} \bigvee_{j\in J} A_j$ *and* $r^{\to} \bigcap_{j\in J} A_j \subseteq \bigcap_{j\in J} r^{\to} A_j$.

Proof It is immediate by Lemma 4.2.14.

A further remarkable result on direlations is as follows:

Theorem 4.2.19 *Let (r, R) be a direlation from $(U, \mathcal{U})$ to $(V, \mathcal{V})$. Then*

(i) $\forall A \in \mathcal{U},\ r^{\to}(r^{\leftarrow}(r^{\to} A)) = r^{\to} A$.

(ii) $\forall A \in \mathcal{U},\ R^{\to}(R^{\leftarrow}(R^{\to} A)) = R^{\to} A$.

(iii) $\forall B \in \mathcal{V},\ r^{\leftarrow}(r^{\to}(r^{\leftarrow} B)) = r^{\leftarrow} B$.

(iv) $\forall B \in \mathcal{V},\ R^{\leftarrow}(R^{\to}(R^{\leftarrow} B)) = R^{\leftarrow} B$.

Proof (i) By Lemma 4.2.15 (1)(ii), we have $A \subseteq r^{\leftarrow}(r^{\to} A)$ and so $r^{\to} A \subseteq r^{\to}(r^{\leftarrow}(r^{\to} A))$. For the reverse inclusion, we know that $r^{\to}(r^{\leftarrow} B) \subseteq B$ for all $B \in \mathcal{V}$. Then we get $r^{\to}(r^{\leftarrow}(r^{\to} A)) \subseteq r^{\to} A$, for $r^{\to} A = B$.

(iv) By Lemma 4.2.15(2)(ii), we have $B \subseteq R^{\to}(R^{\leftarrow} B)$ and then $R^{\leftarrow} B \subseteq R^{\leftarrow}(R^{\to}(R^{\leftarrow} B))$. For the reverse inclusion, we know that $R^{\leftarrow}(R^{\to} A) \subseteq A$ for all $A \in \mathcal{U}$. Then we obtain $R^{\leftarrow}(R^{\to}(R^{\leftarrow} B)) \subseteq R^{\leftarrow} B$ for $R^{\leftarrow} B = A$.

Complemented direlations

Theorem 4.2.20 *Let $(U, \mathcal{U}, c_1)$ and $(V, \mathcal{V}, c_2)$ be complemented texture spaces and (r, R) be a direlation from $(U, \mathcal{U}, c_1)$ to $(V, \mathcal{V}, c_2)$. Then the pair (R', r') is also a direlation from $(U, \mathcal{U}, c_1)$ to $(V, \mathcal{V}, c_2)$ where*

$$R' = \bigvee\{\overline{P}_{(u,v)} \mid \exists w, z \text{ with } \overline{P}_{(w,z)} \not\subseteq R,\ P_w \not\subseteq c_1(P_u) \text{ and } c_2(Q_v) \not\subseteq Q_z\}$$

and

$$r' = \bigcap\{\overline{Q}_{(u,v)} \mid \exists w, z \text{ with } r \not\subseteq \overline{Q}_{(w,z)},\ c_1(Q_u) \not\subseteq Q_w \text{ and } P_z \not\subseteq c_2(P_v)\}.$$

Proof Let us show that r' is a corelation. For the condition CR1, suppose that $\overline{P}_{(u,v)} \not\subseteq r'$ and $P_u \not\subseteq Q_{u'}$. Then we have $u_1 \in U$ and $v_1 \in V$ with $\overline{P}_{(u,v)} \not\subseteq \overline{Q}_{(u_1,v_1)}$ such that for some $w \in U$ and $z \in V$,

$$c_1(Q_{u_1}) \not\subseteq Q_w \text{ and } P_z \not\subseteq c_2(P_{v_1}).$$

By Lemma 4.1.2, $\overline{P}_{(u,v)} \not\subseteq \overline{Q}_{(u_1,v_1)}$ implies that $u = u_1$ and $P_v \not\subseteq Q_{v_1}$. On the other hand, since $P_u \not\subseteq Q_{u'}$, by Theorem 2.3.1 (i), $Q_{u'} \subseteq Q_u$ and hence, $c_1(Q_u) \subseteq c_1(Q_{u'})$. As a result, we conclude that $c_1(Q_{u'}) \not\subseteq Q_w$. However, this means that $r' \subseteq \overline{Q}_{(u',v_1)}$. Finally, since $\overline{P}_{(u',v)} \not\subseteq \overline{Q}_{(u',v_1)}$, $\overline{P}_{(u',v)} \not\subseteq r'$ and so CR1 holds.

Now let us show that r' satisfies the condition CR2. If $\overline{P}_{(u,v)} \not\subseteq r'$, then Theorem 2.3.6 (ii), for some $u_2 \in U$,

$$c_1(Q_{u_1}) \not\subseteq c_1(P_{u_2}) \text{ and } c_1(Q_{u_2}) \not\subseteq Q_w.$$

Then clearly, $r' \subseteq \overline{Q}_{(u_2,v_1)}$ and $P_{u_2} \not\subseteq Q_{u_1}$. However, we also have $\overline{P}_{(u_2,v)} \not\subseteq \overline{Q}_{(u_2,v_1)}$ and so we get $\overline{P}_{(u_2,v)} \not\subseteq r'$. As a result, CR2 holds and hence, r' is a corelation. Using a similar way, we can show that R' is a relation.

The above result leads us to the following definition.

Definition 4.2.21 We denote the complement of (r, R) by $(r, R)'$ and write $(r, R)' = (R', r')$. A direlation (r, R) is called *complemented* if $r = R'$ and $R = r'$.

Example 4.2.22 Consider the textures $(U, \mathcal{U})$ and $(V, \mathcal{V})$ where

$$U = \{a, b, c, d\},\ \mathcal{U} = \{U, \emptyset, \{a\}, \{b\}, \{a, b\}, \{a, d\}, \{b, c\}, \{a, b, c\}, \{a, b, d\}\}$$

and

$$V = \{1, 2, 3\},\ \mathcal{V} = \{V, \emptyset, \{2\}, \{3\}, \{1, 2\}, \{2, 3\}\}.$$

The p-sets and q-sets of the texturings $\mathcal{U}$ and $\mathcal{V}$ are

$$P_a = \{a\},\ \ P_b = \{b\},\ \ P_c = \{b, c\},\ \ P_d = \{a, d\},$$

$$Q_a = \{b, c\},\ Q_b = \{a, d\},\ Q_c = \{a, b, d\},\ Q_d = \{a, b, c\}$$

and

$$P_1 = \{1, 2\},\ \ P_2 = \{2\},\ \ P_3 = \{3\},$$

$$Q_1 = \{2, 3\},\ \ Q_2 = \{3\},\ \ Q_3 = \{1, 2\},$$

respectively. The mapping $c_U : \mathcal{U} \to \mathcal{U}$ is a complementation defined by

$$c_U(\{a\}) = \{a, b, c\},\ \ c_U(\{b\}) = \{a, b, d\},\ \ c_U(\{a, b\}) = \{a, b\},$$

$$c_U(\{a, d\}) = \{b, c\},\ \ c_U(\{b, c\}) = \{a, d\},\ \ c_U(\{a, b, c\}) = \{a\},$$

$$c_U(\{a, b, d\}) = \{b\}, \ \ c_U(\emptyset) = U, \ \ c_U(U) = \emptyset.$$

Further, the mapping $c_V : \mathcal{U} \to \mathcal{V}$ is also complementation mapping on $\mathcal{V}$ where

$$c_V(\{2\}) = \{2, 3\}, \ \ c_V(\{3\}) = \{1, 2\},$$

$$c_V(\{1, 2\}) = \{3\}, \ \ c_V(\{2, 3\}) = \{2\},$$

$$c_V(\emptyset) = V, \ \ c_V(V) = \emptyset.$$

The complementations of p-sets and q-sets can be easily evaluated as follows:

$c_U(Q_a) = c_U(\{b, c\}) = \{a, d\}, \ \ c_U(Q_b) = c_U(\{a, d\}) = \{b, c\}$,

$c_U(Q_c) = c_U(\{a, b, d\}) = \{b\}, \ \ c_U(Q_d) = c_U(\{a, b, c\}) = \{a\},$

$c_V(P_1) = c_V(\{1, 2\}) = \{3\}, \ c_V(P_2) = c_V(\{2\}) = \{2, 3\},$

$c_V(P_3) = c_V(\{3\}) = \{1, 2\},$

$c_U(P_a) = c_U(\{a\}) = \{a, b, c\}, \ \ c_U(P_b) = c_U(\{b\}) = \{a, b, d\}$,

$c_U(P_c) = c_U(\{b, c\}) = \{a, d\}, \ \ c_U(P_d) = c_U(\{a, d\}) = \{b, c\}$,

$c_V(Q_1) = c_V(\{2, 3\}) = \{2\}, \ \ c_V(Q_2) = c_V(\{3\}) = \{1, 2\},$

$c_V(Q_3) = c_V(\{1, 2\}) = \{3\}.$

Let us consider the direlation (r, R) from $(U, \mathcal{U})$ to $(V, \mathcal{V})$ where

$$r = \{(a, 3), (b, 2), (c, 1), (c, 2), (d, 2), (d, 3)\},$$

$$R = \{(a, 2), (b, 3), (c, 2), (c, 3), (d, 1), (d, 2)\}.$$

We have the following cases with respect to the p-sets and q-sets of the product $\mathcal{P}(U) \otimes \mathcal{V}$ and the direlation (r, R):

$\overline{P}_{(a,1)} = \{a\} \times P_1 = \{a\} \times \{1, 2\} = \{(a, 1), (a, 2)\} \not\subseteq R,$

$\overline{P}_{(a,3)} = \{a\} \times P_3 = \{a\} \times \{3\} = \{(a, 3)\} \not\subseteq R$

$\overline{P}_{(b,1)} = \{b\} \times P_1 = \{b\} \times \{1, 2\} = \{(b, 1), (b, 2)\} \not\subseteq R$

$\overline{P}_{(b,2)} = \{b\} \times P_2 = \{b\} \times \{2\} = \{(b, 2)\} \not\subseteq R$

$$\overline{P}_{(c,1)} = \{c\} \times P_1 = \{c\} \times \{1, 2\} = \{(c, 1), (c, 2)\} \nsubseteq R,$$

$$\overline{P}_{(d,3)} = \{d\} \times P_3 = \{d\} \times \{3\} = \{(d, 3)\} \nsubseteq R$$

and

$$r \nsubseteq \overline{Q}_{(a,3)} = (\{b, c, d\} \times V) \cup (U \times Q_3)$$
$$= (\{b, c, d\} \times V) \cup (U \times \{1, 2\})$$
$$= \{(b, 1), (b, 2), (b, 3), (c, 1), (c, 2), (c, 3), (d, 1), (d, 2), (d, 3), (a, 1), (a, 2)\},$$

$$r \nsubseteq \overline{Q}_{(b,2)} = (\{a, c, d\} \times V) \cup (U \times Q_2)$$
$$= (\{a, c, d\} \times V) \cup (U \times \{3\})$$
$$= \{(a, 1), (a, 2), (a, 3), (c, 1), (c, 2), (c, 3), (d, 1), (d, 2), (d, 3), (b, 3)\},$$

$$r \nsubseteq \overline{Q}_{(c,1)} = (\{a, b, d\} \times V) \cup (U \times Q_1)$$
$$= (\{a, b, d\} \times V) \cup (U \times \{2, 3\})$$
$$= \{(a, 1), (a, 2), (a, 3), (b, 1), (b, 2), (b, 3), (d, 1), (d, 2), (d, 3), (c, 2), (c, 3)\},$$

$$r \nsubseteq \overline{Q}_{(c,2)} = (\{a, b, d\} \times V) \cup (U \times Q_2)$$
$$= (\{a, b, d\} \times V) \cup (U \times \{3\})$$
$$= \{(a, 1), (a, 2), (a, 3), (b, 1), (b, 2), (b, 3), (d, 1), (d, 2), (d, 3), (c, 3)\},$$

$$r \nsubseteq \overline{Q}_{(d,2)} = (\{a, b, c\} \times V) \cup (U \times Q_2)$$
$$= (\{a, b, c\} \times V) \cup (U \times \{3\})$$
$$= \{(a, 1), (a, 2), (a, 3), (b, 1), (b, 2), (b, 3), (c, 1), (c, 2), (c, 3), (d, 3)\},$$

$$r \nsubseteq \overline{Q}_{(d,3)} = (\{a, b, c\} \times V) \cup (U \times Q_3)$$
$$= (\{a, b, c\} \times V) \cup (U \times \{1, 2\})$$
$$= \{(a, 1), (a, 2), (a, 3), (b, 1), (b, 2), (b, 3), (c, 1), (c, 2), (c, 3), (d, 1), (d, 2)\}.$$

Now, let us determine the complement r' of r. Recall that r' is defined by

$$r' = \bigcap\{\overline{Q}_{(u,v)} \mid \exists w, z \text{ with } r \not\subseteq \overline{Q}_{(w,z)}, c_U(Q_u) \not\subseteq Q_w \text{ and } P_z \not\subseteq c_V(P_v)\}.$$

For instance, the q-set $\overline{Q}_{(a,1)}$ satisfies the condition due to definition of r'. Indeed, we have

$$c_U(Q_a) = \{a, d\} \not\subseteq Q_d,\ P_1 \not\subseteq c_V(P_1) = Q_2 \text{ and } r \not\subseteq \overline{Q}_{(d,2)}.$$

Likewise, it is easy to see that the q-sets $\overline{Q}_{(a,3)}, \overline{Q}_{(b,1)}, \overline{Q}_{(b,2)}, \overline{Q}_{(c,1)}, \overline{Q}_{(d,3)}$ also hold the condition given in definition of r'. Therefore, we obtain that

$$r' = \overline{Q}_{(a,1)} \cap \overline{Q}_{(a,3)} \cap \overline{Q}_{(b,1)} \cap \overline{Q}_{(b,2)} \cap \overline{Q}_{(c,1)} \cap \overline{Q}_{(d,3)}$$

$$= \{(b, 3), (c, 2), (c, 3), (a, 2), (d, 1), (d, 2)\} = R.$$

On the other hand, the p-sets satisfying the condition with respect to complement R' are as follows:

$$\overline{P}_{(a,3)}, \overline{P}_{(b,2)}, \overline{P}_{(c,1)}, \overline{P}_{(c,2)}, \overline{P}_{(d,2)} \text{ and } \overline{P}_{(d,3)}.$$

Hence,

$$R' = \overline{P}_{(a,3)} \cup \overline{P}_{(b,2)} \cup \overline{P}_{(c,1)} \cup \overline{P}_{(c,2)} \cup \overline{P}_{(d,2)} \cup \overline{P}_{(d,3)}$$

$$= \{(a, 3), (b, 2), (c, 1), (c, 2), (d, 2), (d, 3)\} = r.$$

Therefore, (r, R) is a complemented direlation.

Theorem 4.2.23 *If $(U, \mathcal{U}, c)$ is a complemented texture, then the identity direlation on $(U, \mathcal{U})$ is complemented, that is, $(i, I)' = (i, I)$.*

Proof Let us show that $i' = I$. Suppose that $i' \neq I$. If $i' \not\subseteq I$, then for some $u, u' \in U$, we have $i' \not\subseteq \overline{Q}_{(u,u')}$ and $\overline{P}_{(u,u')} \not\subseteq I$. Note that

$$i' = \bigcap\{\overline{Q}_{(u,u')} \mid \exists u_1, u_2 \text{ with } i \not\subseteq \overline{Q}_{(u_1,u_2)}, c(Q_u) \not\subseteq Q_{u_1} \text{ and } P_{u_2} \not\subseteq c(P_{u'})\}.$$

Since $i' \not\subseteq \overline{Q}_{(u,u')}$, for all $u_1, u_2 \in U$, the implication

$$i \not\subseteq \overline{Q}_{(u_1,u_2)} \text{ and } c(Q_u) \not\subseteq Q_{u_1} \implies P_{u_2} \subseteq c(P_{u'})$$

holds. By definition of the relation i, we also have that for all $u_1, u_2 \in U$,

$$P_{u_1} \not\subseteq Q_{u_2} \text{ and } c(Q_u) \not\subseteq Q_{u_1} \implies P_{u_2} \subseteq c(P_{u'}). \qquad (*)$$

On the other hand, $\overline{P}_{(u,u')} \not\subseteq I$ implies that $\overline{P}_{(u,u')} \not\subseteq \overline{Q}_{(u_1,u_1)}$ for some $u_1 \in U^\flat$. Hence, $u = u_1$ and $P_{u'} \not\subseteq Q_{u_1}$ whence $P_{u'} \not\subseteq Q_u$. This gives that $c(Q_u) \not\subseteq c(P_{u'})$. Now, let us choose $u_2 \in U$ such that $c(Q_u) \not\subseteq Q_{u_2}$ and $P_{u_2} \not\subseteq c(P_{u'})$. Further, take

$u'' \in U$ such that $c(Q_u) \not\subseteq Q_{u''}$ and $P_{u''} \not\subseteq Q_{u_2}$. Then by (*), we obtain $P_{u_2} \subseteq c(P_{u'})$ which is an immediate contradiction. Hence, we obtain $i' \subseteq I$. For the reverse inclusion, suppose that $I \not\subseteq i'$. Let us take $u_3, u_4 \in U$ such that $I \not\subseteq \overline{Q}_{(u_3,u_4)}$ and $\overline{P}_{(u_3,u_4)} \not\subseteq i'$. By definition of i', there exist some $u, u', u_1, u_2 \in U$ such that $\overline{P}_{(u_3,u_4)} \not\subseteq \overline{Q}_{(u,u')}$ and

$$i \not\subseteq \overline{Q}_{(u_1,u_2)}, c(Q_u) \not\subseteq Q_{u_1} \text{ and } P_{u_2} \not\subseteq c(P_{u'}).$$

Then $u = u_3$ and $P_{u_4} \not\subseteq Q_{u'}$ and so

$$P_{u_1} \not\subseteq Q_{u_2},\ c(Q_{u_3}) \not\subseteq Q_{u_1} \text{ and } P_{u_2} \not\subseteq c(P_{u_4}).$$

Since $Q_{u_2} \subseteq Q_{u_1}$, we also have $c(Q_{u_3}) \not\subseteq Q_{u_2}$ and so we get $c(Q_{u_3}) \not\subseteq c(P_{u_4})$. That is, $P_{u_4} \not\subseteq Q_{u_3}$ whence $Q_{u_3} \subseteq Q_{u_4}$. Note that $u_3 \in U^\flat$ and hence,

$$\begin{aligned} I \subseteq \overline{Q}_{(u_3,u_3)} &= (U \setminus \{u_3\} \times U) \cup (U \times Q_3) \\ &\subseteq (U \setminus \{u_3\} \times U) \cup (U \times Q_4) \\ &= \overline{Q}_{(u_3,u_4)} \end{aligned}$$

which is a contradiction since $I \not\subseteq \overline{Q}_{(u_3,u_4)}$.
The proof of the equality $i = I'$ is left to the interested reader.

Corollary 4.2.24 *If (r, R) is a direlation from $(U, \mathcal{P}(U))$ to $(V, \mathcal{P}(V))$, then*

$$r' = (U \times V) \setminus r \text{ and } R' = (U \times V) \setminus R.$$

Proof For discrete textures, we clearly have

$$\begin{aligned} r' &= \bigcap\{\overline{Q}_{(u,v)} \mid \exists w, z \text{ with } r \not\subseteq \overline{Q}_{(w,z)}, c_1(Q_u) \not\subseteq Q_w \text{ and } P_z \not\subseteq c_2(P_v)\} \\ &= \bigcap\{(U \times V) \setminus \{(u,v)\} \mid \exists w, z \text{ with } (w,z) \in r, U \setminus (U \setminus \{u\}) \not\subseteq U \setminus \{w\} \text{ and } \{z\} \not\subseteq V \setminus \{v\}\} \\ &= \bigcap\{(U \times V) \setminus \{(u,v)\} \mid \exists w, z \text{ with } (w,z) \in r, u = w \text{ and } z = v\} \\ &= \bigcap\{(U \times V) \setminus \{(u,v)\} \mid (u,v) \in r\} \\ &= (U \times V) \setminus \{(u,v) \mid (u,v) \in r\} = (U \times V) \setminus r \end{aligned}$$

and similarly, we may show that $R' = (U \times V) \setminus R$.

Corollary 4.2.25 *If (r, R) is a complemented direlation from $(U, \mathcal{P}(U))$ to $(V, \mathcal{P}(V))$, then $R = (U \times V) \setminus r$.*

Corollary 4.2.26 *If (r, R) is complemented a direlation from $(U, \mathcal{P}(U))$ to $(V, \mathcal{P}(V))$, then for all $A \in \mathcal{U}$ and $B \in \mathcal{V}$, we have the following equalities.*

(i) $r^{\to}(A) = r(A)$.

$R^{\rightarrow}A = V \setminus r(U \setminus A)$.

(ii) $r^{\leftarrow}(B) = U \setminus r^{-1}(V \setminus B)$.

$R^{\leftarrow}B = r^{-1}(B)$.

Lemma 4.2.27 *Let $(U, \mathcal{U}, c_U)$ and $(V, \mathcal{V}, c_V)$ be complemented textures and (r, R) be a direlation from $(U, \mathcal{U})$ to $(V, \mathcal{V})$. Then for all $A \in \mathcal{U}$ and $B \in \mathcal{V}$, we have the following equalities.*

(i) $(r')^{\rightarrow}A = c_V(r^{\rightarrow}c_U(A))$.

$(R')^{\rightarrow}A = c_V(R^{\rightarrow}c_U(A))$.

(ii) $(r')^{\leftarrow}B = c_U(r^{\leftarrow}c_V(B))$.

$(R')^{\leftarrow}B = c_U(R^{\leftarrow}c_V(B))$.

Proof We only prove the first equality of (i) leaving the proofs of the other equalities to the reader. Suppose that $(r')^{\rightarrow}A \not\subseteq c_V(r^{\rightarrow}c_U(A))$. Note that r' is a corelation and by definition of a section with respect to a corelation, there exists a $v \in V$ such that $P_v \not\subseteq c_V(r^{\rightarrow}c_U(A))$ and

$$\forall u \in U, \overline{P}_{(u,v)} \not\subseteq r' \implies P_u \subseteq A.$$

Now, let us take a $u' \in U$ such that $P_v \not\subseteq Q_{v'}$ and $P_{v'} \not\subseteq c_V(r^{\rightarrow}c_U(A))$. Applying the complementation to the second statement, we get $r^{\rightarrow}c_U(A) \not\subseteq c_V(P_{v'})$. We take again a $v'' \in V$ such that $r^{\rightarrow}c_U(A) \not\subseteq Q_{v''}$ and $P_{v''} \not\subseteq c_V(P_{v'})$. By definition of a section with respect to relation, for some $u \in U$, we conclude that $r \not\subseteq \overline{Q}_{(u,v'')}$ and $c_U(A) \not\subseteq Q_u$. By Theorem 2.3.6 (ii), we can choose a $u' \in U$ such that $c_U(A) \not\subseteq c_U(P_{u'})$ and $c_U(Q_{u'}) \not\subseteq Q_u$. Consequently, note that we have

$$r \not\subseteq \overline{Q}_{(u,v'')}, c_U(Q_u) \not\subseteq Q_{u'}, \text{ and } P_{v''} \not\subseteq c_V(P_{v'}).$$

Then by definition of r', we get $r' \subseteq \overline{Q}_{(u',v')}$. Hence we obtain $\overline{P}_{(u',v)} \not\subseteq r'$. Therefore, the above implication gives us that $P_{u'} \subseteq A$. However, $c_U(A) \not\subseteq c_U(P_{u'})$ gives the contradiction $P_{u'} \not\subseteq A$.

Now, assume that $c_V(r^{\rightarrow}c_U(A)) \not\subseteq (r')^{\rightarrow}A$. Choose a $v \in V$ such that $c_V(r^{\rightarrow}c_U(A)) \not\subseteq Q_v$ and $P_v \not\subseteq (r')^{\rightarrow}A$. Note that

$$\begin{aligned} c_V(r^{\rightarrow}c_U(A)) &= c_V(\bigcap\{Q_v \mid \forall u, r \not\subseteq \overline{Q}_{(u,v)} \Rightarrow c_u(A) \subseteq Q_u\}) \\ &= \bigvee\{c_V(Q_v) \mid \forall u, r \not\subseteq \overline{Q}_{(u,v)} \Rightarrow c_u(A) \subseteq Q_u\}). \end{aligned}$$

Therefore, for some $v' \in V$, $c_V(Q_{v'}) \not\subseteq Q_v$ such that

$$\forall u \in U, r \not\subseteq \overline{Q}_{(u,v')} \implies c_U(A) \subseteq Q_u \qquad (*)$$

for some $u \in U$. Further, by definition of a section with respect to corelation r', for some $u' \in U$, we also have that

$$\overline{P}_{(u',v)} \not\subseteq r' \text{ and } P_{u'} \not\subseteq A.$$

By definition of r', there exists a $v'' \in V$ such that $\overline{P}_{(u',v)} \not\subseteq \overline{Q}_{(u',v'')}$ and

$$r \not\subseteq \overline{Q}_{(u'',v''')},\ c_U(Q_{u'}) \not\subseteq Q_{u''} \text{ and } P_{v'''} \not\subseteq c_V(P_{v''})$$

for some $u'' \in U$ and $v''' \in V$. Since $c_V(Q_{v'}) \not\subseteq Q_v$, $P_v \not\subseteq Q_{v''}$ gives that $c_V(Q_{v'}) \not\subseteq Q_{v''}$ and so we obtain $c_V(Q_{v'}) \not\subseteq c_V(P_{v'''})$, since $P_{v''} \not\subseteq c_V(P_{v'''})$. Finally, applying the complementation, we find $P_{v'''} \not\subseteq Q_{v'}$. Hence, $r \not\subseteq \overline{Q}_{(u'',v''')}$ implies that $r \not\subseteq \overline{Q}_{(u'',v')}$. Therefore, by (*), we conclude that $c_U(A) \subseteq Q_{u''}$. However, $c_U(Q_{u'}) \not\subseteq Q_{u''}$ and $P_{u'} \not\subseteq A$ implies the contradiction $c_U(A) \not\subseteq Q_{u''}$.

The following is an immediate result of Lemma 4.2.27:

Corollary 4.2.28 *Let $(U, \mathcal{U}, c_U)$ and $(V, \mathcal{V}, c_V)$ be complemented textures and (r, R) be a complemented direlation from $(U, \mathcal{U})$ to $(V, \mathcal{V})$. Then*

(i) $R^{\to} A = c_V(r^{\to} c_U(A))$,

$r^{\to} A = c_V(R^{\to} c_U(A))$.

(ii) $R^{\leftarrow} B = c_U(r^{\leftarrow} c_V(B))$,

$r^{\leftarrow} B = c_U(R^{\leftarrow} c_V(B))$.

Proposition 4.2.29 *Let (r, R) be a direlation from $(U, \mathcal{U})$ to $(V, \mathcal{V})$. Then we have the following equalities.*

(i) $(r')' = r$ *and* $(R')' = R$.

(ii) $(r')^{\leftarrow} = (r^{\leftarrow})'$ *and* $(R')^{\leftarrow} = (R^{\leftarrow})'$.

Proof Note that r' is a corelation and hence, $(r')^{\leftarrow}$ is a relation. Let $B \in \mathcal{V}$. Then we have

$$((r')^{\leftarrow})^{\to} B = (r')^{\leftarrow} B = c_U(r^{\leftarrow} c_V(B)).$$

On the other hand, $(r^{\leftarrow})'$ is also a relation. Hence, for $B \in \mathcal{V}$,

$$(r^{\leftarrow})'^{\to} B = c_U((r^{\leftarrow})^{\to} c_V(B)) = c_U(r^{\leftarrow} c_V(B)).$$

Therefore, $((r')^{\leftarrow})^{\to} B = c_U(r^{\leftarrow} c_V(B))$ and by Lemma 4.2.7 (i), we obtain $(r')^{\leftarrow} = (r^{\leftarrow})'$ as desired.

4.3 Composition of Direlations

Let r and h be ordinary relations where $r \subseteq U \times V$ and $h \subseteq V \times W$. Then we know that r and h are (textural) relations from $(U, \mathcal{P}(U))$ to $(V, \mathcal{P}(V))$ and from $(V, \mathcal{P}(V))$ to $(W, \mathcal{P}(W))$, respectively where $\mathcal{P}(U)$, $\mathcal{P}(V)$ and $\mathcal{P}(W)$ are discrete texturings. Then the ordinary composition of r and h is defined by

$$h \circ r = \{(u, w) \in U \times W \mid \exists v \in V \text{ with } (u, v) \in r \text{ and } (v, w) \in h\}.$$

If we consider this definition in a textural sense, then we observe that

$$\begin{aligned} h \circ r &= \bigcup \{\{(u, w)\} \mid \exists\, v \in V \text{ with } (u, v) \in r \text{ and } (v, w) \in h\} \\ &= \{(u, w) \mid \exists\, v \in V \text{ with } (u, v) \in r \text{ and } (v, w) \in h\} \\ &= \bigcup \{\{u\} \times \{w\} \mid \exists\, v \in V \text{ with } (u, v) \in r \text{ and } (v, w) \in h\} \\ &= \bigcup \{P_u \times P_w \mid \exists\, v \in V \text{ with } r \not\subseteq \overline{Q}_{((u,v)} \text{ and } h \not\subseteq \overline{Q}_{((v,w)}\}. \end{aligned}$$

Now, the above equality leads us to the following concepts.

Definition 4.3.1 Let (r, R) and (h, H) be direlations from $(U, \mathcal{U})$ to $(V, \mathcal{V})$ and from $(V, \mathcal{V})$ to $(W, \mathcal{W})$, respectively. The *composition* of (r, R) and (h, H) is defined by $(h, H) \circ (r, R) = (h \circ r, R \circ H)$ where

$$h \circ r = \bigvee \{\overline{P}_{(u,w)} \mid \exists\, v \in V \ \text{ with } \ r \not\subseteq \overline{Q}_{(u,v)} \text{ and } h \not\subseteq \overline{Q}_{(v,w)}\}$$

and

$$H \circ R = \bigcap \{\overline{Q}_{(u,w)} \mid \exists\, v \in V \text{ with } \overline{P}_{(u,v)} \not\subseteq R \text{ and } \overline{P}_{(v,w)} \not\subseteq H\}.$$

In the above definition, we see that for discrete textures, the (textural) composition of r and h is nothing but the ordinary composition of r and h. However, since $H \circ R$ is dually defined, compositions of corelations are not the same as the ordinary composition of relations:

$$\begin{aligned} H \circ R &= \bigcap \{\overline{Q}_{(u,w)} \mid \exists\, v \in V \text{ with } \overline{P}_{(u,v)} \not\subseteq R \text{ and } \overline{P}_{(v,w)} \not\subseteq H\} \\ &= \bigcap \{(U \setminus \{u\}) \times W \cup (U \times W \setminus \{w\}) \mid \exists\, v \in V \ (u, v) \notin R \text{ and } (v, w) \notin H\} \\ &= (U \times W) \setminus (\bigcup \{(\{u\} \times W) \cap (U \times \{w\}) \mid \exists\, v \in V \ (u, v) \notin R \text{ and } (v, w) \notin H\}) \\ &= (U \times W) \setminus \{(u, v) \mid \exists\, v \in V \ (u, v) \in (U \times V) \setminus R \text{ and } (v, w) \in (V \times W) \setminus H\}) \\ &= (U \times W) \setminus \{(u, v) \mid \exists\, v \in V \ (u, v) \in ((U \times V) \setminus R) \circ ((V \times W) \setminus H)\}) \\ &= (U \times W) \setminus (((U \times V) \setminus R) \circ ((V \times W) \setminus H)). \end{aligned}$$

Now, we have

Theorem 4.3.2 *The pair $(h \circ r, H \circ R)$ is a direlation from $(U, \mathcal{U})$ to $(W, \mathcal{W})$.*

Proof For $R1$, let $h \circ r \not\subseteq \overline{Q}_{(u,w)}$ and $P_{u'} \not\subseteq Q_u$. Then for some $u'' \in U$ and $w'' \in W$, we have $\overline{P}_{(u'',w'')} \not\subseteq \overline{Q}_{(u,w)}$ and

$$\exists v_1 \in V,\ r \not\subseteq \overline{Q}_{(u'',v_1)} \text{ and } h \not\subseteq \overline{Q}_{(v_1,w'')}.$$

By Lemma 4.1.2, $u = u''$ and $P_{w''} \not\subseteq Q_w$. This gives that $h \not\subseteq \overline{Q}_{(u,v_1)}$, that is, $\overline{P}_{(u',w'')} \subseteq h \circ r$. If we consider Lemma 4.1.2, we obtain $h \circ r \not\subseteq \overline{Q}_{(u',w)}$. We showed that $h \circ r$ satisfies the condition R1. The rest of the proof is left to the interested reader. □

Definition 4.3.3 Let (r, R) be a direlation on a texture $(U, \mathcal{U})$. Then

(i) r is *transitive* if $r \circ r \subseteq r$.

(ii) R is *transitive* if $R \subseteq R \circ R$.

(iii) (r, R) is called *transitive* if r and R are transitive.

(iv) If (r, R) is reflexive, symmetric and transitive, then it is called *an equivalence direlation.*

Theorem 4.3.4 *Let (r, R) be a complemented direlation on a discrete texture $(U, \mathcal{P}(U))$. Then the following statements are equivalent:*

(i) *(r, R) is an equivalence direlation on $(U, \mathcal{P}(U))$.*

(ii) *r is an equivalence relation on U.*

Proof For the proof, it is enough to consider the following observations:

$i \subseteq r \Longleftrightarrow \{(u, u) \mid u \in U\} \subseteq r.$

$r^{\leftarrow} = R \Longleftrightarrow (U \times U) \setminus r^{-1} = (U \times U) \setminus r \Longleftrightarrow r = r^{-1}.$

The textural composition is the same as the ordinary composition:

$$\begin{aligned} r \circ r &= \bigvee\{\overline{P}_{(u,v)} \mid \exists w \in U \text{ with } r \not\subseteq \overline{Q}_{(u,v)} \text{ and } r \not\subseteq \overline{Q}_{(v,w)}\} \\ &= \{(u, v) \mid \exists w \in U \text{ with } (u, v) \in r \text{ and } (v, w) \in r\}. \end{aligned}$$

The composition of direlations holds many desirable properties:

Lemma 4.3.5 *Let $A \in \mathcal{U}$ and $B \in \mathcal{W}$. Then*

(i) $(h \circ r)^{\rightarrow} A = h^{\rightarrow}(r^{\rightarrow} A)$.

(ii) $(H \circ R)^{\rightarrow} A = H^{\rightarrow}(R^{\rightarrow} A)$.

(iii) $(h \circ r)^{\leftarrow} B = r^{\leftarrow}(h^{\leftarrow} B)$.

(iv) $(H \circ R)^{\leftarrow} B = R^{\leftarrow}(H^{\leftarrow} B)$.

Proof We prove (i) leaving the items (ii)-(iv) to the interested reader. Let $(h \circ r)^{\rightarrow} A \nsubseteq h^{\rightarrow}(r^{\rightarrow} A)$. Then there exists a $w \in W$ such that

$$(h \circ r)^{\rightarrow} A \nsubseteq Q_w \text{ and } P_w \nsubseteq h^{\rightarrow}(r^{\rightarrow} A).$$

By definition of a section, $h \circ r \nsubseteq \overline{Q}_{(u,w)}$ and $A \nsubseteq Q_u$ for some $u \in U$. Further, by Theorem 4.2.13 (ii), for some $w' \in W$, the inclusion $r^{\rightarrow} A \subseteq h^{\leftarrow} Q_{w'}$ holds where $P_w \nsubseteq Q_{w'}$. Clearly, we also have $r^{\rightarrow} A \subseteq h^{\leftarrow} Q_w$. Because, note that $Q_{w'} \subseteq Q_w$ and so by Lemma 4.2.14 (ii), $h^{\leftarrow} Q_{w'} \subseteq h^{\leftarrow} Q_w$. By definition of composition, for some $v' \in V$,

$$r \nsubseteq \overline{Q}_{(u,v')} \text{ and } h \nsubseteq \overline{Q}_{(v',w'')}$$

where $P_{w''} \nsubseteq Q_w$. By the second statement above, $h \nsubseteq \overline{Q}_{(v',w)}$ and hence, by Lemma 4.2.3 (i), we get $h^{\rightarrow} P_{v'} \nsubseteq Q_w$. First, by Lemma 4.2.3 (i), $r \nsubseteq \overline{Q}_{(u,v')}$ gives that $r^{\rightarrow} P_u \nsubseteq Q_{v'}$. Therefore, $r^{\rightarrow} A \nsubseteq Q_{v'}$, since $P_u \subseteq A$ implies $r^{\rightarrow} P_u \subseteq r^{\rightarrow} A$. On the other hand, by Lemma 4.1.23 (i), $\overline{P}_{w,v')} \nsubseteq h^{\leftarrow}$ and since $h^{\leftarrow}$ is a corelation, by Lemma 4.2.3 (ii), $P_{v'} \nsubseteq (h^{\leftarrow})^{\rightarrow} Q_w = h^{\leftarrow} Q_w$. Then we obtain $r^{\leftarrow} A \nsubseteq h^{\leftarrow} Q_w$ which is an immediate contradiction.

For the reverse inclusion, suppose that $h^{\rightarrow}(r^{\rightarrow} A) \nsubseteq (h \circ r)^{\rightarrow} A$. Choose a $w \in W$ such that

$$h^{\rightarrow}(r^{\rightarrow} A) \nsubseteq Q_w \text{ and } P_w \nsubseteq (h \circ r)^{\rightarrow} A.$$

By the first statement $r^{\rightarrow} A \nsubseteq h^{\leftarrow} Q_w$ and by the second, we get

$$\forall u \in U,\ h \circ r \nsubseteq \overline{Q}_{(u,w')} \implies A \subseteq Q_u \quad (*)$$

for some $w' \in W$ with $P_w \nsubseteq Q_{w'}$. We clearly have that $Q_{w'} \subseteq Q_w$ and then $h^{\rightarrow} Q_{w'} \subseteq h^{\rightarrow} Q_w$. This also implies that $r^{\rightarrow} A \nsubseteq h^{\leftarrow} Q_{w'}$. Let us take a $v' \in V$ such that $r^{\rightarrow} A \nsubseteq Q_{v'}$ and $P_{v'} \nsubseteq h^{\leftarrow} Q_{w'}$. By definition of a section, we have $r \nsubseteq \overline{Q}_{(u,v')}$ and $A \nsubseteq Q_u$ for some $u \in U$. Note that by (*), we have $h \circ r \subseteq \overline{Q}_{(u,w')}$. Further, $P_{v'} \nsubseteq h^{\leftarrow} Q_{w'}$ gives that $h \nsubseteq \overline{Q}_{(v',w')}$. Hence, we conclude that $\overline{P}_{(u,w')} \subseteq h \circ r$. However, this is impossible since $\overline{P}_{(u,w)} \nsubseteq \overline{Q}_{(u,w')}$. □

Theorem 4.3.6 *Let (r, R) be a direlation from $(U, \mathcal{U})$ to $(V, \mathcal{V})$, and (i_U, I_U) and (i_V, I_V) be the identity direlations on $(U, \mathcal{U})$ and $(V, \mathcal{V})$, respectively. Then*

(i) $(r, R) \circ (i_U, I_U) = (r, R)$.

(ii) $(i_V, I_V) \circ (r, R) = (r, R)$.

(iii) *Let (h, H) be a direlation from $(V, \mathcal{V})$ to $(W, \mathcal{W})$ and (k, K) be a direlation from $(W, \mathcal{W})$ to $(Z, \mathcal{Z})$. Then the composition of direlations is associative, that is,*

$$((k, K) \circ (h, H)) \circ (r, R) = (k, K) \circ ((h, H) \circ (r, R)).$$

Proof (i) Since $(r, R) \circ (i_U, I_U) = (r \circ i_U, R \circ I_U)$, we must show that $r \circ i_U = r$ and $R \circ I_U = R$. Let $A \in \mathcal{U}$. Note that for the first equality, by Theorem 4.2.5 and Lemma 4.3.5 (i), we have

$$(r \circ i_U)^{\rightarrow} A = r^{\rightarrow}(i_U^{\rightarrow} A) = r^{\rightarrow} A.$$

Thus, by Lemma 4.2.7 (i), we obtain that $r \circ i_U = r$. The proof of the second equality is similar.

(ii) The proof is left to the reader.

(iii) Let $A \in \mathcal{U}$. Then by Lemma 4.3.5 (i), we conclude that

$$\begin{aligned}(k \circ (h \circ r))^{\rightarrow} A &= k^{\rightarrow}(h \circ r)^{\rightarrow} A = k^{\rightarrow}(h^{\rightarrow} r^{\rightarrow} A) \\ &= (k \circ h)^{\rightarrow} r^{\rightarrow} A = ((k \circ h) \circ r)^{\rightarrow} A.\end{aligned}$$

Hence, by Theorem 4.2.5, we obtain $k \circ (h \circ r) = (k \circ h) \circ r$. Following a similar way, one can show that $K \circ (H \circ R) = (K \circ H) \circ R$. As a result, the desired equality holds.

Theorem 4.3.7 *Let $(U, \mathcal{U}, c_U)$, $(V, \mathcal{V}, c_V)$ and $(W, \mathcal{W}, c_W)$ be complemented textures. If (r, R) and (h, H) are direlations from $(U, \mathcal{U})$ to $(V, \mathcal{V})$ and from $(V, \mathcal{V})$ to $(W, \mathcal{W})$, respectively, then we have the following propositions:*

(i) $(h \circ r)' = h' \circ r'$.

(ii) $(H \circ R)' = H' \circ R'$.

Proof Let $A \in \mathcal{U}$. By Lemmas 4.2.3 (i) and 4.2.27 (i), we easily observe that (i)

$$\begin{aligned}(h' \circ r')^{\rightarrow} A &= (h')^{\rightarrow}((r')^{\rightarrow} A) \\ &= (h')^{\rightarrow} c_V(r^{\rightarrow} c_U(A)) \\ &= c_W(h^{\rightarrow} c_V(c_V r^{\rightarrow} c_U(A))) \\ &= c_W h^{\rightarrow}(r^{\rightarrow} c_U A)) \\ &= c_W((h \circ r)^{\rightarrow}(c_U(A)) \\ &= ((h \circ r)')^{\rightarrow} A.\end{aligned}$$

Finally, by Lemma 4.2.7 (i), we obtain the equality as desired. The proof of (ii) is similar.

Theorem 4.3.8 *If (r, R) is a reflexive and transitive direlation, then*

$$\forall A \in \mathcal{U}, r^{\leftarrow} r^{\leftarrow} A = r^{\leftarrow} A \text{ and } R^{\leftarrow} R^{\leftarrow} A = R^{\leftarrow} A.$$

Proof By Theorem 4.2.16 (i), $r^{\leftarrow} A \subseteq A$ and so by Lemma 4.2.14 (ii) we obtain $r^{\leftarrow} r^{\leftarrow} A \subseteq r^{\leftarrow} A$. Further, since r is transitive, $r \circ r \subseteq r$ and so by Lemma 4.2.7 (ii), $r^{\leftarrow} \subseteq (r \circ r)^{\leftarrow}$. This implies that

$$r^{\leftarrow}A \subseteq (r \circ r)^{\leftarrow}A \subseteq r^{\leftarrow}r^{\leftarrow}A.$$

The proof of the second equality is dual. □

Theorem 4.3.9 *If* (r, R) *is a symmetric direlation on* $(U, \mathcal{U})$*, then*

$$\forall A \in \mathcal{U},\ A \subseteq r^{\leftarrow}R^{\leftarrow}A, \text{ and } R^{\rightarrow}r^{\rightarrow}A \subseteq A.$$

Proof For the first inclusion, suppose that $A \not\subseteq r^{\leftarrow}R^{\leftarrow}A$. Let us take a $u \in U$ such that $A \not\subseteq Q_u$ and $P_u \not\subseteq r^{\leftarrow}R^{\leftarrow}A$. By Theorem 4.2.13 (i), $r^{\rightarrow}P_u \not\subseteq R^{\leftarrow}A$. Since (r, R) is symmetric,

$$r^{\rightarrow}P_u \not\subseteq R^{\leftarrow}A = (R^{\leftarrow})^{\rightarrow}A = r^{\rightarrow}A,$$

that is, $r^{\rightarrow}P_u \not\subseteq r^{\rightarrow}A$. By Lemma 4.2.14 (i), this implies that $P_u \not\subseteq A$ whence $A \subseteq Q_u$ which is an immediate contradiction.

4.4 Seriality and Injectivity

As is known that the seriality and compatibility of a relation play an important role in rough set theory. Let us start with the seriality. Recall that a relation r from a set U to a set V is called *serial* if

for all $u \in U$, there exists a $v \in V$ such that $(u, v) \in r$.

Now let (r, R) be a direlation from $(U, \mathcal{P}(U))$ to $(V, \mathcal{P}(V))$ where $R = (U \times V) \setminus r$. For discrete textures, the seriality of r can be restated as follows:

r is serial if for all $u \in U$, there exists a $v \in V$ such that $r \not\subseteq \overline{Q}_{(u,v)}$.

Note that since $R = (U \times V) \setminus r$, we also have that $\overline{P}_{(u,v)} \not\subseteq R$. For discrete textures, if $u \in U$, we always have $P_u \not\subseteq Q_u$. However, for $u \neq u'$ where $u' \in U$, we have $P_u \subseteq Q_{u'}$. For general textures, this may not be necessarily true. If we consider the fuzzy texture $(M_{[0,1]}, \mathcal{M}_{[0,1]})$ given in Example 2.1.4 (ii), we see that for $u, u' \in M_{[0,1]}$ with $u' < u$, we have $P_u = (0, u] \not\subseteq (0, u'] = Q_{u'}$. Therefore, for many concepts of textures, $P_u \not\subseteq Q_{u'}$ is case where $u \neq u'$.

The above argument leads us to the following definition in textures:

Definition 4.4.1 Let (r, R) be a direlation from $(U, \mathcal{U})$ to $(V, \mathcal{V})$. Then we call (r, R) *t-serial* if

$$\forall u, u' \in U, P_u \not\subseteq Q_{u'} \implies \exists\, v \in V \text{ with } r \not\subseteq \overline{Q}_{(u,v)} \text{ and } \overline{P}_{(u',v)} \not\subseteq R.$$

Theorem 4.4.2 *Let* (r, R) *be a direlation from* $(U, \mathcal{U})$ *to* $(U, \mathcal{U})$*.*

(i) *If* (r, R) *is reflexive, then it is t-serial.*

(ii) *If* (r, R) *is t-serial, symmetric and transitive, then it is reflexive.*

Proof (i) Let $u, u' \in U$ and $P_u \not\subseteq Q_{u'}$. Then by definition of i, we get $i \not\subseteq \overline{Q}_{(u,u')}$. Take $v \in U$ such that $i \not\subseteq \overline{Q}_{(u,v)}$ and $\overline{P}_{(u,v)} \not\subseteq \overline{Q}_{(u,u')}$. This gives that $P_v \not\subseteq Q_{u'}$. By definition of I, we have $\overline{P}_{(u',v)} \not\subseteq I$ and by reflexivity of R, we obtain that $\overline{P}_{(u',v)} \not\subseteq R$. That is (r, R) is t-serial.

(ii) Suppose that $i \not\subseteq r$. Take $u, v \in U$ such that $i \not\subseteq \overline{Q}_{(u,v)}$ and $\overline{P}_{(u,v)} \not\subseteq r$. Then we have $P_u \not\subseteq Q_v$ and by t-seriality of (r, R), we get $r \not\subseteq \overline{Q}_{(u,v')}$ and $\overline{P}_{(v,v')} \not\subseteq R$. This gives that $R^{\leftarrow} \not\subseteq \overline{Q}_{(u,v')}$ and $\overline{P}_{(v,v')} \not\subseteq r^{\leftarrow}$ and hence, $\overline{P}_{(v',u)} \not\subseteq R$ and $r \not\subseteq \overline{Q}_{(v',v)}$. Hence, by definition of composition, we conclude that $\overline{P}_{(u,v)} \subseteq r \circ r$ and then we find $\overline{P}_{(u,v)} \subseteq r$ which is an immediate contradiction. Using a similar argument we may show that $R \subseteq I$.

Note that a relation $r \subseteq U \times V$ is *inverse serial* if for all $v \in V$ there exists a $u \in U$ such that $(u, v) \in r$. For textures, the inverse seriality of a direlation from $(U, \mathcal{U})$ to $(V, \mathcal{V})$ is the following:

Definition 4.4.3 Let (r, R) be a direlation from $(U, \mathcal{U})$ to $(V, \mathcal{V})$. Then we call (r, R) *t-inverse serial* if

$$\forall v, v' \in V, P_v \not\subseteq Q_{v'} \implies \exists u \in U, r \not\subseteq \overline{Q}_{(u,v')} \text{ and } \overline{P}_{(u,v)} \not\subseteq R.$$

Theorem 4.4.4 *Let (r, R) be a direlation from $(U, \mathcal{U})$ to $(V, \mathcal{V})$. Then*

(i) *(r, R) is t-serial iff $(r, R)^{\leftarrow}$ is t-inverse serial.*

(ii) *(r, R) is t-inverse serial iff $(r, R)^{\leftarrow}$ is t-serial.*

Proof (i) For the necessity, let $P_u \not\subseteq Q_{u'}$ for some $u, u' \in U$. Since (r, R) is t-serial, we have a $v \in V$ with $r \not\subseteq \overline{Q}_{(u,v)}$ and $\overline{P}_{(u',v)} \not\subseteq R$. Then by Lemma 4.1.23 (i), we obtain $\overline{P}_{(v,u)} \not\subseteq r^{\leftarrow}$ and $R^{\leftarrow} \not\subseteq \overline{Q}_{(v,u')}$. The sufficiency and the rest of the proof is similar.

Corollary 4.4.5 *Let (r, R) be a direlation from $(U, \mathcal{P}(U))$ to $(V, \mathcal{P}(V))$. Then we have the following implication.*

(i) *If (r, R) is t-serial, then r and $(U \times V) \setminus R$ are serial.*

(ii) *If (r, R) is t-inverse serial, then r and $(U \times V) \setminus R$ are inverse serial.*

Proof (i) Let $u \in U$. Clearly, we have $\{u\} \not\subseteq U \setminus \{u\}$, that is, $P_u \not\subseteq Q_u$. Then by the assumption, $r \not\subseteq \overline{Q}_{(u,v)}$ and $P_{(u,v)} \not\subseteq R$ for some $v \in V$. Hence, for the discrete textures, we get $(u, v) \in r$ and $(u, v) \notin R$, that is, $(u, v) \in (U \times V) \setminus R$. (ii) The proof is similar.

In particular, if (r, R) is complemented, then the above implication turn into an equivalence:

Corollary 4.4.6 *Let (r, R) be a complemented direlation from $(U, \mathcal{P}(U))$ to $(V, \mathcal{P}(V))$. Then*

(i) *(r, R) is t-serial if and only if r is serial.*

(ii) *(r, R) is t-inverse serial if and only if r is inverse serial.*

Proof (i) Let r be serial and $P_u \not\subseteq Q_{u'}$. Then we have $\{u\} \not\subseteq U \setminus \{u'\}$ and so $u = u'$. By the assumption, we get $(u, v) \in r$ for some $v \in V$. Hence, $(u, v) \notin (U \times V) \setminus r$, that is, $r \not\subseteq \overline{Q}_{(u,v)}$ and $\overline{P}_{(u',v)} \notin (U \times V) \setminus r = R$. Sufficiency follows from Corollary 4.4.5.

(ii) The proof is similar to (i).

Finally, the compatibility of a direlation can be given as

Definition 4.4.7 A direlation (r, R) is called *t-compatible* if it is t-serial and t-inverse serial.

Definition 4.4.8 A direlation (r, R) is called *injective* if

$$\forall u, u' \in U, v \in V, \ r \not\subseteq \overline{Q}_{(u,v)} \text{ and } \overline{P}_{(u',v)} \not\subseteq R \implies P_u \not\subseteq Q_{u'}.$$

Example 4.4.9 Consider the textures $(U, \mathcal{U})$ and $(V, \mathcal{V})$ where

$U = \{a, b, c, d\}$,
$\mathcal{U} = \{\emptyset, \{a\}, \{b\}, \{a, b\}, \{a, b, c\}, \{b, c\}, \{a, d\}, \{a, b, d\}, U\}$

and

$V = \{1, 2, 3\}$,
$\mathcal{V} = \{\emptyset, \{2\}, \{3\}, \{1, 2\}, \{2, 3\}, V\}$.

Let us consider the direlation (r, R) from $(U, \mathcal{U})$ to $(V, \mathcal{V})$ with

$r = \{(c, 1), (b, 2), (c, 2), (d, 2), (a, 3), (d, 3)\}$

and

$R = \{(d, 1), (a, 2), (c, 2), (d, 2), (b, 3), (c, 3)\}$.

Note that for the texturing $\mathcal{U}$, p-sets and q-sets are

$P_a = \{a\}$, $P_b = \{b\}$, $P_c = \{b, c\}$, $P_d = \{a, d\}$, and

$Q_a = \{b, c\}$, $Q_b = \{a, d\}$, $Q_c = \{a, b, d\}$, $Q_d = \{a, b, c\}$.

For the texturing $\mathcal{V}$, p-sets and q-sets are

$P_1 = \{1, 2\}$, $P_2 = \{2\}$, $P_3 = \{3\}$, and

$Q_1 = \{2, 3\}$, $Q_2 = \{3\}$, $Q_3 = \{1, 2\}$.

It is easy to see that (r, R) is t-serial. For instance, note that we have $P_a \not\subseteq Q_a$. In this case,

$$\begin{aligned}\overline{Q}_{(a,3)} &= (U \setminus \{a\} \times V) \cup (U \times Q_3)\\ &= (\{b, c, d\} \times V) \cup (U \times \{1, 2\})\\ &= \{(b, 1), (b, 2), (b, 3), (c, 1), (c, 2), (c, 3), (d, 1),\\ &(d, 2), (d, 3), (a, 1), (a, 2)\}\end{aligned}$$

and $\overline{P}_{(a,3)} = \{a\} \times P_3 = \{(a, 3)\}$. Clearly, we have $r \nsubseteq \overline{Q}_{(a,3)}$ and $\overline{P}_{(a,3)} \nsubseteq R$. That is, for $P_a \nsubseteq Q_a$, the condition of t-seriality holds. Likewise, for $P_c \nsubseteq Q_b$, since

$$\begin{aligned}\overline{Q}_{(c,1)} &= (U \setminus \{c\} \times V) \cup (U \times Q_1)\\ &= (\{a, b, d\} \times V) \cup (U \times \{2, 3\})\\ &= \{(a, 1), (a, 2), (a, 3), (b, 1), (b, 2), (b, 3),\\ &(d, 1), (d, 2), (d, 3), (c, 2), (c, 3)\}\end{aligned}$$

and $\overline{P}_{(b,1)} = \{b\} \times P_1 = \{(b, 1), (b, 2)\}$, we clearly have $r \nsubseteq \overline{Q}_{(c,1)}$ and $\overline{P}_{(b,1)} \nsubseteq R$. This means that for $P_c \nsubseteq Q_b$, the condition of t-seriality holds. Similarly, for the cases $P_b \nsubseteq Q_b$, $P_c \nsubseteq Q_c$, $P_d \nsubseteq Q_d$ and $P_d \nsubseteq Q_a$, we may show that the condition of t-seriality holds. Hence, the direlation (r, R) is t-serial. Further, (r, R) is t-inverse serial. Indeed, for $P_1 \nsubseteq Q_1$, we have $r \nsubseteq \overline{Q}_{(c,1)}$ and $\overline{P}_{(c,1)} \nsubseteq R$. Similarly, for the cases $P_1 \nsubseteq Q_2$, $P_2 \nsubseteq Q_2$ and $P_3 \nsubseteq Q_3$, the condition of t-inverse seriality also holds. As a result, (r, R) is t-compatible.

Corollary 4.4.10 *Let (r, R) be a complemented direlation from $(U, \mathcal{P}(U))$ to $(V, \mathcal{P}(V))$. Then*

(i) *(r, R) is injective if and only if r is injective.*

(ii) *If $(R^{\leftarrow}, r^{\leftarrow})$ is injective if and only if r^{-1} is injective.*

Proof We give proof of (i) leaving the proof of (ii) to the interested reader. Let (r, R) be injective. We show that r is injective. Let us take $v \in V$, $u, u' \in U$ where $(u, v) \in r$ and $(u', v) \in r$. Then we have $r \nsubseteq \overline{Q}_{(u,v)}$ and $\overline{P}_{(u',v)} \nsubseteq R = (U \times V) \setminus r$. By the assumption, $\{u\} = P_u \nsubseteq Q_{u'} = U \setminus \{u'\}$ and then we get $u = u'$ as desired. Now let r be injective. If $r \nsubseteq \overline{Q}_{(u,v)}$ and $\overline{P}_{(u',v)} \nsubseteq R = (U \times V) \setminus r$, then $(u, v) \in r$ and $(u', v) \in r$. Since r is injective, $u = u'$ and hence, $P_u = \{u\} \nsubseteq U \setminus \{u'\} = Q_{u'}$.

For t-seriality, we have the following equivalences:

Theorem 4.4.11 *Let (r, R) be a direlation from $(U, \mathcal{U})$ to $(V, \mathcal{V})$. The following statements are equivalent:*

(i) *(r, R) is t-serial.*

(ii) $r^{\leftarrow} \circ R \subseteq I_U$.

(iii) $\forall A \in \mathcal{U},\ r^{\leftarrow}(R^{\rightarrow} A) \subseteq A$.

(iv) $\forall B \in \mathcal{V},\ r^{\leftarrow} B \subseteq R^{\leftarrow} B$.

(v) $\forall A_1, A_2 \in \mathcal{U},\ r^{\rightarrow} A_1 \subseteq R^{\rightarrow} A_2 \implies A_1 \subseteq A_2$.

Proof (i) $\Longrightarrow$ (ii): Suppose that $r^{\leftarrow} \circ R \not\subseteq I_U$. Then there exist some $u_1, u_2 \in U$ such that $r^{\leftarrow} \circ R \not\subseteq \overline{Q}_{(u_1,u_2)}$ and $\overline{P}_{(u_1,u_2)} \not\subseteq I_U$. From definition of I_U, we have $P_{u_1} \not\subseteq Q_{u_2}$. By the assumption, there exists a $v \in U$ with $r \not\subseteq \overline{Q}_{(u_1,v)}$ and $R \not\subseteq \overline{Q}_{(u_2,v)}$. Note that by Lemma 4.1.23 (i), we also have $\overline{P}_{(v,u_1)} \not\subseteq r^{\leftarrow}$. Hence, from definition of the composition of corelations, we get $r^{\leftarrow} \circ R \subseteq \overline{Q}_{(u_1,u_2)}$. However, this is an immediate contradiction.

(ii) $\Longrightarrow$ (iii): The proof is immediate by Lemma 4.2.7 (ii) and Theorem 4.2.5.

(iii) $\Longrightarrow$ (iv): From Lemma 4.2.15 (2)(ii), for all $B \in \mathcal{V}$, we have $B \subseteq R^{\rightarrow}(R^{\leftarrow}B)$. if we take $A = R^{\leftarrow}B$, then by the assumption, we get $r^{\leftarrow}(R^{\rightarrow}R^{\leftarrow}B) \subseteq R^{\leftarrow}B$. On the other hand, by Lemma 4.2.14 (ii), we already have $r^{\leftarrow}B \subseteq r^{\leftarrow}(R^{\rightarrow}R^{\leftarrow}B)$. This gives the desired inclusion $r^{\leftarrow}B \subseteq R^{\leftarrow}B$.

(iv) $\Longrightarrow$ (v): Suppose that $r^{\rightarrow}A_1 \subseteq R^{\rightarrow}A_2$. Then by Lemma 4.2.14 (ii) and Lemma 4.2.15 (1) (ii), the inclusions $A_1 \subseteq r^{\leftarrow}r^{\rightarrow}A_1 \subseteq r^{\leftarrow}R^{\rightarrow}A_2$ hold. By the assumption and by Lemma 4.2.15 (1) (ii), we conclude that $r^{\leftarrow}R^{\rightarrow}A_2 \subseteq R^{\leftarrow}R^{\rightarrow}A_2 \subseteq A_2$, that is, $A_1 \subseteq A_2$ as desired.

(v) $\Longrightarrow$ (i): Let $u, u' \in U$ and $P_u \not\subseteq Q_{u'}$. By the assumption, we have $r^{\rightarrow}P_u \not\subseteq R^{\rightarrow}Q_{u'}$. Let us choose a $v \in V$ such that $r^{\rightarrow}P_u \not\subseteq Q_v$ and $P_v \not\subseteq R^{\rightarrow}Q_{u'}$. Then by Lemma 4.2.3, we get $r \not\subseteq \overline{Q}_{(u,v)}$ and $\overline{P}_{(u',v)} \not\subseteq R$, that is, (r, R) is t-serial.

The following theorem is a dual form of Theorem 4.4.11 and the proof is similar.

Theorem 4.4.12 *Let (r, R) be a direlation from $(U, \mathcal{U})$ to $(V, \mathcal{V})$. The following statements are equivalent:*

(i) *(r, R) is t-inverse serial.*

(ii) $r \circ R^{\leftarrow} \subseteq I_V$.

(iii) $\forall B \in \mathcal{V},\ B \subseteq r^{\rightarrow}(R^{\leftarrow}B)$.

(iv) $\forall A \in \mathcal{U},\ R^{\rightarrow}A \subseteq r^{\rightarrow}A$.

(v) $\forall B_1, B_2 \in \mathcal{V},\ R^{\leftarrow}B_1 \subseteq r^{\leftarrow}B_2 \implies B_1 \subseteq B_2$.

Theorem 4.4.13 *Let (r, R) be a t-inverse serial direlation. Then*

(i) $\forall v \in V,\ R^{\leftarrow}P_v \neq \emptyset$.

(ii) *If $\emptyset \neq B \in \mathcal{V}$, then $R^{\leftarrow}B \neq \emptyset$.*

(iii) *If $V \neq B \in \mathcal{V}$, then $r^{\leftarrow}B \neq U$.*

Proof (i) By Theorem 4.4.12 (iv), we have $R^{\rightarrow}(R^{\leftarrow}P_v) \subseteq r^{\rightarrow}(R^{\leftarrow}P_v)$ for $A = R^{\leftarrow}P_v$. Again by Theorem 4.4.12 (iv), we also get $P_v \subseteq r^{\rightarrow}(R^{\leftarrow}P_v)$. Suppose that $R^{\leftarrow}P_v = \emptyset$. Then by Lemma 4.2.15 (1)(i), $r^{\rightarrow}(R^{\leftarrow}P_v) = r^{\rightarrow}\emptyset = \emptyset$ gives that $P_v \subseteq \emptyset$. But this is a contradiction.

(ii) Since $B \neq \emptyset$, we may choose a $v \in B$. Then we have $P_v \subseteq B$ and by (i), $R^{\leftarrow} P_v \neq \emptyset$. By Lemma 4.2.14 (iv), we obtain $R^{\leftarrow} P_v \subseteq R^{\leftarrow} B$. This gives that $R^{\leftarrow} B \neq \emptyset$.

(iii) Take $B \in \mathcal{V}$. Suppose that $V \neq B$ and $r^{\leftarrow} B = U$. By Theorem 4.4.12 (iv), we have $R^{\rightarrow}(r^{\leftarrow} B) \subseteq r^{\rightarrow}(r^{\leftarrow} B)$ for $A = r^{\leftarrow} B$. By Lemma 4.2.15 (1)(iii), we get $r^{\rightarrow}(r^{\leftarrow} B) \subseteq B$ and so $R^{\rightarrow}(r^{\leftarrow} B) \subseteq B$. This implies that $R^{\rightarrow} U \subseteq B$. However, by Lemma 4.2.15 (2)(i), we know that the equality $R^{\rightarrow} U = V$ holds and this is a contradiction since $V \neq B$.

Theorem 4.4.14 *Let* (r, R) *be a t-serial direlation. Then*

(i) $\forall u \in U, \; R^{\rightarrow} P_u \neq \emptyset$.

(ii) *If* $\emptyset \neq A \in \mathcal{U}$, *then* $R^{\rightarrow} A \neq \emptyset$.

(iii) *If* $U \neq A \in \mathcal{U}$, *then* $r^{\rightarrow} A \neq V$.

Theorem 4.4.15 *Let* (r, R) *be a direlation from* $(U, \mathcal{U})$ *to* $(V, \mathcal{V})$.

(1) If (r, R) is t-serial, then for all $A \in \mathcal{U}$,

$A \subseteq R^{\leftarrow}(r^{\rightarrow} A)$.

(2) If (r, R) is injective, then for all $A \in \mathcal{U}$,

(i) $R^{\leftarrow}(r^{\rightarrow} A) \subseteq A$.

(ii) $r^{\rightarrow} A \subseteq R^{\rightarrow} A$.

(iii) $A \subseteq r^{\leftarrow}(R^{\rightarrow} A)$.

(3) If (r, R) is t-inverse serial and injective, then for all $A \in \mathcal{U}$,

$r^{\rightarrow} A = R^{\rightarrow} A$.

(4) If (r, R) is injective and t-serial, then for all $A \in \mathcal{U}$, we have the following equalities.

(i) $r^{\leftarrow}(r^{\rightarrow} A) = A$.

(ii) $R^{\leftarrow}(R^{\rightarrow} A) = A$.

Proof (1) By Lemma 4.2.15 (1)(ii) and Theorem 4.4.11 (iv), we get $A \subseteq r^{\leftarrow}(r^{\rightarrow} A) \subseteq R^{\leftarrow}(r^{\rightarrow} A)$.

(2) (i) Suppose that $R^{\leftarrow}(r^{\rightarrow} A) \not\subseteq A$. Take a $u \in U$ such that $R^{\leftarrow}(r^{\rightarrow} A) \not\subseteq Q_u$ and $P_u \not\subseteq A$. By definition of a presection, there exists a $v \in V$ such that $\overline{P}_{(u,v)} \not\subseteq R$ and $r^{\rightarrow} A \not\subseteq Q_v$. Then by definition of a section, we have $r \not\subseteq \overline{Q}_{(u',v)}$ and $A \not\subseteq Q_{u'}$ for some $u' \in U$. Since (r, R), is injective, we find $P_{u'} \not\subseteq Q_u$ and so $A \not\subseteq Q_u$. That is, $P_u \subseteq A$ and this gives a contradiction.

(ii) Suppose that $r^{\rightarrow} A \not\subseteq R^{\rightarrow} A$. Take a $v \in V$ such that $r^{\rightarrow} A \not\subseteq Q_v$ and $P_v \not\subseteq R^{\rightarrow} A$. Then we have $r \not\subseteq \overline{Q}_{(u,v)}$ and $A \not\subseteq Q_u$ for some $u \in U$. On the other hand, we get

$\overline{P}_{(u',v)} \not\subseteq R$ and $P_{u'} \not\subseteq A$ for some $u' \in U$. Since (r, R) is injective, $P_u \not\subseteq Q_{u'}$ implies that $P_u \not\subseteq A$. But this is a contradiction.

(iii) By (2) (ii) and Lemmas 4.2.14(ii) and 4.2.15(1)(ii), we have $A \subseteq r^{\leftarrow}(r^{\rightarrow}A) = r^{\leftarrow}(R^{\rightarrow}A)$.

(3) By the item (2) (ii) and by Theorem 4.4.12 (iv), we have $R^{\rightarrow}A \subseteq r^{\rightarrow}A$. Then we obtain $r^{\rightarrow}A = R^{\rightarrow}A$.

(4) (i) By the items (1) and (2)(i), and Theorem 4.4.11 (iii), we get $A = r^{\leftarrow}(R^{\rightarrow}A) = r^{\leftarrow}(r^{\rightarrow}A)$.

(ii) By (2) (iii), Theorem 4.4.11 (iv) and Lemma 4.2.15 (2) (iii), we obtain

$$A = r^{\leftarrow}(R^{\rightarrow}A) \subseteq R^{\leftarrow}(R^{\rightarrow}A) \subseteq A.$$

Therefore, we conclude that $A = R^{\leftarrow}(R^{\rightarrow}A)$.

Dually, we have the following result:

Theorem 4.4.16 *Let* (r, R) *be a direlation from* $(U, \mathcal{U})$ *to* $(V, \mathcal{V})$.

(1) If (r, R) is t-inverse serial, then for all $B \in \mathcal{V}$,

$$R^{\rightarrow}(r^{\leftarrow}B) \subseteq B.$$

(2) If $(R^{\leftarrow}, r^{\leftarrow})$ is injective, then for all $B \in \mathcal{V}$,

(i) $B \subseteq R^{\rightarrow}(r^{\leftarrow}B)$.

(ii) $R^{\leftarrow}B \subseteq r^{\leftarrow}B$.

(iii) $r^{\rightarrow}(R^{\leftarrow}B) \subseteq B$.

(3) If (r, R) is t-serial and $(R^{\leftarrow}, r^{\leftarrow})$ is injective, then for all $B \in \mathcal{V}$,

$$r^{\leftarrow}B = R^{\leftarrow}B.$$

(4) If (r, R) is t-inverse serial and $(R^{\leftarrow}, r^{\leftarrow})$ is injective, then for all $B \in \mathcal{V}$, we have the following equalities.

(i) $r^{\rightarrow}(r^{\leftarrow}B) = B$.

(ii) $R^{\rightarrow}(R^{\leftarrow}B) = B$.

Proof It is similar to the proof of Theorem 4.4.15.

Theorem 4.4.17 *Let* $(U, \mathcal{U})$ *and* $(V, \mathcal{V})$ *be any two textures and* (r, R) *be a direlation from* $(U, \mathcal{U})$ *to* $(V, \mathcal{V})$. *Then*

(1) $r^{\leftarrow}V = U$ *and* $R^{\leftarrow}\emptyset = \emptyset$.

(2) *For all* $B \in \mathcal{V}$, $R^{\leftarrow}(r^{\rightarrow}(r^{\leftarrow}(B)) \subseteq R^{\leftarrow}B$.

(3) *If* (r, R) *is t- serial, then* $r^{\leftarrow}\emptyset = \emptyset$ *and* $R^{\leftarrow}V = U$.

(4) *If* (r, R) *is t- serial, then* $R^{\leftarrow}B \subseteq R^{\leftarrow}(r^{\rightarrow}(R^{\leftarrow}B))$.

(5) *If* (r, R) *is t-inverse serial, then* $r^{\rightarrow}U = V$ *and* $R^{\rightarrow}\emptyset = \emptyset$.

Proof (1) Suppose that $r^{\leftarrow}V \neq U$. Take a point $u \in U$ with $U \nsubseteq Q_u$ and $P_u \nsubseteq r^{\leftarrow}V$. By definition of presection, there exists a $v \in V$ such that $r \nsubseteq \overline{Q}_{(u,v)}$ and $P_v \nsubseteq V$. However, this is a contradiction since V is a domain of discourse. Hence, we get $r^{\leftarrow}V = U$. For the second equality, note that by Corollary 2.2.4,

$$R^{\leftarrow}\emptyset = \bigcap\{Q_u \mid \forall v \in V,\ \overline{P}_{(u,v)} \nsubseteq R \implies \emptyset \subseteq Q_v\} = \bigcap\{Q_u \mid u \in U\} = \emptyset.$$

(2) By Lemma 4.2.15 (1) (iii), we have $r^{\rightarrow}(r^{\leftarrow}B) \subseteq B$. If we apply presection operations on both sides of the inclusion, by Lemma 4.2.14 (iv), we obtain $R^{\leftarrow}(r^{\rightarrow}(r^{\leftarrow}(B)) \subseteq R^{\leftarrow}B$.

(3) By Theorem 4.4.11 (iv), we have $r^{\leftarrow}\emptyset \subseteq R^{\leftarrow}\emptyset$. By (1), $R^{\leftarrow}\emptyset = \emptyset$ gives that $r^{\leftarrow}\emptyset = \emptyset$. For the proof of second equality, again by Theorem 4.4.11 (iv), we have $r^{\leftarrow}V \subseteq R^{\leftarrow}V$. By (1), $U \subseteq R^{\leftarrow}V$ and so $R^{\leftarrow}V = U$.

(4) By Lemma 4.2.15 (1) (ii), we have $A \subseteq r^{\leftarrow}(r^{\rightarrow}A)$. Then for $A = R^{\leftarrow}B$, we get $R^{\leftarrow}B \subseteq r^{\leftarrow}(r^{\rightarrow}(R^{\leftarrow}B))$. By Theorem 4.4.11 (iv), we obtain $r^{\leftarrow}(r^{\rightarrow}(R^{\leftarrow}B)) \subseteq R^{\leftarrow}(r^{\rightarrow}(R^{\leftarrow}B))$ and so $R^{\leftarrow}B \subseteq R^{\leftarrow}(r^{\rightarrow}(R^{\leftarrow}B))$.

(5) By Theorem 4.4.13 (ii), we have $R^{\rightarrow}U \subseteq r^{\rightarrow}U$. By Theorem 4.2.15 (2)(i), the equality $R^{\rightarrow}U = V$ gives that $r^{\rightarrow}U = V$. The proof of the second equality is similar.

Theorem 4.4.18 *Let (r, R) be a direlation from $(U, \mathcal{U})$ to $(V, \mathcal{V})$.*

(i) (a) *If (r, R) is injective, then $R^{\leftarrow}B \subseteq r^{\leftarrow}(r^{\rightarrow}(R^{\leftarrow}B))$.*

(b) *If $(R^{\leftarrow}, r^{\leftarrow})$ is injective, then $r^{\leftarrow}(r^{\rightarrow}(R^{\leftarrow}B)) \subseteq R^{\leftarrow}B$.*

(ii) *If $(R^{\leftarrow}, r^{\leftarrow})$ is injective and (r, R) is t-serial, then*

(a) $R^{\leftarrow}(r^{\rightarrow}(R^{\leftarrow}B)) = R^{\leftarrow}B$.

(b) $r^{\leftarrow}(R^{\rightarrow}(r^{\leftarrow}B)) = r^{\leftarrow}B$.

(c) $R^{\leftarrow}(R^{\rightarrow}(r^{\leftarrow}B)) = r^{\leftarrow}B$.

(d) $r^{\leftarrow}(R^{\rightarrow}(R^{\leftarrow}B)) = R^{\leftarrow}B$.

(iii) *If (r, R) is t-serial, then $R^{\leftarrow}(r^{\rightarrow}(r^{\leftarrow}B)) = r^{\leftarrow}B$.*

Proof (i) (a) By Lemma 4.2.15 (ii), $A \subseteq r^{\leftarrow}r^{\rightarrow}A$. If we take $A = R^{\leftarrow}B$, we obtain

$$R^{\leftarrow}B \subseteq r^{\leftarrow}(r^{\rightarrow}(R^{\leftarrow}B)).$$

(b) Note that from Theorem 4.4.16 (2) (iii), $r^{\rightarrow}(R^{\leftarrow}B) \subseteq B$ and so finally, if we apply Theorem 4.4.16 (2) (ii) and Lemma 4.2.14 (ii), we obtain that

$$r^{\leftarrow}r^{\rightarrow}(R^{\leftarrow}B) \subseteq r^{\leftarrow}B \subseteq R^{\leftarrow}B.$$

(ii) (a) By Theorem 4.4.16 (2) (iii), we have $r^{\rightarrow}(R^{\leftarrow}B) \subseteq B$. Then from Lemma 4.2.14 (iv), we get

$$R^{\leftarrow}r^{\rightarrow}(R^{\leftarrow}B) \subseteq R^{\leftarrow}B.$$

For the reverse inclusion, by Theorem 4.4.15 (1), we have $A \subseteq R^{\leftarrow}(r^{\rightarrow}A)$ and hence, if we take $A = R^{\leftarrow}B$, we conclude that

$$R^{\leftarrow}B \subseteq R^{\leftarrow}r^{\rightarrow}(R^{\leftarrow}B).$$ □

Theorem 4.4.19 *Let (r, R) be a direlation from $(U, \mathcal{U})$ to $(V, \mathcal{V})$.*

(i) (a) *If $(R^{\leftarrow}, r^{\leftarrow})$ is injective, then $R^{\rightarrow}A \subseteq r^{\rightarrow}(r^{\leftarrow}(R^{\rightarrow}A))$.*

(b) *If (r, R) is injective, then $r^{\rightarrow}(r^{\leftarrow}(R^{\rightarrow}A)) \subseteq R^{\rightarrow}A$.*

(ii) *If (r, R) is injective and (r, R) is t-inverse serial, then*

(a) $R^{\rightarrow}(r^{\leftarrow}(R^{\rightarrow}A)) = R^{\rightarrow}A$.

(b) $r^{\rightarrow}(R^{\rightarrow}(r^{\rightarrow}A)) = r^{\rightarrow}A$.

(c) $R^{\rightarrow}(R^{\leftarrow}(r^{\rightarrow}A)) = r^{\rightarrow}A$.

(d) $r^{\rightarrow}(R^{\leftarrow}(R^{\rightarrow}A)) = R^{\rightarrow}A$.

(iii) *If (r, R) is t-inverse serial, then $R^{\rightarrow}(r^{\leftarrow}(r^{\rightarrow}A)) = r^{\rightarrow}A$.*

Chapter 5
Rough Sets

5.1 Pawlak's Approximation Spaces

As is known that Pawlak's rough set theory arised from the fundamental requirements about the data mining due to information systems. An information system is a table which consists of some objects and attributes obtained from data. In rough set theory, an indiscernibility relation is the first step for the discovery of the basic knowledge hidden inside the information system. In this respect, as the main tools of the rough set theory, approximation operators and definable sets are purely mathematical notions. An indiscernibility relation is in fact, an equivalence relation, that is, a reflexive, symmetric and transitive relation on a given domain of discourse. An equivalence relation provides some equivalent classes which are called granules containing information due to objects with the common attributes. Approximation operators play an important role in determining the definable sets via granules. Essentially, granules are also definable sets. The approximations, called lower and upper approximation of sets, respectively, give a way to specify the target set of the objects containing the necessary information.

Formally, *an information system (or information table)* is a quadruple $S = (U, AT, V, f)$ where U is a set of objects, AT is a set of attributes, $V = \bigcup_{a \in AT} V_a$ is a set of *values* of attributes and V_a is the domain of a where $f : U \times AT \to V$ is a description function such that $f(x, a) \in V_a$ for every $x \in U$ and $a \in AT$ [1]. Take $B \subseteq AT$ and $x, y \in U$. Then the relation

$$IND(B) = \{(x, y) \mid \forall a \in B,\ f(x, a) = f(y, a)\}$$

is called *B-indiscernibility relation*. Clearly, $IND(B)$ is an equivalence relation on U and it exactly represents the information system $S = (U, AT, V, f)$ for the attribute set $B \subseteq AT$. In other words, the pair $(U, IND(B))$ is an abstract description of the system S and it is called *an approximation space* corresponding to the information system $S = (U, AT, V, f)$ for some $B \subseteq AT$. This argument can be extended to a

M. Diker, *Texture Spaces*, Studies in Fuzziness and Soft Computing 411,
https://doi.org/10.1007/978-3-031-39748-6_5

more general framework taking a domain of discourse and any equivalence relation on this domain as follows:

Definition 5.1.1 Let U be domain of discourse and r be an equivalence relation on U. Then the pair (U, r) is called *a Pawlak approximation space.*

An equivalence relation r on U provides us two operations on the power set $\mathcal{P}(U)$ as follows:

Definition 5.1.2 Let (U, r) be a Pawlak approximation space. Let us consider the mappings

$$\underline{apr}, \ \overline{apr} : \mathcal{P}(U) \to \mathcal{P}(U)$$

defined by

$\forall A \in \mathcal{P}(U), \ \underline{apr}(A) = \{u \in U \mid [u]_r \subseteq A\}$, and

$\forall A \in \mathcal{P}(U), \ \overline{apr}(A) = \{u \in U \mid [u]_r \cap A \neq \emptyset\}$,

respectively, where $[u]_r = \{v \in U \mid (u, v) \in r\}$ is an equivalence class for each $u \in U$. The operator $\underline{apr}$ is called *the lower approximation operator* and $\overline{apr}$ is called *the upper approximation operator.*

It is easy to see that the lower and upper approximation operators are dual with respect to ordinary set complementation, that is,

$$U \setminus \underline{apr}(A) = \overline{apr}(U \setminus A).$$

Alternatively, the lower and upper approximation of a set $A \in \mathcal{P}(U)$ can be also stated as follows:

$\underline{apr}(A) = \bigcup\{[u]_r \mid [u]_r \subseteq A\}$,

$\overline{apr}(A) = \bigcup\{[u]_r \mid [u]_r \cap A \neq \emptyset\}$.

In other words, the lower and upper approximations of a set A can be stated as the union of equivalence classes. For every subset A of U, the inclusions

$$\underline{apr}(A) \subseteq A \subseteq \overline{apr}(A)$$

hold. Pawlak's idea starts with the above inclusions that provide us where we do search the stored information with respect to target set A. These inclusions determine three different regions for the domain of discourse. The sets $\underline{apr}(A)$ and $U \setminus \overline{apr}(A)$ are called *the positive region* and *the negative region* (*exterior*), respectively. The third region is $\overline{apr}(A) \setminus \underline{apr}(A)$ which is called *the border*. For $A \in \mathcal{P}(U)$, if the border of A is non-empty set, then A is called *a rough set*. Note that if the border is empty, then A is definable.

The approximations $\underline{apr}(A)$ and $\overline{apr}(A)$ can be easily obtained in terms of granules $[u]_r$. In a Pawlak's approximation space (U, r), if a set $A \subseteq U$ can be written as the unions of some granules, then A is called *definable* in the space (U, r). In other words, a set $A \subseteq U$ is definable if for some $B \subseteq U$,

$$A = \bigcup_{u \in B} [u]_r.$$

Clearly, the lower and upper approximations of a set A are definable sets. Definable sets can be easily determined in terms of granules.

5.2 Generalized Approximation Spaces

Let U be a domain of discourse and r be any relation on U. Then the pair (U, r) is called *an approximation space*. Note that for an approximation space, the relation may not be necessarily equivalence relation on U. In a Pawlak's space, a remarkable result is due to definability:

$\forall A \subseteq U$, A is definable if and only if $\underline{apr}(A) = \overline{apr}(A)$.

For non-equivalence relations, this equivalence is not true in general. Since symmetry and transitivity lead us to inefficient observations with respect to data in informations systems, an equivalence relation is not always convenient for the real-life applications. However, we may extend the above arguments to arbitrary non-equivalence relations. Note that if r is not symmetric, then we trivially see that for some $u \in U$, the sets

$$\{v \in U \mid (u, v) \in r\}, \quad \{v \in U \mid (v, u) \in r\}$$

may be different. In other words, if we remove the symmetry condition of the relation r, then for some $u \in U$, the granule $[u]_r$ may turn into the two parts. This leads us to the following two operators which we call *the successor* and *predecessor neighborhood operators*, respectively:

Let (U, r) be an approximation space where r is any relation. Let us consider the mappings

$$\begin{aligned} &r_s : U \to \mathcal{P}(U), \\ &\forall u \in U, \ r_s(u) = \{v \in U \mid (u, v) \in r\} \end{aligned}$$

and

$$\begin{aligned} &r_p : U \to \mathcal{P}(U), \\ &\forall u \in U, \ r_p(u) = \{v \in U \mid (v, u) \in r\}. \end{aligned}$$

Note that if r is reflexive relation, then we guarantee that the neighborhoods $r_s(u)$ and $r_p(u)$ are non-empty sets. More generally, if r is a non-equivalence relation, then for each $u \in U$, the neighborhoods $r_s(u)$ and $r_p(u)$ above may not be disjoint anymore. In this case, the similarity plays a dominant role than that of indiscernibility of objects. Then we may revisit the approximations called *singleton approximations* as follows:

$$\underline{apr}_{r_p}(A) = \{u \mid r_p(u) \subseteq A\},$$
$$\overline{apr}_{r_p}(A) = \{u \mid r_p(u) \cap A \neq \emptyset\}$$

and

$$\underline{apr}_{r_s}(A) = \{u \mid r_s(u) \subseteq A\},$$
$$\overline{apr}_{r_s}(A) = \{u \mid r_s(u) \cap A \neq \emptyset\}.$$

It is easy to see that the singleton approximations are still dual. Indeed, for instance, for the r_p-approximation observe that

$$\begin{aligned} u \in U \setminus \underline{apr}_{r_p}(A) \text{ iff } & u \notin \underline{apr}_{r_p}(A) \\ \text{iff } & r_p(u) \nsubseteq A \\ \text{iff } & \exists u' \in U, (u', u) \in r \text{ and } u' \notin A \\ \text{iff } & r_p(u) \cap (U \setminus A) \neq \emptyset \\ \text{iff } & u \in \overline{apr}_{r_p}(U \setminus A). \end{aligned}$$

Moreover, as in Pawlak approximation space, we may give the definability due to operators r_s and r_p. A non-equivalence relation r gives us two different definabilities:

A subset A of U is called

predecessor definable if $A = \bigcup_{u \in B} r_p(u)$ for some $B \subseteq U$,

successor definable if $A = \bigcup_{u \in B} r_s(u)$ for some $B \subseteq U$,

r_p-*definable* if $\underline{apr}_{r_p}(A) = \overline{apr}_{r_p}(A)$,

r_s-*definable* if $\underline{apr}_{r_s}(A) = \overline{apr}_{r_s}(A)$.

The following propositions give an easy characterization for successor and predecessor definable sets:

(i) $A \subseteq U$ is successor definable if and only if $A = r(B)$ for some $B \subseteq U$.

(ii) $A \subseteq U$ is predecessor definable if $A = r^{-1}(B)$ for some $B \subseteq U$.

The successor and predecessor definabilites (Pawlak's definability for non-equivalence relations) are weaker than r_p-definability and r_s-definability, respectively. In other words, if r holds the reflexivity, then for every subset A of U, we have the following implications:

(i) A is r_p-definable $\Longrightarrow$ A is predecessor definable.

(ii) A is r_s-definable $\Longrightarrow$ A is successor definable.

Example 5.2.1 Take the approximation space (U, r) where

$$U = \{a, b, c, d\} \text{ and } r = \{(a, a), (d, d), (a, b), (c, d), (d, b), (b, c)\}.$$

Note that all successor neighborhoods of (U, r) are

$r_s(a) = \{a, b\}, r_s(b) = \{c\}, r_s(c) = \{d\}, r_s(d) = \{d, b\}$

and all predecessor neighborhoods are

$r_p(a) = \{a\}, \quad r_p(b) = \{a, d\}, r_p(c) = \{b\}, r_p(d) = \{c, d\}.$

Take $A = \{a, b, c\} \subseteq U$. Clearly, $A = r_s(a) \cup r_s(b)$ and therefore, A is successor definable. However A is not r_s-definable, that is, $\underline{apr}_{r_s}(A) \neq \overline{apr}_{r_s}(A)$. Indeed, we have

$\underline{apr}_{r_s}(A) = U \setminus r^{-1}(U \setminus A) = U \setminus r^{-1}(\{d\}) = U \setminus \{c, d\} = \{a, b\}$

and

$\overline{apr}_{r_s}(A) = r^{-1}(A) = \{a, d, b\}.$

For the subset $B = \{a, d\} \subseteq U$, we clearly have $B = r_p(a) \cup r_p(b)$ and hence, B is predecessor definable. But, B is not r_p-definable successor since

$\underline{apr}_{r_p}(B) = U \setminus r(\{b, c\}) = U \setminus \{a, b, d\} = \{c\}$

and

$\overline{apr}_{r_p}(A) = r(A) = \{a, d, c\}.$

Note that here, we do not loose the r_p and r_s-definabilities of the approximations of sets, that is, the sets $\underline{apr}_{r_p}(A)$, $\overline{apr}_{r_p}(A)$ are r_p-definable while $\underline{apr}_{r_s}(A)$ and $\overline{apr}_{r_s}(A)$ are r_s-definable sets:

$\underline{apr}_{r_p}(\underline{apr}_{r_p}(A)) = \overline{apr}_{r_p}(\underline{apr}_{r_p}(A)).$

$\underline{apr}_{r_p}(\overline{apr}_{r_p}(A)) = \overline{apr}_{r_p}(\overline{apr}_{r_p}(A)).$

$\underline{apr}_{r_s}(\overline{apr}_{r_s}(A)) = \overline{apr}_{r_s}(\overline{apr}_{r_s}(A)).$

$\underline{apr}_{r_s}(\underline{apr}_{r_s}(A)) = \overline{apr}_{r_s}(\underline{apr}_{r_s}(A)).$

Hence, by the implications given before Example 5.2.1, they are also successor and predecessor definable sets, respectively. On the other hand, the successor and predecessor neighborhood operators also give alternative approximations, called *the subset approximations*:

$\underline{apr}'_{r_p}(A) = \bigcup\{r_p(u) \mid r_p(u) \subseteq A\},$

$\overline{apr}''_{r_p}(A) = \bigcup\{r_p(u) \mid r_p(u) \cap A \neq \emptyset\}$

and

$\underline{apr}'_{r_s}(A) = \bigcup\{r_s(u) \mid r_s(u) \subseteq A\},$

$\overline{apr}''_{r_s}(A) = \bigcup\{r_s(u) \mid r_s(u) \cap A \neq \emptyset\}.$

The subset approximations satisfy the inclusion property, that is, for all $A \subseteq U$, the inclusions

$\underline{apr}'_{r_p}(A) \subseteq A \subseteq \overline{apr}''_{r_p}(A),$

and

$\underline{apr}'_{r_s}(A) \subseteq A \subseteq \overline{apr}''_{r_s}(A)$

hold. They are not dual as Pawlak approximations. However, taking the set theoretical complement, we can determine the corresponding duals of these approximations. For instance, note that we may also express the r_p-subset lower approximation of a set $A \subseteq U$ as

$\underline{apr}'_{r_p}(A) = \{u \in U \mid \exists u' \in U, u \in r_p(u') \ \wedge \ r_p(u') \subseteq A\}.$

Then

$$\begin{aligned} U \setminus \underline{apr}'_{r_p}(U \setminus A) &= U \setminus \{u \in U \mid \exists u' \in U, u \in r_p(u') \ \wedge \ r_p(u') \subseteq U \setminus A\} \\ &= U \setminus \{u \in U \mid \exists u' \in U, u \in r_p(u') \ \wedge \ r_p(u') \cap A = \emptyset\} \\ &= \{u \in U \mid \forall u' \in U, \ u \in r_p(u') \implies r_p(u') \cap A \neq \emptyset\}. \end{aligned}$$

Now, if we say

$$\overline{apr}'_{r_p}(A) = \{u \in U \mid \forall u' \in U, \ u' \in r_p(u) \implies r_p(u) \cap A \neq \emptyset\},$$

then the approximations $\underline{apr}'_{r_p}$ and $\overline{apr}'_{r_p}$ will be dual. Further, if r is reflexive, then it is reasonable to say that we also have

$$\underline{apr}'_{r_p}(A) \subseteq A \subseteq \overline{apr}'_{r_p}(A).$$

This once, we get the upper approximation $\overline{apr}''_{r_p}(A)$ to obtain dual one with respect to set complementation. Note that

$$\begin{aligned} U \setminus \overline{apr}''_{r_p}(U \setminus A) &= U \setminus (\bigcup\{r_p(u) \mid r_p(u) \cap (U \setminus A) \neq \emptyset\}) \\ &= U \setminus \{u \in U \mid \exists u' \in U, u \in r_p(u') \text{ and } r_p(u') \cap (U \setminus A) \neq \emptyset\} \\ &= \{u \in U \mid \forall u' \in U, \ u \notin r_p(u') \ \text{ or } \ r_p(u') \subseteq A\} \\ &= \{u \in U \mid \forall u' \in U, \ u \in r_p(u') \implies r_p(u') \subseteq A\}. \end{aligned}$$

If we take

$\underline{apr}''_{r_p}(A) = \{u \in U \mid \forall u' \in U, \ u \in r_p(u') \implies r_p(u') \subseteq A\},$

then the approximations $\underline{apr}''_{r_p}(A)$ and $\overline{apr}''_{r_p}(A)$ will be dual and moreover, if r is reflexive, then we have

$$\underline{apr}''_{r_p}(A) \subseteq A \subseteq \overline{apr}''_{r_p}(A).$$

We may use a similar way to obtain the corresponding dual approximations to $\underline{apr}'_{r_s}(A)$ and $\overline{apr}'_{r_s}(A)$.

5.3 Generalized Approximation Spaces with Two Domains of Discourse

Rough Set Models with Two Domains of Discourse

Let U and V be any two domains of discourse and r be a relation from U to V, that is, let $r \subseteq U \times V$. Then the triple (U, V, r) is called *a generalized approximation space*. Recall that *an interval structure* is a pair of functions defined between two Boolean algebras whose elements are objects and descriptions, respectively [2]. Let $f, g : \mathcal{P}(U) \to \mathcal{P}(V)$ be any two mappings. The pair (f, g) is called *an interval structure* if for every $A, B \in \mathcal{P}(U)$, the following conditions hold:

$$f(A) \cup f(B) \subseteq f(A \cup B),$$

$$f(A) \cap f(B) = f(A \cap B),$$

$$f(\emptyset) = \emptyset \text{ and } f(U) = V,$$

and

$$g(A) \cup g(B) = g(A \cup B),$$

$$g(A \cap B) \subseteq g(A) \cap g(B),$$

$$g(\emptyset) = \emptyset \text{ and } g(U) = V,$$

$$f(U \setminus A) = V \setminus g(A),\ g(U \setminus A) = U \setminus f(A).$$

Note that Pawlak's approximation operators $(\underline{apr}, \overline{apr})$ clearly constitute an interval structure. For the generalized approximation space (U, V, r), the compatibility notion guarantees a connection between the elements of sets U and V. In the framework of interval structures, the compatibility is needed for the analysis of the uncertain information. This also gives a motivation for the generalization of rough set approximations with two domains of discourse.

Now, let (U, V, r) be a generalized approximation space. Then r is called

serial if for all $u \in U$ there exists a $v \in V$ such that $(u, v) \in r$,

inverse serial if for all $v \in V$ there exists a $u \in U$ such that $(u, v) \in r$,

compatible if it is serial and inverse serial.

To define the rough set approximations in (U, V, r), we need the following mappings:

$$F : U \to \mathcal{P}(V),$$

$$\forall u \in U,\ F(u) = \{v \in V \mid (u, v) \in r\},$$

and

$$G : V \to \mathcal{P}(U),$$

$$\forall v \in V,\ G(v) = \{u \in U \mid (u, v) \in r\}.$$

In some sense, the sets $F(u)$ and $G(v)$ correspond to the successor and predecessor neighborhood operators given for the approximation spaces with a single universe, respectively. In other words, note that

$$F(u) = r_s(u) \text{ and } G(v) = r_p(v)$$

for $u, v \in U = V$. Further, note that if r is a compatible relation, then for all $u \in U$ and $v \in V$, we guarantee that $F(u)$ and $G(v)$ are non-empty. In this case, using the mappings F and G, the lower and upper approximations can be defined by

$$\underline{apr}_F(B) = \{u \in U \mid F(u) \subseteq B\}, \qquad \underline{apr}_G(A) = \{v \in V \mid G(v) \subseteq A\},$$

$$\overline{apr}_F(B) = \{u \in U \mid F(u) \cap B \neq \emptyset\}, \qquad \overline{apr}_G(A) = \{v \in V \mid G(v) \cap A \neq \emptyset\}.$$

for all subset $A \subseteq U$ and $B \subseteq V$.

Approximations and Set-Valued Mappings

The approximations of a set can be restated using set-valued functions. Note that F and G provide us the following mappings $\mathcal{F}$ and $\mathcal{G}$ defined by

$$\mathcal{F} : \mathcal{P}(U) \to \mathcal{P}(V), \qquad \mathcal{G} : \mathcal{P}(V) \to \mathcal{P}(U),$$

$$\forall A \in \mathcal{P}(U),\ \mathcal{F}(A) = \bigcup_{u \in A} F(u), \qquad \forall B \in \mathcal{P}(V),\ \mathcal{G}(B) = \bigcup_{v \in B} G(v),$$

respectively.

Dually, we may also consider the mappings

$$F' : U \to \mathcal{P}(V), \qquad G' : V \to \mathcal{P}(U),$$

$$\forall u \in U,\ F'(u) = \{v \in V \mid (u, v) \notin r\}, \qquad \forall v \in V,\ G'(v) = \{u \in U \mid (u, v) \notin r\}.$$

Likewise, we define

$$\mathcal{F}' : \mathcal{P}(U) \to \mathcal{P}(V), \qquad \mathcal{G}' : \mathcal{P}(V) \to \mathcal{P}(U),$$

$$\forall A \in \mathcal{P}(U),\ \mathcal{F}'(A) = \bigcap_{u \notin A} F'(u), \qquad \forall B \in \mathcal{P}(V),\ \mathcal{G}'(B) = \bigcap_{v \notin B} G'(v).$$

In fact, the mappings $\mathcal{F}$ and $\mathcal{G}$ are the upper rough set approximation operators determined by the mappings G and F, respectively. Further, $\mathcal{F}'$ is the lower rough set approximation operators determined by G while $\mathcal{G}'$ is the lower rough set approximation operator determined by F.

Theorem 5.3.1 *Let (U, V, r) be a generalized approximation space. Then*

(i) $\mathcal{F}(A) = \overline{apr}_G(A)$ *and* $\mathcal{F}'(A) = \underline{apr}_G(A)$,

(ii) $\mathcal{G}(B) = \overline{apr}_F(B)$ *and* $\mathcal{G}'(B) = \underline{apr}_F(B)$

for all $A \subseteq U$ *and* $B \subseteq V$.

Proof (i) Suppose that $\mathcal{F}'(A) \nsubseteq \underline{apr}_G(A)$. Take $v \in V$ such that $v \in \mathcal{F}'(A)$ and $v \notin \underline{apr}_G(A)$. Then $G(v) \nsubseteq A$ and so $(u, v) \in r$ and $u \notin A$ for some $u \in U$. By definition of $\mathcal{F}'(A)$,

$$\forall u \in U,\ u \notin A \implies (u, v) \notin r.$$

However, this is an immediate contradiction. For the reverse inclusion, suppose that $\underline{apr}_G(A) \nsubseteq \mathcal{F}'(A)$. Then there is a $v \in V$ such that $v \in \underline{apr}_G(A)$ and $v \notin \mathcal{F}'(A)$. This gives that $G(v) \subseteq A$ and $v \notin F'(u)$ for some $u \in U \setminus A$. That is, $(u, v) \in r$ and hence, $u \in G(v)$. Therefore, $u \notin A$ is a contradiction, since $G(v) \subseteq A$. The proof of $\mathcal{F}(A) = \overline{apr}_G(A)$ is similar.

(ii) We show that $\mathcal{G}(B) = \overline{apr}_F(B)$ leaving the second inequality to the reader. Let $u \in \overline{apr}_F(B)$ and $u \notin \mathcal{G}(B)$. Then $F(u) \cap B \neq \emptyset$ and

$$\forall v \in V,\ v \in B \implies u \notin G(v).$$

Now, there exists a $v \in V$ such that $v \in F(u)$ and $v \in B$. Since we also have $u \in G(v)$, it is an immediate contradiction. The proof of the reverse inclusion is similar.

5.4 Textural Rough Sets

As we know, a direlation (r, R) between any two textures $(U, \mathcal{U})$ and $(V, \mathcal{V})$ is actually a generalization of the ordinary relation between the sets U and V. This is due to the fact that if the textures are discrete, then the pair (r, R) where $R = (U \times V) \setminus r$ automatically holds the conditions given in Definitions 4.1.4 and 4.1.5, and so it is a (complemented) direlation.

Now, let us give the following result:

Theorem 5.4.1 *Let r be inverse serial. Then for all $A \subseteq U$, we have the equalities*

$$\overline{apr}_G(A) = \{v \in V \mid G(v) \cap A \neq \emptyset\} = r(A).$$

$$\underline{apr}_G(A) = \{v \in V \mid G(v) \subseteq A\} = V \setminus r(U \setminus A),$$

where the mapping $G : V \to \mathcal{P}(U)$ is defined by

$$\forall v \in V,\ G(v) = \{u \in U \mid (u, v) \in r\}.$$

as in the Sect. 5.3.

Proof Let us prove first equality, leaving the second to the interested reader. Let $v \in V \setminus r(U \setminus A)$ and $v \notin \underline{apr}_G(A)$. By definiton of the lower approximation, $G(v) \nsubseteq A$ and hence, $u \in G(v)$ and $u \notin A$ for some $u \in U$. That is, $(u, v) \in r$ and $u \in U \setminus A$. This implies that $v \in r(U \setminus A)$ whence $v \notin V \setminus r(U \setminus A)$. However, this gives a contradiction. As a result, we conclude that $V \setminus r(U \setminus A) \subseteq \underline{apr}_G(A)$. The reverse inclusion is similar. □

The equalities presented in Theorem 5.4.1 surprisingly provide a connection with the approximations of $A \subseteq U$ in generalized approximation space (U, V, r) as follows:

Corollary 5.4.2 *Let (r, R) be a direlation from $(U, \mathcal{P}(U))$ to $(V, \mathcal{P}(V))$ where $(U, \mathcal{P}(U))$ and $(V, \mathcal{P}(V))$ are any two discrete textures and $R = (U \times V) \setminus r$. Then for all $A \subseteq U$, we have the following statements:*

(i) $r^{\rightarrow} A = r(A)$.

(ii) $R^{\rightarrow} A = V \setminus r(U \setminus A)$.

Proof By Theorem 4.2.3, the proof is straightforward.

Theorem 5.4.3

$$\overline{apr}_F(B) = \{u \in U \mid F(u) \cap B \neq \emptyset\} = r^{-1}(B).$$

$$\underline{apr}_F(B) = \{u \in U \mid F(u) \cap B \neq \emptyset\} = U \setminus r^{-1}(V \setminus B),$$

where the mapping $F : U \to \mathcal{P}(V)$ is defined by

$$\forall u \in U, \;\; F(u) = \{v \in V \mid (u, v) \in r\}$$

as in the Sect. 5.3.

Likewise, by Theorem 4.2.10, for the presections, for all $B \subseteq V$, we also have that

(i) $r^{\leftarrow} B = U \setminus r^{-1}(V \setminus B)$.

(ii) $R^{\leftarrow} B = r^{-1}(B)$.

As a result, we obtain the following reasonable equalities for discrete textures:

Corollary 5.4.4

$$r^{\rightarrow} A = \overline{apr}_G(A), \quad R^{\rightarrow} A = \underline{apr}_G(A),$$

$$r^{\leftarrow} B = \underline{apr}_F(B), \quad R^{\leftarrow} B = \overline{apr}_F(B).$$

Theorem 5.4.1 says that for discrete textures, the pair $(R^{\rightarrow} A, r^{\rightarrow} A)$ constitutes nothing but the lower and upper approximations of a subset $A \subseteq U$ in V for the generalized approximation space (U, V, r). Theorem 5.4.3 states that the pair $(r^{\leftarrow} B, R^{\leftarrow} B,)$ consists of the lower and upper approximations of a subset $B \subseteq V$ in U for the generalized approximation space (U, V, r). Note that for $U = V$, $(r^{\rightarrow} A, R^{\rightarrow} A)$ is the pair of approximations of A in generalized approximation space (U, r) with a single domain of discourse.

More generally, the above observations lead us to the following definition:

Definition 5.4.5 Let $\mathbf{U} = (U, \mathcal{U})$ and $\mathbf{V} = (V, \mathcal{V})$ be any two textures and (r, R) be a direlation from $(U, \mathcal{U})$ to $(V, \mathcal{V})$. Then the quadruple $(\mathbf{U}, \mathbf{V}, r, R)$ is called *a generalized textural approximation space*, or in brief *a generalized t-approximation space*. For the sake of simplicity, for a generalized t-approximation space, we use the notation (U, V, r, R).

Note that by Theorem 4.2.16, if (r, R) is t-serial, then for all $B \in \mathcal{V}$, we have the inclusion

$$r^{\leftarrow} B \subseteq R^{\leftarrow} B.$$

Further, again by Theorem 4.2.16, if (r, R) is a t-inverse serial direlation, then for all $A \in \mathcal{U}$, we have

$$R^{\rightarrow} A \subseteq r^{\rightarrow} A.$$

If (r, R) is a reflexive direlation on the texture $(U, \mathcal{U})$, then for all $A \in \mathcal{U}$, we also have the inclusions

$$r^{\leftarrow} A \subseteq A \subseteq R^{\leftarrow} A$$

and

$$R^{\rightarrow} A \subseteq A \subseteq r^{\rightarrow} A.$$

This means that the pairs

$$(r^{\leftarrow} B, R^{\leftarrow} B) \text{ and } (R^{\rightarrow} A, r^{\rightarrow} A)$$

are the natural generalizations of the approximations of Pawlak's rough set theory. Then we call the sets A and B *textural rough sets* or in brief, *t-rough sets* if the conditions

$$r^{\rightarrow} A \neq R^{\rightarrow} A \text{ and } R^{\leftarrow} B \neq r^{\leftarrow} B$$

hold.

As we have seen in Chap. 1, the notions of p-set and q-set are basic tools in texture space theory and they play a vital role in the development of textures just as the points of classical set theory. For discrete textures, the sections and presections of a p-set and a q-set correspond to the neighborhoods or the complement of neighborhoods of the points in (U, V, r):

Proposition 5.4.6 *Let (r, R) be a complemented direlation from $(U, \mathcal{P}(U))$ to $(V, \mathcal{P}(V))$. Then for all $u \in U$ and $v \in V$, we have the following equalities.*

(i) $r^{\rightarrow} P_u = F(u), \quad R^{\rightarrow} Q_u = F'(u) = V \setminus F(u).$
(ii) $R^{\leftarrow} P_v = G(v), \quad r^{\leftarrow} Q_v = G'(v) = V \setminus G(v).$

Proof As we have discussed in Chap. 1, note that for a discrete texture, we have the following statements

$$P_u = \{u\},$$

$U \setminus \{u\} = Q_u,$

$(u, v) \in r \iff r \not\subseteq \overline{Q}_{(u,v)}$ and

$(u, v) \in r \iff \overline{P}_{(u,v)} \not\subseteq R.$

Then we easily observe that

$$\begin{aligned} r^{\to} P_u &= \bigcap \{Q_v \mid \forall u' \in U,\ r \not\subseteq \overline{Q}_{(u',v)} \implies P_u \subseteq Q_{u'}\} \\ &= \bigcap \{V \setminus \{v\} \mid \forall u' \in U,\ (u', v) \in r \implies \{u\} \subseteq U \setminus \{u'\}\} \\ &= \{v \mid \exists u' \in U,\ (u', v) \in r \text{ and } u = u'\} \\ &= \{v \mid (u, v) \in r\} \\ &= F(u) \end{aligned}$$

and

$$\begin{aligned} R^{\to} Q_u &= \bigvee \{P_v \mid \forall u' \in U,\ \overline{P}_{(u',v)} \not\subseteq R \implies P_{u'} \subseteq Q_u\} \\ &= \bigcup \{\{v\} \mid \forall u' \in U,\ (u', v) \in r \implies \{u'\} \subseteq U \setminus \{u\}\} \\ &= \{v \mid \forall u' \in U,\ (u', v) \in r \implies u' \neq u\} \\ &= \{v \mid (u, v) \notin r\} \\ &= V \setminus F(u). \end{aligned}$$

The proof of (ii) is similar. □

Remark 5.4.7 The set $F(u)$ can be regarded as a neigbourhood of u in V. By the same way, $G(v)$ can also be considered as a neighborhood of v in U. Therefore, we say the set $r^{\to} P_u$ of u is *a textural neighborhood* or in brief *t-neighborhood* in $\mathcal{V}$ while $R^{\leftarrow} P_v$ is *a t-neighborhood* of v in $\mathcal{U}$.

Corollary 5.4.8 *Let* (r, R) *be a complemented direlation on the discrete texture* $(U, \mathcal{P}(U))$*. Then for all* $w \in U$*, we have the following equalities.*

(i) $r^{\to} P_w = r_s(w),\quad R^{\to} Q_w = U \setminus r_s(w).$
(ii) $R^{\leftarrow} P_w = r_p(w),\quad r^{\leftarrow} Q_w = U \setminus r_p(w).$
(iii) $r^{\to} Q_w = r(U \setminus \{w\}),\quad R^{\to} P_w = U \setminus r(U \setminus \{w\}).$
(iv) $r^{\leftarrow} P_w = U \setminus r^{-1}(U \setminus \{w\}),\quad R^{\leftarrow} Q_w = r^{-1}(U \setminus \{w\}).$

Proof (i) By Proposition 5.4.6 (i), we have

$$r^{\to} P_w = F(w) = r_s(w)$$

and

$$R^{\to} Q_w = F'(w) = U \setminus F(u) = U \setminus r_s(w).$$

The proof of the rest of items is similar.

Set-Valued Mappings for Textural Rough Sets

In view of Proposition 5.4.6, we can also obtain the approximations for textures considering the set-valued mappings as in generalized t-approximation spaces with two domains of discourse. Let us consider the following mappings:

$$F_t : U \to \mathcal{V}, \qquad F_t' : U \to \mathcal{V},$$
$$\forall u \in U,\ F_t(u) = r^{\to} P_u, \qquad \forall u \in U,\ F_t'(u) = R^{\to} Q_u,$$

$$\mathcal{F}_t : \mathcal{U} \to \mathcal{V}, \qquad \mathcal{F}_t' : \mathcal{U} \to \mathcal{V},$$
$$\forall A \in \mathcal{U},\ \mathcal{F}_t(A) = \bigvee_{A \not\subseteq Q_u} F_t(u) \qquad \forall A \in \mathcal{U},\ \mathcal{F}_t'(A) = \bigcap_{P_u \not\subseteq A} F_t'(u)$$

and

$$G_t : V \to \mathcal{U}, \qquad G_t' : V \to \mathcal{U},$$
$$\forall v \in V,\ G_t(v) = R^{\leftarrow} P_v, \qquad \forall v \in V,\ G_t'(v) = r^{\leftarrow} Q_v,$$

$$\mathcal{G}_t : \mathcal{V} \to \mathcal{U}, \qquad \mathcal{G}_t' : \mathcal{V} \to \mathcal{U}$$
$$\forall B \in \mathcal{V},\ \mathcal{G}_t(B) = \bigvee_{B \not\subseteq Q_v} G(v), \qquad \forall\ B \in \mathcal{V},\ \mathcal{G}_t'(B) = \bigcap_{P_v \not\subseteq B} G_t'(v).$$

Corollary 5.4.9 *Let (U, V, r, R) be a generalized t-approximation space. Then we have*

(i) *For all $A \in \mathcal{U}, \mathcal{F}_t(A) = r^{\to} A$ and $\mathcal{F}_t'(A) = R^{\to} A$.*

(ii) *For all $B \in \mathcal{V}$, $\mathcal{G}_t(B) = R^{\leftarrow} B$ and $\mathcal{G}_t'(B) = r^{\leftarrow} B$.*

Proof (i) By Theorem 4.2.17 (ii) and Theorem 2.3.1(2), we have

$$\mathcal{F}_t(A) = \bigvee_{A \not\subseteq Q_u} F_t(u) = \bigvee_{A \not\subseteq Q_u} r^{\to} P_u = r^{\to} \bigvee_{A \not\subseteq Q_u} P_u = r^{\to} A.$$

(ii) By Theorem 4.2.17(ii) and Theorem 2.3.1(2), we get

$$\mathcal{G}_t(B) = \bigvee_{B \not\subseteq Q_v} G_t(v) = \bigvee_{B \not\subseteq Q_v} R^{\leftarrow} P_v = R^{\leftarrow} \bigvee_{B \not\subseteq Q_v} P_v = R^{\leftarrow} B.$$

The proofs of the other equalities are similar.

Remark 5.4.10 In view of Theorem 5.3.1 and the Proposition 5.4.1, the set-valued mappings for the discrete textures, turn in to the classical set-valued mappings. For instance, if $\mathcal{P}(V) = \mathcal{V}$, then we clearly obtain that

$$F = F_t : U \to \mathcal{V},\ \forall u \in U,\ F_t(u) = r^{\to} P_u = F(u),$$

and

$$\mathcal{F}_t : \mathcal{U} \to \mathcal{V}, \forall A \subset \mathcal{U},\ \mathcal{F}_t(A) = \bigvee_{A \not\subseteq Q_u} F_t(u) = \bigcup_{u \in A} F(u) = \mathcal{F}(A).$$

5.5 Textural Definability

Definition 5.5.1 Let (U, V, r, R) be a generalized t-approximation space. Then

(1) $B \in \mathcal{V}$ is called *presectional definable*, if $r^{\leftarrow}B = R^{\leftarrow}B$.

(2) $A \in \mathcal{U}$ is called *sectional definable*, if $r^{\rightarrow}A = R^{\rightarrow}A$.

We know that a direlation is a suitable morphism between textures generalizing the ordinary relation between sets [3–5]. In texture space theory, it is possible to define a more strong concept (morphism) corresponding to ordinary functions using direlations.

A direlation (r, R) from a texture $(U, \mathcal{U})$ to a texture $(V, \mathcal{V})$ is called *a difunction* [6] if it satisfies the conditions

DF1 For $u, u' \in U$, $P_u \not\subseteq Q_{u'} \implies \exists v \in V$ with $r \not\subseteq \overline{Q}_{(u,v)}$ and $\overline{P}_{(u',v)} \not\subseteq R$,

DF2 For $v, v' \in V$ and $u \in U$, $r \not\subseteq \overline{Q}_{(u,v)}$ and $\overline{P}_{(u,v')} \not\subseteq R \implies P_{v'} \not\subseteq Q_v$.

The condition DF1 is in fact the t-seriality condition of a direlation. The injectivity of $(R^{\leftarrow}, r^{\leftarrow})$ implies the condition DF2:

Corollary 5.5.2 (i) *Let (r, R) be a t-serial direlation from a texture $(U, \mathcal{U})$ to a texture $(V, \mathcal{V})$. If $(R^{\leftarrow}, r^{\leftarrow})$ is injective, then (r, R) is a difunction.*

(ii) *Let (r, R) be a t-inverse serial direlation from a texture $(U, \mathcal{U})$ to a texture $(V, \mathcal{V})$. If (r, R) is injective, then $(R^{\leftarrow}, r^{\leftarrow})$ is a difunction.*

Proof (i) Let $(R^{\leftarrow}, r^{\leftarrow})$ be injective. Take $v, v' \in V$ and $u \in U$ such that

$$r \not\subseteq \overline{Q}_{(u,v)} \text{ and } \overline{P}_{(u,v')} \not\subseteq R.$$

By Lemma 4.1.23 (i) and (ii), we also have that

$$R^{\leftarrow} \not\subseteq \overline{Q}_{(v',u)} \text{ and } \overline{P}_{(v,u)} \not\subseteq r^{\leftarrow}.$$

Since $(R^{\leftarrow}, r^{\leftarrow})$ is injective, we get $P_{v'} \not\subseteq Q_v$ as desired, that is (r, R) satisfies the condition DF2. The condition DF1 is nothing but the t-seriality of (r, R). Hence, (r, R) is a difunction.

(ii) The proof is omitted.

Let (r, R) be a difunction from $(U, \mathcal{P}(U))$ to $(V, \mathcal{P}(V))$. Then r is a function from U to V since DF1 and DF2 clearly correspond to the conditions

(F1) For any $u \in U$ we have $v \in V$ such that $(u, v) \in r$,

(F2) If for some $u \in U$, $(u, v) \in r$ and $(u, v') \in r$ where $v, v' \in V$, then $v = v'$.

To see this, it is enough to consider the following observations.

Indeed, for DF1,

$$P_u = \{u\} \not\subseteq Q_{u'} = U \setminus \{u'\} iff u = u',$$

$$r \not\subseteq \overline{Q}_{(u,v)} \text{ iff } r \not\subseteq ((U \setminus \{u\}) \times V) \cup (U \times (V \setminus \{v\})) \text{ iff } (u, v) \in r$$

and

$$\overline{P}_{(u',v)} \not\subseteq R \text{ iff } \{u'\} \times \{v\} \not\subseteq R \text{ iff } (u', v) \notin R \text{ iff } (u, v) \notin R.$$

For DF2,

$$r \not\subseteq \overline{Q}_{(u,v)} \text{ iff } (u, v) \in r,$$

$$\overline{P}_{(u,v')} \not\subseteq R \text{ iff } (u, v') \notin R,$$

and

$$P_v \not\subseteq Q_{v'} \text{ iff } v = v'.$$

By the above argument, we also have that $R = (U \times V) \setminus r$.

Remark 5.5.3 Note that for an ordinary function, the counterpart of Corollary 5.5.2 is straightforward. For instance, if r is a serial and injective relation, then r is a function.

Just as a direlation, a difunction is also defined in terms of p-sets and q-sets of the textures. However, it is interesting that there is a close relation between the notions of (textural) definability and difunction.

Theorem 5.5.4 *Let* (r, R) *be a direlation from a texture* $(U, \mathcal{U})$ *to a texture* $(V, \mathcal{V})$. *Then the following conditions are equivalent:*

(i) (r, R) *is a difunction from* $(U, \mathcal{U})$ *to* $(V, \mathcal{V})$.

(ii) *Every set* $B \in \mathcal{V}$ *is presectional definable that is,* $r^{\leftarrow} B = R^{\leftarrow} B$.

Proof (i)$\Longrightarrow$(ii): Note that the condition DF1 is infact t-seriality of (r, R). Therefore by Theorem 4.4.11 (iv), for all $B \in \mathcal{V}$, the inclusion $r^{\leftarrow} B \subseteq R^{\leftarrow} B$ holds. For the reverse inclusion, assume that $R^{\leftarrow} B \not\subseteq r^{\leftarrow} B$. By Corollary 2.3.5, there exists a $u \in U$ such that

$$R^{\leftarrow} B \not\subseteq Q_u \text{ and } P_u \not\subseteq r^{\leftarrow} B.$$

Using definitions of presections, we conclude that

$$\overline{P}_{(u,v)} \not\subseteq R \text{ and } B \not\subseteq Q_v$$

and

$$r \not\subseteq \overline{Q}_{(u,v')} \text{ and } P_{v'} \not\subseteq B$$

for some $v, v' \in V$. In this case, by DF2, we get that $P_v \not\subseteq Q_{v'}$. By Theorem 2.3.1 (i), $B \subseteq Q_{v'}$ and this implies that $Q_{v'} \not\subseteq Q_v$, that is $P_{v'} \subseteq Q_v$ which is an immediate contradiction. Hence we obtain the inclusion $R^{\leftarrow}B \subseteq r^{\leftarrow}B$ as desired. That is the implication (i)$\Longrightarrow$(ii) holds.

(ii)$\Longrightarrow$(i) First, let us show that for all $A_1, A_2 \in \mathcal{U}$,

$$r^{\rightarrow}A_1 \subseteq R^{\rightarrow}A_2 \implies A_1 \subseteq A_2. \qquad (*)$$

Let $r^{\rightarrow}A_1 \subseteq R^{\rightarrow}A_2$. Then by Lemma 4.2.15 (1)(ii) and (2)(iii) and by the assumption, we find that

$$A_1 \subseteq r^{\leftarrow}r^{\rightarrow}A_1 \subseteq r^{\leftarrow}R^{\rightarrow}A_2 = R^{\leftarrow}R^{\rightarrow}A_2 \subseteq A_2.$$

Now, let $P_u \not\subseteq Q_{u'}$. By the implication (*), we get $r^{\rightarrow}P_u \not\subseteq R^{\rightarrow}Q_{u'}$ and so for some $v \in V$, $r^{\rightarrow}P_u \not\subseteq Q_v$ and $P_v \not\subseteq R^{\rightarrow}Q_{u'}$, that is, $r \not\subseteq \overline{Q}_{(u,v)}$ and $\overline{P}_{(u',v)} \not\subseteq R$. Then the condition DF1 holds. Now let us prove the condition DF2. Note that by Lemma 4.2.3(i), for all $B \in \mathcal{V}$,

$$r^{\rightarrow}r^{\leftarrow}B \subseteq B$$

and so by the assumption for all $B \in \mathcal{V}$, we conclude that

$$r^{\rightarrow}R^{\leftarrow}B \subseteq B.$$

Now let $v, v' \in V$ and $u \in U$, $\quad r \not\subseteq \overline{Q}_{(u,v)}$ and $\overline{P}_{(u,v')} \not\subseteq R$.

Then by Lemma 4.2.15 (1)(iii)

$$r^{\rightarrow}P_u \not\subseteq Q_v \text{ and } P_{v'} \not\subseteq R^{\rightarrow}Q_u.$$

By the assumption, we have $r^{\rightarrow}R^{\leftarrow}Q_v = Q_v$ and therefore we get $r^{\rightarrow}P_u \not\subseteq r^{\rightarrow}R^{\leftarrow}Q_v$. Then by Lemma 4.2.14 $P_u \not\subseteq R^{\leftarrow}Q_v$. Finally, note that

$$Q_v \subseteq R^{\rightarrow}R^{\leftarrow}Q_v \subseteq R^{\rightarrow}Q_u.$$

This gives that $P_{v'} \not\subseteq Q_v$ as desired.

As it is seen from Theorem 5.5.4, if all subsets of a domain of discourse are presectional definable with respect to a direlation (r, R), then (r, R) is a difunction and vice versa. Dually, we have the following result:

Theorem 5.5.5 *Let (r, R) be a direlation from a texture $(U, \mathcal{U})$ to a texture $(V, \mathcal{V})$. Then the following conditions are equivalent:*

(i) *$(R^{\leftarrow}, r^{\leftarrow})$ is a difunction from $(V, \mathcal{V})$ to $(U, \mathcal{U})$.*

(ii) *Every set $A \in \mathcal{U}$ is sectional definable that is, $r^{\rightarrow}A = R^{\rightarrow}A$.*

Theorem 5.5.6 *Let $(\mathbf{U}, \mathbf{V}, r, R)$ be a generalized t-approximation space where (r, R) is t-compatible. Then we have the following equivalences:*

(i) *$B \in \mathcal{V}$ is presectional definable $\iff \exists A \in \mathcal{U},\ r^{\rightarrow}A = B$ and $R^{\leftarrow}B = A$.*

(ii) *$A \in \mathcal{U}$ is sectional definable $\iff \exists B \in \mathcal{V},\ R^{\rightarrow}A = B$ and $r^{\leftarrow}B = A$.*

Proof (i) Suppose $A = r^{\leftarrow}B$. Since B is presectional definable, we have $r^{\leftarrow}B = R^{\leftarrow}B$. By Lemma 4.2.15 (1) (iii),

$$r^{\rightarrow}A = r^{\rightarrow}(r^{\leftarrow}B) = r^{\rightarrow}(R^{\leftarrow}B) \subseteq B.$$

Since (r, R) is t-inverse serial, by Theorem 4.4.13 (i), we have $B \subseteq r^{\rightarrow}(R^{\leftarrow}B)$ and this implies that $r^{\rightarrow}A = B$. Further, we already have that $A = r^{\leftarrow}B = R^{\leftarrow}B$. Conversely, for some $A \in \mathcal{U}$, if $r^{\rightarrow}A = B$ and $R^{\leftarrow}B = A$, then by Theorem 4.2.15 (1)(ii) and Theorem 4.4.11(iv), we conclude that

$$A \subseteq r^{\leftarrow}(r^{\rightarrow}A)) = r^{\leftarrow}B \subseteq R^{\leftarrow}B = A.$$

Hence, we obtain $r^{\leftarrow}B = R^{\leftarrow}B$ and so B is presectional definable.

(ii) The proof is dual to (i). □

Definition 5.5.7 Let (r, R) be a direlation from a texture $(U, \mathcal{U})$ to a texture $(V, \mathcal{V})$. Then

(i) $A \in \mathcal{U}$ is called *textural predecessor definable*, in brief, *t-predecessor definable* in $\mathcal{U}$ if for some $B \in \mathcal{V}$,

$$A = \bigvee_{B \not\subseteq Q_v} R^{\leftarrow}P_v.$$

(ii) $B \in \mathcal{V}$ is called *textural successor definable* in brief, *t-predecessor definable* in $\mathcal{V}$ if for some $A \in \mathcal{U}$,

$$B = \bigvee_{A \not\subseteq Q_u} r^{\rightarrow}P_u.$$

Proposition 5.5.8 *Let (U, V, r, R) be a generalized t-approximation space. Then we have the following equivalences:*

(i) *$A \in \mathcal{U}$ is t-predecessor definable if and only if $A = R^{\leftarrow}B$ for some $B \in \mathcal{V}$.*

(ii) *$B \in \mathcal{V}$ is t-successor definable if and only if $B = r^{\rightarrow}A$ for some $A \in \mathcal{U}$.*

(iii) *$r^{\rightarrow}A$ and $R^{\leftarrow}B$ are t-successor definable and t-predecessor definable, respectively.*

Proof (i) If A is t-predecessor definable, then by Theorem 4.2.17 (i), for some $B \in \mathcal{V}$,

$$A = \bigvee_{B \not\subseteq Q_v} R^{\leftarrow}P_v = R^{\leftarrow} \bigvee_{B \not\subseteq Q_v} P_v = R^{\leftarrow}B.$$

Because, by Theorem 2.3.5 (ii), every set $B \in \mathcal{V}$ can be written in terms of p-sets as

$$B = \bigvee_{B \not\subseteq Q_v} P_v.$$

Now, the sufficiency is clear.

(ii) The proof is similar.

(iii) Note that by Theorem 2.3.5 (ii) and Theorem 4.2.17 (i),

$$r^{\rightarrow}A = r^{\rightarrow}\bigvee_{A\not\subseteq Q_u} P_u = \bigvee_{A\not\subseteq Q_u} r^{\rightarrow}P_u$$

and if we say $B = r^{\rightarrow}A$, then B, that is, $r^{\rightarrow}A$ is t-successor definable. Similarly, one can show that $R^{\leftarrow}B$ is t-predecessor definable.

Corollary 5.5.9 *Let (U, V, r, R) be a generalized t-approximation space where (r, R) is t-serial and injective.*

(i) *If $A \in \mathcal{U}$ is sectional definable, then it is also t-predecessor definable.*

(ii) *If $B \in \mathcal{V}$ is presectional definable, then it is also t-successor definable.*

Proof (i) If $A \in \mathcal{U}$ is sectional definable, then $r^{\rightarrow}A = R^{\rightarrow}A$. Thus by Theorem 4.4.15 (4)(ii),

$$R^{\leftarrow}r^{\rightarrow}A = R^{\leftarrow}R^{\rightarrow}A = A.$$

If we say $r^{\rightarrow}A = B$, we obtain that $R^{\leftarrow}B = A$, that is, A is t-predecessor definable.

(ii) Omitted.

We may also discuss on definabilities for generalized approximation spaces with a single domain of discourse:

Theorem 5.5.10 *Let (r, R) be a direlation on $(U, \mathcal{U})$ and $A \in \mathcal{U}$.*

(1) *If (r, R) is reflexive, then*

(i) $r^{\rightarrow}A = R^{\rightarrow}A \implies$ *A is t-successor definable.*

(ii) $r^{\leftarrow}A = R^{\leftarrow}A \implies$ *A is t-predecessor definable.*

(2) *If (r, R) is a symmetric direlation on $(U, \mathcal{U})$, then we have the following equivalences:*

(i) *A is t-successor definable $\iff$ A is t-predecessor definable.*

(ii) $r^{\rightarrow}A = R^{\rightarrow}A \iff r^{\leftarrow}A = R^{\leftarrow}A$.

(3) *If (r, R) is an equivalence direlation on $(U, \mathcal{U})$, then the following conditions are equivalent.*

(i) *A is t-successor definable.*

(ii) *A is t-predecessor definable.*

(iii) $r^{\leftarrow}A = R^{\leftarrow}A$.

(iv) $r^{\rightarrow}A = R^{\rightarrow}A$.

Proof (1) (i) Let $r^{\rightarrow}A = R^{\rightarrow}A$. Since (r, R) is reflexive, by Theorem 4.2.16 (1), it is easy to see that $R^{\rightarrow}A \subseteq A \subseteq r^{\rightarrow}A$. Therefore, we have $R^{\rightarrow}A = A = r^{\rightarrow}A$. By Proposition 5.5.8 (ii), A is textural successor definable.

(ii) The proof is dual to (i).

(2) (i) Let A be textural successor definable. Since (r, R) is symmetric, we have $r = R^{\leftarrow}$ and hence, by Theorem 4.2.11, for some $B \in \mathcal{U}$,

$$A = \bigvee_{B \not\subseteq Q_u} r^{\rightarrow}P_u = \bigvee_{B \not\subseteq Q_u} (R^{\leftarrow})^{\rightarrow}P_u = \bigvee_{B \not\subseteq Q_u} R^{\leftarrow}P_u.$$

(ii) Since (r, R) is reflexive, we have $r = R^{\leftarrow}$ and hence, again by Theorem 4.2.11,

$$r^{\rightarrow}A = R^{\rightarrow}A \Longleftrightarrow (R^{\leftarrow})^{\rightarrow}A = (r^{\leftarrow})^{\rightarrow}A \Longleftrightarrow R^{\leftarrow}A = r^{\leftarrow}A.$$

(3) (i) $\Longrightarrow$ (ii): Immediate by (2) (i).

(ii) $\Longrightarrow$ (iii): Since (r, R) is reflexive and transitive, by Theorem 4.2.17 and Lemma 4.3.5,

$$R^{\leftarrow}A = R^{\leftarrow}\bigvee_{B \not\subseteq Q_u} R^{\leftarrow}P_u = \bigvee_{B \not\subseteq Q_u} R^{\leftarrow}R^{\leftarrow}P_u = \bigvee_{B \not\subseteq Q_u} (R \circ R)^{\leftarrow}P_u = \bigvee_{B \not\subseteq Q_u} R^{\leftarrow}P_u = A.$$

Further, since (r, R) is symmetric, by Lemma 4.2.15 (2)(ii) and by Corollary 4.1.24, we also have $A \subseteq r^{\leftarrow}R^{\leftarrow}A$. Hence, we obtain $A \subseteq r^{\leftarrow}A$. On the other hand, the reflexivity of (r, R) implies that $r^{\leftarrow}A \subseteq A$.

(iii) $\Longrightarrow$ (iv): Let $r^{\leftarrow}A = R^{\leftarrow}A$. Since (r, R) is symmetric, by Corollary 4.1.24, $r^{\leftarrow} = R$ and $r = R^{\leftarrow}$ and then by Lemma 4.3.5, we conclude that

$$R^{\rightarrow}A = (r^{\leftarrow})^{\rightarrow}A = r^{\leftarrow}A = R^{\leftarrow}A = (R^{\leftarrow})^{\rightarrow}A = r^{\rightarrow}A.$$

(iv) $\Longrightarrow$ (i): Immediate by (1) (i).

References

1. J.W. Grzymala-Busse, Algebraic properties of knowledge representation systems, in *Proceedings of the ACM SIGART International Symposium on Methodologies for Intelligent Systems* (Knoxville, 1986), pp. 432–440
2. S.K.M. Wong, L.S. Wang, Y.Y. Yao, On modeling uncertainty with interval structures. Comput. Intell. **11**(2), 406–426 (1995)
3. M. Diker, A category approach to relation preserving functions in rough set theory. Int. J. Approx. Reason. **56**(1), 71–86 (2015)
4. M. Diker, Categories of rough sets and textures. Theor. Comput. Sci. **488**, 46–65 (2013)

5. M. Diker, Categories of direlations and rough set approximation operators, in *Rough Sets and Current Trends in Computing. RSCTC 2010. Lecture Notes in Computer Science*, vol. 6086, eds. by M. Szczuka, M. Kryszkiewicz, S. Ramanna, R. Jensen, Q. Hu (Springer, Berlin, Heidelberg, 2010), pp. 288–298
6. L.M. Brown, R. Ertürk, Ş Dost, Ditopological texture spaces and fuzzy topology. I. Basic concepts. Fuzzy Sets Syst. **147**, 171–199 (2004)

Chapter 6
Basic Results in Rough Set Theory via Textures

Check for updates

6.1 Generalized Approximation Spaces and Discrete Textures

As is known, the lower and upper approximations provide reasonable information about the target set in information systems (tables). Therefore, most of the researchers also focused on the discussions about the basic properties of the approximation operators. From the theoretical point of view, textures can provide sound and precise analysis for the results due to approximations. Here, we fairly give almost all basic results as a corollary of the textural arguments given in Chap. 3. Note that while some of the results presented here are already known, there are many properties that are not noticed before. We know that every generalized approximation space (U, V, r) with two domains of discourse can be regarded as a t-generalized approximation space (U, V, r, R) where $\mathcal{U} = \mathcal{P}(U)$ and $\mathcal{V} = \mathcal{P}(V)$ are the discrete textures. Further, note that by Corollaries 4.2.24 and 4.2.25, the pair (r, R) is a direlation from $(U, \mathcal{P}(U))$ to $(V, \mathcal{P}(V))$ where $R = (U \times V) \setminus r$. Recall that from Corollary 5.4.4, for all $A \subseteq U$ and $B \subseteq V$, we also have

$$r^{\to} A = \overline{apr}_G(A), \quad R^{\to} A = \underline{apr}_G(A),$$

$$r^{\leftarrow} B = \underline{apr}_F(B), \quad R^{\leftarrow} B = \overline{apr}_F(B).$$

Therefore, for the proof of the corollaries, we use the same technic and so we mostly leave the proofs to the interested reader.

We start with the duality.

Corollary 6.1.1 *If (U, V, r) is a generalized approximation space, then we have the following statements:*

(i) $\underline{apr}_G(A) = V \setminus (\overline{apr}_G(U \setminus A))$.

(ii) $\overline{apr}_G(A) = V \setminus (\underline{apr}_G(U \setminus A))$.

M. Diker, *Texture Spaces*, Studies in Fuzziness and Soft Computing 411,
https://doi.org/10.1007/978-3-031-39748-6_6

(iii) $\overline{apr}_F(B) = U \setminus (\underline{apr}_F(V \setminus B))$.

(iv) $\underline{apr}_F(B) = U \setminus (\overline{apr}_F(V \setminus B))$.

Proof (i) By Corollary 4.2.28, for any generalized t-approximation space (U, V, r, R), we already have that $R^{\to} A = c_V(r^{\to} c_U(A))$. Since the textures are discrete, by Corollary 5.4.4, we obtain that

$$\underline{apr}_G(A) = R^{\to} A = c_V(r^{\to} c_U(A)) = V \setminus (\overline{apr}_G(U \setminus A)).$$

Corollary 6.1.2 *If (U, V, r) is a generalized approximation space, then we have the following statements:*

(i) $\overline{apr}_G(\emptyset) = \emptyset$.

(ii) *For all $A \subseteq U$, $A \subseteq \underline{apr}_F(\overline{apr}_G(A))$.*

(iii) *For all $B \subseteq V$, $\overline{apr}_G(\underline{apr}_F(B)) \subseteq B$.*

(iv) $\underline{apr}_G(U) = V$.

(v) *For all $B \subseteq V$, $B \subseteq \underline{apr}_G(\overline{apr}_F(B))$.*

(vi) *For all $A \subseteq U$, $\overline{apr}_F(\underline{apr}_G(A)) \subseteq A$.*

Proof (iii) From Lemma 4.2.15, for any generalized t-approximation space (U, V, r, R), we have the inclusion

$$\forall B \in \mathcal{V},\ r^{\to}(r^{\leftarrow} B) \subseteq B.$$

Since the textures are discrete, we obtain

$$\overline{apr}_G(\underline{apr}_F(B)) = r^{\to}(r^{\leftarrow} B) \subseteq B.$$

(iv) Note that for discrete textures, by Corollary 5.4.4, we immediately conclude that $V = R^{\to} U = \underline{apr}_G(U)$.

Corollary 6.1.3 *If (U, V, r) is a generalized approximation space, then the following statements are equivalent:*

(i) *r is serial.*

(ii) $\forall A \subseteq U$, $\underline{apr}_F(\underline{apr}_G(A)) \subseteq A$.

(iii) $\forall B \subseteq V$, $\underline{apr}_F(B) \subseteq \overline{apr}_F(B)$.

(iv) $\forall A_1, A_2 \subseteq U$, $\overline{apr}_G(A_1) \subseteq \underline{apr}_G(A_2) \implies A_1 \subseteq A_2$.

Proof We prove the equivalence (i)$\iff$(iii). By Theorem 4.4.11, for a t-generalized approximation space (U, V, r, R), we have that

$$(r, R) \text{ is t-serial if and only if } \forall B \in \mathcal{V},\quad r^{\leftarrow} B \subseteq R^{\leftarrow} B.$$

Further, by Corollary 4.4.6, for discrete textures, we know that (r, R) is t-serial if and only if r is serial. Then by Corollary 5.4.4, we automatically conclude that

$$r \text{ is serial if and only if } r^{\leftarrow} B = \underline{apr}_F(B) \subseteq R^{\leftarrow} B = \overline{apr}_F(B).$$

Corollary 6.1.4 *Let r be an inverse serial relation from U to V. Then we have the following statements:*

(i) $\forall B \subseteq V$, $\underline{apr}_G(\underline{apr}_F(B)) \subseteq B \subseteq \overline{apr}_G(\overline{apr}_F(B))$.

(ii) $\forall A \subseteq U$, $\underline{apr}_G(A) \subseteq \overline{apr}_G(A)$.

(iii) $\forall v \in V$, $\overline{apr}_F(\{v\}) \neq \emptyset$.

(iv) $\forall \emptyset \neq B \in \mathcal{V}$, $\overline{apr}_F(B) \neq \emptyset$.

(v) $\forall V \neq B \in \mathcal{V}$, $\underline{apr}_F(B) \neq U$.

Proof We prove (i). Note that r is inverse serial if and only if (r, R) is t-inverse serial where $R = (U \times V) \setminus r$. On the other hand, by Theorem 4.4.13 we have that

$$\forall B \in \mathcal{V},\ R^{\to}(r^{\leftarrow}B) \subseteq B \subseteq r^{\to}(R^{\leftarrow}B).$$

Therefore, from Corollary 5.4.4, we obtain that

$$\underline{apr}_G(\underline{apr}_F(B)) = R^{\to}(r^{\leftarrow}B) \subseteq B \subseteq r^{\to}(R^{\leftarrow}B) = \overline{apr}_G(\overline{apr}_F(B)).$$

Corollary 6.1.5 *Let r be a relation from U to V. Then for all $B, B_1, B_2 \subseteq V$, we have the following assertions:*

(i) *If r is serial and r^{-1} injective, then* $\underline{apr}_F(B) = \overline{apr}_F(B)$.

(ii) *If r is inverse serial and injective, then* $\underline{apr}_G(\underline{apr}_F(B)) = B$.

(iii) *If r^{-1} is injective, then* $\overline{apr}_G(\overline{apr}_F(B) \subseteq B$.

(iv) *If r is inverse serial and r^{-1} is injective, then*

$$\overline{apr}_G(\underline{apr}_F(B)) = B = \underline{apr}_G(\overline{apr}_F(B)).$$

(v) $B_1 \subseteq B_2 \Longrightarrow \underline{apr}_F(B_1) \subseteq \underline{apr}_F(B_2)$.

(vi) $B_1 \subseteq B_2 \Longrightarrow \overline{apr}_F(B_1) \subseteq \overline{apr}_F(B_2)$.

Proof (i) By Theorem 4.4.16 (3), $r^{\leftarrow}B = R^{\leftarrow}B$. Then we easily see that

$$\underline{apr}_F(B) = r^{\leftarrow}B = R^{\leftarrow}B = \overline{apr}_F(B).$$

(iii) Since r is injective, by Theorem 4.4.10 (i), (r, R) is t-injective where $R = (U \times V) \setminus r$ and therefore, by Theorem 4.4.16 (2)(iii), $r^{\to}R^{\leftarrow}B \subseteq B$. Hence, we find that

$$\overline{apr}_G(\overline{apr}_F(B) = r^{\to}R^{\leftarrow}B \subseteq B.$$

(iv) By Theorem 4.4.6 (i) and Theorem 4.4.10, the corresponding direlation (r, R) is inverse serial and injective. Therefore, by Theorem 4.4.16 (4), we have $r^{\to}(r^{\leftarrow}B) = B = R^{\to}(R^{\leftarrow}B)$. This gives that

$$\overline{apr}_G(\underline{apr}_F(B)) = r^{\to}(r^{\leftarrow}B) = B = R^{\to}(R^{\leftarrow}B) = \underline{apr}_G(\overline{apr}_F(B)).$$

Corollary 6.1.6 *Let r be a relation from U to V. Then for all $A, A_1, A_2 \subseteq U$, we have the following assertions:*

(i) *If r is inverse serial and injective, then $\underline{apr}_G(A) = \overline{apr}_G(A)$.*

(ii) *If r is serial and injective, then $\underline{apr}_F(\underline{apr}_G(A)) = A$.*

(iii) *If r is injective, then $A \subseteq \overline{apr}_F(\overline{apr}_G(A)$.*

(iv) *If r is serial and injective, then*

$$\overline{apr}_F(\underline{apr}_G(A)) = A = \underline{apr}_F(\overline{apr}_G(A)).$$

(v) $A_1 \subseteq A_2 \Longrightarrow \underline{apr}_G(A_1) \subseteq \underline{apr}_G(A_2)$.

(vi) $A_1 \subseteq A_2 \Longrightarrow \overline{apr}_G(A_1) \subseteq \overline{apr}_G(A_2)$.

Proof The proof is similar to the proof of Corollary 5.5.10 and is left to the interested reader.

Corollary 6.1.7 *Let (U, V, r) be a generalized approximation space. Then for all $\{A_j \mid j \in J\} \subseteq \mathcal{P}(U)$ and $\{B_j \mid j \in J\} \subseteq \mathcal{P}(V)$, we have the following statements:*

(i) $\underline{apr}_F(\bigcap_{j\in J} B_j) = \bigcap_{j\in J} \underline{apr}_F(B_j)$ *and*

$$\underline{apr}_G(\bigcap_{j\in J} A_j) = \bigcap_{j\in J} \underline{apr}_G(A_j).$$

(ii) $\overline{apr}_G(\bigcup_{j\in J} A_j) = \bigcup_{j\in J} \overline{apr}_G(A_j)$ *and*

$$\overline{apr}_F(\bigcup_{j\in J} B_j) = \bigcup_{j\in J} \overline{apr}_F(B_j).$$

Proof (i) We give the proof of the first equality. From Theorem 4.2.17, we have $r^{\leftarrow} \bigcap_{j\in J} B_j = \bigcap_{j\in J} r^{\leftarrow} B_j$.

Therefore, by Corollary 5.4.4, we conclude that

$$\underline{apr}_F(\bigcap_{j\in J} B_j) = r^{\leftarrow} \bigcap_{j\in J} B_j = \bigcap_{j\in J} r^{\leftarrow} B_j = \bigcap_{j\in J} \underline{apr}_F(A_j).$$

(ii) For the first equality, as in the proof of (i), note that

$$r^{\rightarrow} \bigvee_{j\in J} A_j = \overline{apr}_G(\bigcup_{j\in J} A_j) = \bigvee_{j\in J} r^{\rightarrow} A_j = \bigcup_{j\in J} \overline{apr}_G(A_j).$$

Corollary 6.1.8 *Let (U, V, r) be a generalized approximation space. Then we have the following statements:*

(i) $\bigcup_{j\in J} \underline{apr}_F B_j \subseteq \underline{apr}_F(\bigcup_{j\in J} B_j)$ *and*

$\overline{apr}_F(\bigcap_{j\in J} B_j) \subseteq \bigcap_{j\in J} \overline{apr}_F(B_j)$.

(ii) $\bigcup_{j\in J} \underline{apr}_G A_j \subseteq \underline{apr}_G(\bigcup_{j\in J} A_j)$ *and*

$\overline{apr}_G(\bigcap_{j\in J} A_j) \subseteq \bigcap_{j\in J} \overline{apr}_G(A_j)$.

Proof We left to the reader.

Corollary 6.1.9 *If (U, V, r) is a generalized approximation space, then we have the following statements:*

(i) $\underline{apr}_F(V) = U$ *and* $\overline{apr}_F(\emptyset) = \emptyset$.

(ii) $\overline{apr}_F(\overline{apr}_G(\underline{apr}_F(B))) \subseteq \overline{apr}_F(B)$.

(iii) *If r is serial, then*

(1) $\underline{apr}_F(\emptyset) = \emptyset$ *and* $\overline{apr}_F(V) = U$.

(2) $\overline{apr}_F(B) \subseteq \overline{apr}_F(\overline{apr}_G(\overline{apr}_F(B)))$.

(iv) *If r is inverse serial, then*

(1) $\underline{apr}_G(\emptyset) = \emptyset$ *and* $\overline{apr}_G(U) = V$.

(2) $\underline{apr}_G(\underline{apr}_F(\underline{apr}_G(A))) \subseteq \underline{apr}_G(A)$.

Proof (iii) (2). Let r be serial. Then by Corollary 4.4.6, (r, R) is t-serial. Hence, by Theorem 4.4.17 (2),

$$R^{\leftarrow}B \subseteq R^{\leftarrow}(r^{\rightarrow}(R^{\leftarrow}B)).$$

Hence, by Corollary 5.4.4, we immediately find that

$$\overline{apr}_F(B) = R^{\leftarrow}B \subseteq R^{\leftarrow}(r^{\rightarrow}(R^{\leftarrow}B)) = \overline{apr}_F(\overline{apr}_G(\overline{apr}_F(B))).$$

Corollary 6.1.10 *If (U, V, r) is a generalized approximation space, then we have the following statements:*

(i) $\forall A \in \mathcal{P}(U)$, $\overline{apr}_G(\underline{apr}_F(\overline{apr}_G(A))) = \overline{apr}_G(A)$.

(ii) $\forall A \in \mathcal{P}(U)$, $\underline{apr}_G(\overline{apr}_F(\underline{apr}_G(A))) = \underline{apr}_G(A)$.

(iii) $\forall B \in \mathcal{P}(V)$, $\underline{apr}_F(\overline{apr}_G(\underline{apr}_F(B))) = \underline{apr}_F(B)$.

(iv) $\forall B \in \mathcal{P}(V)$, $\overline{apr}_F(\underline{apr}_G(\overline{apr}_F(B))) = \overline{apr}_F(B)$.

Proof (ii) From Theorem 4.2.19, for a t-generalized approximation space (U, V, r, R), we have

$$\forall A \in \mathcal{U},\ R^{\rightarrow}(R^{\leftarrow}(R^{\rightarrow}A)) = R^{\rightarrow}A.$$

Then by Corollary 5.4.4, we obtain that

$$\underline{apr}_G(\overline{apr}_F(\underline{apr}_G(A))) = R^{\rightarrow}(R^{\leftarrow}(R^{\rightarrow}A)) = R^{\rightarrow}A = \underline{apr}_G(A).$$

Corollary 6.1.11 *Let (U, V, r) be a generalized approximation space. If r is serial, then*

$$A \subseteq \overline{apr}_F(\overline{apr}_G(A)).$$

Proof By Theorem 4.4.15 (1), for all $A \subseteq U$, we have $A \subseteq R^{\leftarrow}r^{\rightarrow}A$. Then we obtain

$$A \subseteq R^{\leftarrow}r^{\rightarrow}A = \overline{apr}_F(\overline{apr}_G(A)).$$

Corollary 6.1.12 *Let (U, V, r) be a generalized approximation space. Then we have the following statements:*

(1) *Let r be serial and r^{-1} be injective, then*

(i) $\overline{apr}_F(B) = \overline{apr}_F(\overline{apr}_G(\overline{apr}_F(B)))$,

(ii) $\overline{apr}_F(B) = \underline{apr}_F(\underline{apr}_G(\overline{apr}_F(B)))$.

(iii) $\underline{apr}_F(B) = \overline{apr}_F(\overline{apr}_G(\underline{apr}_F(B)))$,

(iv) $\underline{apr}_F(B) = \overline{apr}_F(\underline{apr}_G(\underline{apr}_F(B)))$,

(v) $\underline{apr}_F(B) = \underline{apr}_F(\underline{apr}_G(\underline{apr}_F(B)))$.

(2) *Let r be injective and serial. Then*

(i) $\overline{apr}_F(B) = \underline{apr}_F(\overline{apr}_G(\overline{apr}_F(B))$,

(ii) $\underline{apr}_G(A) = \overline{apr}_G(\underline{apr}_F(\underline{apr}_G(A))$.

(3) *Let r be inverse serial and injective. Then*

(i) $\underline{apr}_G(A) = \underline{apr}_G(\underline{apr}_F(\underline{apr}_G(A)))$.

(ii) $\underline{apr}_G(A) = \overline{apr}_G(\overline{apr}_F(\underline{apr}_G(A)))$.

(iii) $\overline{apr}_G(A) = \underline{apr}_G(\overline{apr}_F(\overline{apr}_G(A))$.

(iv) $\overline{apr}_G(A) = \overline{apr}_G(\overline{apr}_F(\overline{apr}_G(A)))$.

Proof (1) (i) By Theorem 4.4.18 (ii)(a), we already have that $R^{\leftarrow}(r^{\rightarrow}(R^{\leftarrow}B)) = R^{\leftarrow}B$. Hence,

$$\underline{apr}_G(\underline{apr}_F(\underline{apr}_G(A))) = R^{\leftarrow}(r^{\rightarrow}(R^{\leftarrow}B)) = R^{\leftarrow}B = \underline{apr}_G(A).$$

(2)(ii) By Theorem 4.4.19 (ii)(d), we already have that $R^{\rightarrow}A = r^{\rightarrow}R^{\leftarrow}R^{\rightarrow}A$. As a result, we conclude that

$$\overline{apr}_G(\underline{apr}_F(\underline{apr}_G(A)) = r^{\rightarrow} R^{\leftarrow} R^{\rightarrow} A = R^{\rightarrow} A = \overline{apr}_G(A)$$

as desired.

(iii) By Theorem 4.4.19 (ii)(c), for the generalized t-approximation space (U, V, r, R), the equality $R^{\rightarrow} A = r^{\rightarrow} R^{\leftarrow} R^{\rightarrow} A$ holds. Therefore, we get

$$\underline{apr}_G(\overline{apr}_F(\overline{apr}_G(A)) = R^{\rightarrow} R^{\leftarrow} r^{\rightarrow} A = R^{\rightarrow} A = \overline{apr}_G(A).$$

6.2 Definability in Generalized Approximation Spaces with Two Domains of Discourse

If textures are discrete, then by Corollary 5.4.4, the presectional definability and sectional definability given in Sect. 5.5 correspond to the well-known F-definability and G-definability, respectively, in the literature.

Definition 6.2.1 Let (U, V, r) be a generalized approximation space.

(1) $B \subseteq V$ is called *F-definable* if $\underline{apr}_F(B) = \overline{apr}_F(B)$.

(2) $A \subseteq U$ is called *G-definable* if $\underline{apr}_G(A) = \overline{apr}_G(A)$.

Corollary 6.2.2 *Let (U, V, r) be a generalized approximation space where r is compatible.*

(i) *$B \subseteq V$ is F-definable $\Longleftrightarrow \exists A \subseteq U$, $\overline{apr}_G(A) = B$ and $\overline{apr}_F(B) = A$.*

(ii) *$A \subseteq U$ is G-definable $\Longleftrightarrow \exists B \subseteq V$, $\underline{apr}_G(A) = B$ and $\underline{apr}_F(B) = A$.*

Proof Let $B \subseteq V$ be F-definable. Let us consider the corresponding textural approximation space (U, V, r, R) where $R = (U \times V) \setminus r$ (note that textures are discrete, that is, $\mathcal{U} = \mathcal{P}(U)$ and $\mathcal{V} = \mathcal{P}(V)$). Then by Corollary 5.4.4, B is presectional definable, namely $r^{\leftarrow} B = R^{\leftarrow} B$. Hence, by Theorem 5.5.6 (i), there exists a set $A \in \mathcal{U}$ such that

$$r^{\rightarrow} A = B \text{ and } R^{\leftarrow} B = A.$$

Again, by Corollary 5.4.4, we obtain

$$\overline{apr}_G(A) = B \text{ and } \overline{apr}_F(B) = A$$

as desired. Sufficiency and the proof of (ii) are similar.

We have the following characterization.

Theorem 6.2.3 *Let (U, V, r) be a generalized approximation space. Then the following conditions are equivalent:*

(i) *r is a function from U to V.*

(ii) *$\forall B \subseteq V$, V is F-definable.*

(iii) *$\forall v \in V$, $\{v\}$ is F-definable.*

Proof (i) $\Longrightarrow$ (ii) : Let r be a function from U to V. Then (r, R) is a difunction from $(U, \mathcal{P}(U))$ to $(V, \mathcal{P}(V))$ where $R = (U \times V) \setminus r$. Hence, by Theorem 5.5.4, every subset $B \subseteq V$ is presectional definable, that is, $r^{\leftarrow}B = R^{\leftarrow}B$. Now, by Corollary 5.4.4, we obtain $\underline{apr}_F(B) = \overline{apr}_F(B)$.

(ii) $\Longrightarrow$ (iii) : Immediate.

(iii) $\Longrightarrow$ (i) Let $B \in \mathcal{P}(V)$. Since r is a function, it is also serial, and so by Corollary 6.1.3 (iii), we already have the inclusion

$$\underline{apr}_F(B) = \overline{apr}_F(B.$$

Let us show the reverse inclusion. By the assumption, for all $v \in V$, the singleton $\{v\}$ is F-definable, that is, $\underline{apr}_F(\{v\}) = \overline{apr}_F(\{v\})$. Therefore, by Corollaries 6.1.7 (ii) and 6.1.8 (i), we conclude that

$$\begin{aligned}\underline{apr}_F(B) &= \underline{apr}_F\Big(\bigcup_{v \in B}\{v\}\Big) \supseteq \bigcup_{v \in B}\underline{apr}_F(\{v\})\\ &= \bigcup_{v \in B}\overline{apr}_F(\{v\}) = \overline{apr}_F\Big(\bigcup_{v \in B}\{v\}\Big) = \overline{apr}_F(B).\end{aligned}$$

Therefore for all $B \in \mathcal{V}$, we obtain that $r^{\leftarrow}B = R^{\leftarrow}B$. Then by Theorem 5.5.4, (r, R) is a difunction from $(U, \mathcal{P}(U))$ to $(V, \mathcal{P}(V))$, which implies that r is a function from U to V. □

Dually, we may also give the following characterization.

Theorem 6.2.4 *Let (U, V, r) be a generalized approximation space. Then the following conditions are equivalent:*

(i) r^{-1} *is a function from* V *to* U.

(ii) $\forall A \subseteq U,\ A$ *is* G*-definable.*

(iii) $\forall u \in U,\ \{u\}$ *is* G*-definable.*

Successor and predecessor definabilities

Successor and predecessor definabilities are first defined in terms of neighborhoods by Grzymala-Busse and Rzasa in [1] for generalized approximation space with a single domain of discourse. Recall that for a generalized approximation space (U, r), a set $A \subseteq U$ is called predecessor definable if

$$A = \bigcup_{u \in B} r_p(u),$$

and $A \subseteq U$ is called successor definable if

$$A = \bigcup_{u \in B} r_s(u).$$

Let the textures be discrete and $U = V$, then by Proposition 5.4.8, if $A \subseteq U$ is t-predecessor definable, then it is predecessor definable:

$$A = \bigvee_{B \not\subseteq Q_u} R^{\leftarrow} P_u = \bigcup_{u \in B} r_p(u).$$

Likewise, a t-successor definable set is successor definable:

$$A = \bigvee_{B \not\subseteq Q_u} r^{\rightarrow} P_u = \bigcup_{u \in B} r_s(u).$$

In some sense, this concept can be extended to an approximation space with two domains of discourse. Let $(U, \mathcal{U})$ and $(V, \mathcal{V})$ be discrete texture spaces. Then by Proposition 5.5.4, a t-predecessor definable set is predecessor definable:

$$B = \bigvee_{A \not\subseteq Q_u} r^{\rightarrow} P_u = \bigcup_{u \in A} F(u).$$

Similarly, a textural successor definable set is successor definable:

$$A = \bigvee_{B \not\subseteq Q_v} R^{\leftarrow} P_v = \bigcup_{v \in B} G(v).$$

The right sides of the above equalities give us the successor and predecessor definabilities for generalized approximation spaces with two domains of discourse.

Corollary 6.2.5 *Let (U, V, r) be a generalized approximation space. Then*

(i) *$B \subseteq V$ is successor definable if and only if $B = \overline{apr}_G(A)$ for some $A \subseteq U$.*

(ii) *$A \subseteq U$ is predecessor definable if $A = \overline{apr}_F(B)$ for some $B \subseteq V$.*

Proof (i) Let B be successor definable. Then by Theorem 5.3.1 (i), for some $A \subseteq U$, we get

$$B = \bigcup_{u \in A} F(u) = \mathcal{F}(A) = \overline{apr}_G(A).$$

(ii) Let B be predecessor definable. Again, by Theorem 5.3.1 (ii), for some $B \subseteq V$, we get

$$A = \bigcup_{v \in B} G(v) = \mathcal{G}(B) = \overline{apr}_G(B) = \mathcal{G}(B) = \overline{apr}_F(B).$$

Remark 6.2.6 The well-known definabilities of sets are given in Table 6.1. Note that the sets $F(u)$ and $G(v)$ correspond to the successor and predecessor neighborhood operators given of the approximation spaces with a single universe, respectively. In other words, we have the equalities

$$F(u) = r_s(u) \text{ and } G(v) = r_p(v)$$

for $u, v \in U = V$. Therefore, by Theorem 5.4.4, we observe that for every $A \subseteq U$,

$$r^{\rightarrow} A = \overline{apr}_G(A) = \overline{apr}_{r_s}(A), \quad R^{\rightarrow} A = \underline{apr}_G(A) = \underline{apr}_{r_s}(A),$$

$$r^{\leftarrow} A = \underline{apr}_F(A) = \underline{apr}_{r_p}(A), \quad R^{\leftarrow} A = \overline{apr}_F(A) = \overline{apr}_{r_p}(A). \qquad \square$$

Table 6.1 Well-known definabilities

Approximation spaces	Definability with approximations	Definability with neighborhoods
Pawlak's approximation space	$\underline{apr}(A) = \overline{apr}(A)$	$A = \bigcup_{u \in A} [u]_r$
Generalized approximation space with a single domain	$\underline{apr}_{r_p}(A) = \overline{apr}_{r_p}(A)$ $\underline{apr}_{r_s}(A) = \overline{apr}_{r_s}(A)$	$A = \bigcup_{u \in B} r_p(u)$ (predecessor definable) $A = \bigcup_{u \in B} r_s(u)$ (successor definable)
Generalized approximation space with two domains	$\underline{apr}_F(B) = \overline{apr}_F(B)$ (F-definable) $\underline{apr}_G(A) = \overline{apr}_G(A)$ (G-definable)	$\exists A \subseteq U,\ B = \bigcup_{u \in A} F(u)$ $\exists B \subseteq V,\ A = \bigcup_{v \in B} G(v)$
t-Generalized approximation space	$r^{\to} A = R^{\to} A$ (sectional definable) $r^{\leftarrow} B = R^{\leftarrow} B$ (presectional definable)	$\exists B \in \mathcal{V},\ A = \bigvee_{B \not\subseteq Q_v} R^{\leftarrow} P_v$ (t-predecessor definable) $\exists A \in \mathcal{U},\ B = \bigvee_{A \not\subseteq Q_u} r^{\to} P_u$ (t-successor definable)

Theorem 6.2.7 *Let* (U, r) *be a generalized approximation space with a single domain.*

(1) *If* r *is a reflexive relation on* U*, then*

(i) $\underline{apr}_{r_p}(A) = \overline{apr}_{r_p}(A) \implies$ *A is successor definable.*

(ii) $\underline{apr}_{r_s}(A) = \overline{apr}_{r_s}(A) \implies$ *A is predecessor definable.*

(2) *If* r *is a symmetric relation on* U*, then we have the following equivalences:*

(i) *A is successor definable* $\iff$ *A is predecessor definable.*

(ii) $\underline{apr}_{r_s}(A) = \overline{apr}_{r_s}(A) \iff \underline{apr}_{r_p}(A) = \overline{apr}_{r_p}(A)$.

(3) *If* r *is an equivalence relation on* U*, then the following conditions are equivalent:*

(i) *A is successor definable.*

(ii) *A is predecessor definable.*

(iii) $\underline{apr}_{r_s}(A) = \overline{apr}_{r_s}(A)$.

(iv) $\underline{apr}_{r_p}(A) = \overline{apr}_{r_p}(A)$.

Proof We give the proof of (1) and (2) leaving (3) to the interested reader.

(1) (i) Let r be reflexive and $\underline{apr}_{r_p}(A) = \overline{apr}_{r_p}(A)$. By Theorem 4.1.14, ($r$, R) is a reflexive direlation on the generalized t-approximation space (U, r, R). Then by Corollary 5.4.4, $r^{\rightarrow}A = R^{\rightarrow}A$ and so by Theorem 5.5.10 (1)(i), we get that A is t-successor definable. Finally, by the arguments given before Corollary 6.2.5, A is successor definable.

(ii) Let $\underline{apr}_{r_s}(A) = \overline{apr}_{r_s}(A)$. As in item (i), ($r$, R) is a reflexive direlation on the generalized t-approximation space (U, r, R). Then by Corollary 5.4.4, $r^{\leftarrow}A = R^{\leftarrow}A$ and then by Theorem 5.5.10 (1)(ii), A is t-predecessor definable and hence, it is predecessor definable as desired.

(2) If r is symmetric, then by Theorem 4.1.25, (r, R) is also symmetric on the generalized t-approximation space (U, r, R) where $R = (U \times U) \setminus r$.

(i) Let A be successor definable. Then A is t-successor definable in (U, r, R). Then by Theorem 4.1.25 (2)(i), A is t-predecessor definable. Hence, by the arguments given before Corollary 6.2.5, A is predecessor definable. The proof of the necessity is similar.

(ii) We only prove the sufficiency. Let $\underline{apr}_{r_s}(A) = \overline{apr}_{r_s}(A)$. Then $r^{\rightarrow}A = R^{\rightarrow}A$ in (U, r, R). Then by Theorem 5.5.10 (2)(ii), we also have that $r^{\leftarrow}A = R^{\leftarrow}A$ and so we conclude that $\underline{apr}_{r_p}(A) = \overline{apr}_{r_p}(A)$. Necessity is also easy.

Let $S = (U, AT, V, f)$ be an information system. Suppose that for an attribute $a \in AT$, we have $f(x, a) \notin V_a$. Then we say that "the value is lost" and we denote this by $f(x, a) =?$. If for at least one attribute $a \in AT$, we have $f(x, a) =?$, then the quadruple $S = (U, AT, V, f)$ is called *an incomplete information system* [2, 3].

In [4], for any incomplete information system $S = (U, AT, V, f)$ and for any $A \subseteq AT$, the similarity relation is defined as follows:

$$\forall x, y \in U, \quad SIM(A)(x, y) \Longleftrightarrow \forall a \in A : ((f(x, a) =?) \text{ or } (f(x, a) = f(y, a)).$$

The relation $SIM(A)$ is reflexive and transitive, but not necessarily symmetric. Now for any object $x \in U$, let us consider the following sets defined in [4]:

$SIM_A(x) = \{y \in U : (y, x) \in SIM(A)\}$,

$SIM_A^{-1}(x) = \{y \in U : (x, y) \in SIM(A)\}$.

In fact, by definition of predecessor and successor neighborhood, we have

$$SIM_A(x) = r_p(x) \text{ and } SIM_A^{-1}(x) = r_s(x)$$

where $SIM(A) = r$. Now look at the following example.

Table 6.2 Incomplete information system S

	a	b	c	d
x_1	1	1	1	2
x_2	2	?	2	3
x_3	2	2	?	3

Example 6.2.8 Let us consider the incomplete information system in Table 6.2. For $A = \{a, b\}$, we have

$$r = SIM(A) = \{(x_1, x_1), (x_2, x_2), (x_3, x_3), (x_2, x_3)\}.$$

Then the family of all successor definable sets is

$$\mathcal{D}_s = \{r(B) \mid B \subseteq U\} = \{U, \{x_1\}, \{x_3\}, \{x_1, x_3\}, \{x_2, x_3\}\}.$$

Clearly, $\{x_1, x_2\}$ is not successor definable. Similarly, the family of all predecessor definable sets is

$$\mathcal{D}_p = \{r^{-1}(B) \mid B \subseteq U\} = \{U, \{x_1\}, \{x_2\}, \{x_2, x_3\}, \{x_1, x_2\}\}.$$

It is easy to see that $\{x_1, x_3\}$ is not predecessor definable. □

For a generalized approximation space (U, r), the subset approximation operators

$\underline{apr}'_{r_p}(A) = \bigcup\{r_p(u) \mid r_p(u) \subseteq A\}$,

$\overline{apr}'_{r_p}(A) = \{u \in U \mid \forall u' \in U,\ u' \in r_p(u) \implies r_p(u) \cap A \neq \emptyset\}$,

and

$\underline{apr}'_{r_s}(A) = \bigcup\{r_s(u) \mid r_s(u) \subseteq A\}$,

$\overline{apr}'_{r_s}(A) = \{u \in U \mid \forall u' \in U,\ u' \in r_s(u) \implies r_s(u) \cap A \neq \emptyset\}$,

are given by Yao in [7].

Theorem 6.2.9 *Let $(U, \mathcal{U})$ be a discrete texture, that is, $\mathcal{U} = \mathcal{P}(U)$ and (r, R) be a complemented direlation on $(U, \mathcal{P}(U))$. Then*

$$r^{\to} r^{\leftarrow} A = \underline{apr}'_{r_s}(A) \ \text{and}\ R^{\to} R^{\leftarrow} A = \overline{apr}'_{r_s}(A)$$

for all $A \subseteq U$. □

Proof If (r, R) is a complemented direlation on $(U, \mathcal{P}(U))$, then it is easy to see that r is an ordinary relation on U, that is, $r \subseteq U \times U$ where $R = (U \times U) \setminus r$. Now

$$\begin{aligned} r^{\to} r^{\leftarrow} A &= r^{\to} \bigvee\{P_u \mid \forall v \in U, r \not\subseteq \overline{Q}_{(u,v)} \implies P_v \subseteq A\} \\ &= \bigvee\{r^{\to} P_u \mid \forall v \in U, r \not\subseteq \overline{Q}_{(u,v)} \implies P_v \subseteq A\} \\ &= \bigcup\{r_s(u) \mid \forall v \in U, (u,v) \in r \implies v \in A\} \\ &= \bigcup\{r_s(u) \mid r_s(u) \subseteq A\} = \underline{apr'}_{r_s}(A), \end{aligned}$$

and

$$\begin{aligned} R^{\to} R^{\leftarrow} A &= R^{\to} \bigcap\{Q_u \mid \forall v, \overline{P}_{(u,v)} \not\subseteq R \implies A \subseteq Q_v\} \\ &= \bigcap\{R^{\to} Q_u \mid \forall v, \overline{P}_{(u,v)} \not\subseteq R \implies A \subseteq Q_v\} \\ &= \bigcap\{U \setminus r_s(u) \mid \forall v, (u,v) \in r \implies v \notin A\} \\ &= U \setminus (\bigcup\{r_s(u) \mid \forall v, (u,v) \in r \implies v \notin A\}) \\ &= U \setminus (\{w \mid \exists u, w \in r_s(u)[\forall v, v \in r_s(u) \implies v \notin A]\}) \\ &= U \setminus \{w \mid \exists u, w \in r_s(u), r_s(u) \cap A = \emptyset\} \\ &= \{w \mid \forall u, w \in r_s(u) \implies r_s(u) \cap A \neq \emptyset\} \\ &= \overline{apr}'_{r_s}(A). \end{aligned}$$

We also have the following result.

Theorem 6.2.10 *Let $(U, \mathcal{U})$ be a discrete texture, that is, $U = \mathcal{P}(U)$ and (r, R) be a complemented direlation on $(U, \mathcal{P}(U))$. Then*

$$R^{\leftarrow} R^{\to} A = \underline{apr'}_{r_p}(A) \text{ and } r^{\leftarrow} r^{\to} A = \overline{apr}'_{r_p}(A)$$

for all $A \subseteq U$. □

Proof

$$\begin{aligned} r^{\leftarrow} r^{\to} A &= r^{\leftarrow} \bigcap\{Q_v \mid \forall u, r \not\subseteq \overline{Q}_{(u,v)} \implies A \subseteq Q_u\} \\ &= \bigcap\{r^{\leftarrow} Q_v \mid \forall u, (u,v) \in r \implies u \notin A\} \\ &= \bigcap\{U \setminus r_p(v) \mid r_p(v) \cap A = \emptyset\} \\ &= U \setminus (\bigcup\{r_p(v) \mid r_p(v) \cap A = \emptyset\}) \\ &= U \setminus \{w \mid \exists v \in U, w \in r_p(v),\ r_p(v) \cap A = \emptyset\} \\ &= \{w \mid \forall v \in U, w \in r_p(v) \implies r_p(v) \cap A \neq \emptyset\} = \overline{apr}'_{r_p}(A) \end{aligned}$$

and

$$\begin{aligned} R^{\leftarrow} R^{\to} A &= R^{\leftarrow} \bigvee\{P_v \mid \forall u, \overline{P}_{(u,v)} \not\subseteq R \implies P_u \subseteq A\} \\ & \bigvee\{R^{\leftarrow} P_v \mid \forall u, \overline{P}_{(u,v)} \not\subseteq R \implies P_u \subseteq A\}) \\ &= \bigcup\{r_p(v) \mid \forall u, (u,v) \in r \implies u \in A\} \\ &= \bigcup\{r_p(v) \mid r_p(v) \subseteq A\} = \underline{apr'}_{r_p}(A). \end{aligned}$$

□

It is interesting that the subset approximations can be written in terms of single approximations as follows:

Corollary 6.2.11 *(i)* $\underline{apr}'_{r_s}(A) = \overline{apr}_{r_p}(\underline{apr}_{r_s}(A))$.

(ii) $\overline{apr}'_{r_s}(A) = \underline{apr}_{r_p}(\overline{apr}_{r_s}(A))$.

(iii) $\underline{apr}'_{r_p}(A) = \overline{apr}_{r_s}(\underline{apr}_{r_p}(A))$.

(iv) $\overline{apr}'_{r_p}(A) = \underline{apr}_{r_s}(\overline{apr}_{r_p}(A))$.

Proof By Theorem 5.4.4 and Remark 6.2.6, we know that for every $A \subseteq U$,

$$r^{\rightarrow}A = \overline{apr}_{r_s}(A), \quad R^{\rightarrow}A = \underline{apr}_{r_s}(A),$$

$$r^{\leftarrow}A = \underline{apr}_{r_p}(A), \quad R^{\leftarrow}A = \overline{apr}_{r_p}(A).$$

Then by Theorems 5.4.4 and 6.2.9, the proof is immediate.

6.3 Revised Textural Rough Sets

Let (r, R) be a direlation from $(U, \mathcal{U})$ to $(V, \mathcal{V})$. We know that for every set $B \in \mathcal{V}$, the section $r^{\leftarrow}B$ is an element of $\mathcal{U}$. This gives us a mapping

$$r^{\leftarrow} : \mathcal{V} \rightarrow \mathcal{U}$$

defined by

$$\forall B \in \mathcal{V},\ r^{\leftarrow}B = \bigvee\{P_u \mid \forall v, r \nsubseteq \overline{Q}_{(u,v)} \Rightarrow P_v \subseteq B\}.$$

Similarly, we take the mapping

$$r^{\rightarrow} : \mathcal{U} \rightarrow \mathcal{V}$$

defined by

$$\forall A \in \mathcal{U},\ r^{\rightarrow}A = \bigcap\{Q_v \mid \forall u, r \nsubseteq \overline{Q}_{(u,v)} \Rightarrow A \subseteq Q_u\}.$$

Then we may consider the ordinary composition of the mappings $r^{\leftarrow}$ and $r^{\rightarrow}$ as

$$r^{\rightarrow}r^{\leftarrow} : \mathcal{V} \rightarrow \mathcal{V}$$

where

$$\forall B \in \mathcal{V},\ r^{\rightarrow}r^{\leftarrow}B = r^{\rightarrow}(r^{\leftarrow}B).$$

Dually, for the corelation, we also take the mapping

$$R^{\rightarrow}R^{\leftarrow} : \mathcal{V} \rightarrow \mathcal{V},\ \forall B \in \mathcal{V},\ R^{\rightarrow}R^{\leftarrow}B = R^{\rightarrow}(R^{\leftarrow}B).$$

Now we may give the following definition.

Definition 6.3.1 Let $(\mathbf{U}, \mathbf{V}, r, R)$ be a generalized textural approximation space and $B \in \mathcal{V}$.

(i) The operators $r^{\rightarrow}r^{\leftarrow}$ and $R^{\rightarrow}R^{\leftarrow}$ are called *the revised t-lower approximation operator* and *t-upper approximation operator* on $\mathcal{V}$, respectively.

(ii) The operators $R^{\leftarrow}R^{\rightarrow}$ and $r^{\leftarrow}r^{\rightarrow}$ are called *the revised t-lower approximation operator* and *t-upper approximation operator* on $\mathcal{U}$, respectively.

Theorem 6.3.2 *If $(U, \mathcal{U}, c_U)$ and $(V, \mathcal{V}, c_V)$ are complemented textures and (r, R) is a complemented direlation from $(U, \mathcal{U}, c_U)$ to $(V, \mathcal{V}, c_V)$, then for all $B \in \mathcal{P}(V)$,*

$$r^{\rightarrow}r^{\leftarrow}B = c_V(R^{\rightarrow}R^{\leftarrow}c_V(B)) \text{ and}$$

$$R^{\rightarrow}R^{\leftarrow}B = c_V(r^{\rightarrow}r^{\leftarrow}c_V(B)).$$

Proof Since (r, R) is complemented, we have $r' = R$ and $R' = r$. Then by Lemma 4.2.29, we get

$$\begin{aligned} c_V(R^{\rightarrow}R^{\leftarrow}(c_V(B))) &= c_V((r')^{\rightarrow}((r')^{\leftarrow}(c_V(B))) \\ &= c_V((r')^{\rightarrow}c_U(r^{\leftarrow}c_V(c_V(B))) \\ &= c_V((r')^{\rightarrow}c_U(r^{\leftarrow}B) \\ &= c_V(c_V(r^{\rightarrow}c_U(c_U(r^{\leftarrow}B)))) \\ &= r^{\rightarrow}r^{\leftarrow}B. \end{aligned}$$

The proof of the second equality is dual and it is omitted.

We have the following results.

Theorem 6.3.3 *Let (U, V, r, R) be a generalized t-approximation space, $B \in \mathcal{V}$ and $\{B_j \mid j \in J\} \subseteq \mathcal{V}$. Then we have the following properties:*

(1) $r^{\rightarrow}r^{\leftarrow}B \subseteq B \subseteq R^{\rightarrow}R^{\leftarrow}B.$

(2) $B_1 \subseteq B_2 \implies r^{\rightarrow}r^{\leftarrow}B_1 \subseteq r^{\rightarrow}r^{\leftarrow}B_2.$

(3) $B_1 \subseteq B_2 \implies R^{\rightarrow}R^{\leftarrow}B_1 \subseteq R^{\rightarrow}R^{\leftarrow}B_2.$

(4) $\bigvee_{j\in J} r^{\rightarrow}r^{\leftarrow}B_j \subseteq r^{\rightarrow}r^{\leftarrow}\bigvee_{j\in J} B_j.$

(5) $R^{\rightarrow}R^{\leftarrow}(\bigcap_{j\in J} B_j) \subseteq \bigcap_{j\in J} R^{\rightarrow}R^{\leftarrow}B_j.$

(6) $r^{\rightarrow}r^{\leftarrow}(r^{\rightarrow}r^{\leftarrow}B) = r^{\rightarrow}r^{\leftarrow}B.$

(7) $R^{\rightarrow}R^{\leftarrow}(R^{\rightarrow}R^{\leftarrow}B) = R^{\rightarrow}R^{\leftarrow}B.$

Proof (1) It is immediate by Lemmas 4.2.15 (1)(iii) and (2)(ii).

(2) By Lemma 4.2.14 (ii), for $B_1, B_2 \in \mathcal{V}$ with $B_1 \subseteq B_2$, we have $r^{\leftarrow}B_1 \subseteq r^{\leftarrow}B_2$ and then by Lemma 4.2.14 (i), if we apply the section operator on both sides of the inclusion, we conclude that

$$r^{\to}r^{\leftarrow}B_1 \subseteq r^{\to}r^{\leftarrow}B_2$$

as desired.

(3) It is similar to the proof of (2).

(4) By Theorem 4.2.17 (i) and Theorem 4.2.18 (i), we clearly get

$$\bigvee_{j\in J} r^{\to}r^{\leftarrow}B_j = r^{\to}\bigvee_{j\in J} r^{\leftarrow}B_j \subseteq r^{\to}r^{\leftarrow}\bigvee_{j\in J} B_j.$$

(5) By Theorem 4.2.18 (i), $R^{\leftarrow}\bigcap_{j\in J} B_j \subseteq \bigcap_{j\in J} R^{\leftarrow}B_j$ and so by Theorem 4.2.17 (i), we get

$$R^{\to}R^{\leftarrow}\bigcap_{j\in J} B_j \subseteq R^{\to}(\bigcap_{j\in J} R^{\leftarrow}B_j) = \bigcap_{j\in J} R^{\to}R^{\leftarrow}B_j.$$

(6) By Theorem 4.2.19 (i), we already have that $r^{\leftarrow}(r^{\to}r^{\leftarrow}B) = r^{\leftarrow}B$. If we apply the operator $r^{\to}$ on both sides of the equality we find that

$$r^{\to}r^{\leftarrow}(r^{\to}r^{\leftarrow}B) = r^{\to}r^{\leftarrow}B.$$

(7) Similar.

Theorem 6.3.4 *Let (U, V, r, R) be a generalized t-approximation space, $B \in \mathcal{V}$ and $\{B_j \mid j \in J\} \subseteq \mathcal{V}$. Then we have the following properties:*

(1) *If (r, R) is t-serial, then $r^{\to}r^{\leftarrow}\emptyset = \emptyset$ and $R^{\to}R^{\leftarrow}V = V$.*

(2) *If (r, R) is t-inverse serial, then $r^{\to}r^{\leftarrow}V = V$ and $R^{\to}R^{\leftarrow}\emptyset = \emptyset$.*

(3) *Let (r, R) be t-inverse serial and injective.*

(i) $r^{\to}r^{\leftarrow}\bigcap_{j\in J} B_j = \bigcap_{j\in J} r^{\to}r^{\leftarrow}B_j$.

(ii) $R^{\to}R^{\leftarrow}\bigvee_{j\in J} B_j = \bigvee_{j\in J} R^{\to}R^{\leftarrow}B_j$.

(4) *Let (r, R) be t-serial and injective.*

(i) $R^{\to}R^{\leftarrow}\bigcap_{j\in J} B_j = \bigcap_{j\in J} R^{\to}R^{\leftarrow}B_j$.

(ii) $r^{\to}r^{\leftarrow}\bigvee_{j\in J} B_j = \bigvee_{j\in J} r^{\to}r^{\leftarrow}B_j$.

(ii) *If (r, R) is inverse serial and injective, then $R^{\to}R^{\leftarrow}(B) = r^{\to}r^{\leftarrow}(R^{\to}R^{\leftarrow}B)$.*

(5) *Let (r, R) be t-inverse serial and $(R^{\leftarrow}, r^{\leftarrow})$ be injective.*

(i) $R^{\to}R^{\leftarrow}(r^{\to}r^{\leftarrow}B) = B$.

(ii) $B = r^{\to}r^{\leftarrow}(R^{\to}R^{\leftarrow}B)$.

Proof (1) Let (r, R) be t-serial. By Theorem 4.4.17 (3), we already have $r^{\leftarrow}\emptyset = \emptyset$ and $R^{\leftarrow}V = U$. Further, by Lemma 4.2.15 (1)(i) and (2)(i) $r^{\rightarrow}\emptyset = \emptyset$ and $R^{\rightarrow}U = V$. If we apply section operators $r^{\rightarrow}$ and $R^{\rightarrow}$ on both sides of the equalities, we obtain

$$r^{\rightarrow}r^{\leftarrow}\emptyset = r^{\rightarrow}\emptyset = \emptyset \text{ and } R^{\rightarrow}R^{\leftarrow}V = R^{\rightarrow}U = V.$$

(2) Let (r, R) be t-inverse serial. By Theorem 4.4.17 (1), we have $r^{\leftarrow}V = U$ and $R^{\leftarrow}\emptyset = \emptyset$. Further, by Theorem 4.4.17 (5), we already have $r^{\rightarrow}U = V$ and $R^{\rightarrow}\emptyset = \emptyset$. If we apply section operators on both sides of the equalities, we get

$$r^{\rightarrow}r^{\leftarrow}V = r^{\rightarrow}U = V \text{ and } R^{\rightarrow}R^{\leftarrow}\emptyset = R^{\rightarrow}\emptyset = \emptyset.$$

(3) (i) By Theorem 4.2.17 (i)–(ii) and Theorem 4.4.15 (3), we get

$$\begin{aligned} r^{\rightarrow}r^{\leftarrow}\bigcap_{j\in J} B_j &= r^{\rightarrow}(\bigcap_{j\in J} r^{\leftarrow}B_j) = R^{\rightarrow}(\bigcap_{j\in J} r^{\leftarrow}B_j) \\ &= \bigcap_{j\in J} R^{\rightarrow}(r^{\leftarrow}B_j) = \bigcap_{j\in J} r^{\rightarrow}r^{\leftarrow}B_j. \end{aligned}$$

(ii) The proof is similar to (i).

(4) (i) By Theorem 4.4.15 (3), Theorem 4.2.17 (ii),

$$\begin{aligned} R^{\rightarrow}R^{\leftarrow}\bigcap_{j\in J} B_j &= R^{\rightarrow}r^{\leftarrow}\bigcap_{j\in J} B_j = R^{\rightarrow}\bigcap_{j\in J} r^{\leftarrow}B_j \\ &= \bigcap_{j\in J} R^{\rightarrow}r^{\leftarrow}B_j = \bigcap_{j\in J} R^{\rightarrow}R^{\leftarrow}B_j. \end{aligned}$$

(ii) Similar.

(iii) From Lemma 4.2.15 (2) (ii), we already have that $B \subseteq R^{\rightarrow}R^{\leftarrow}B$. Then by Lemma 4.2.14 (ii)

$r^{\leftarrow}B \subseteq r^{\leftarrow}R^{\rightarrow}R^{\leftarrow}B,$

and by Lemma 4.2.14 (i) and Theorem 4.4.15 (4)(i), we obtain

$B = r^{\rightarrow}r^{\leftarrow}B \subseteq r^{\rightarrow}r^{\leftarrow}R^{\rightarrow}R^{\leftarrow}B.$

(iv) Similar.

(5) (i) By Theorem 4.4.15 (4)(i), we know that $r^{\rightarrow}(r^{\leftarrow}B) = B$ for all $B \in \mathcal{V}$. By Lemma 4.2.14 (iv),

$R^{\leftarrow}r^{\rightarrow}(r^{\leftarrow}B) = R^{\leftarrow}B$

and so by Lemma 4.2.14 (iii) and Theorem 4.4.16 (4) (i) and (ii), we find

$R^{\rightarrow}R^{\leftarrow}r^{\rightarrow}(r^{\leftarrow}B) = R^{\rightarrow}R^{\leftarrow}B = B.$

The following theorems are the dual of Theorems 6.3.2, 6.3.3, and 6.3.4, respectively, and the proofs are omitted.

Theorem 6.3.5 *If $(U, \mathcal{U}, c_U)$ and $(V, \mathcal{V}, c_V)$ are complemented textures and (r, R) is a complemented direlation from $(U, \mathcal{U}, c_U)$ to $(V, \mathcal{V}, c_V)$, then for all $B \in \mathcal{V}$,*

$$r^{\leftarrow} r^{\rightarrow} A = c_U(R^{\leftarrow} R^{\rightarrow} c_U(A)) \text{ and}$$

$$R^{\leftarrow} R^{\rightarrow} A = c_U(r^{\leftarrow} r^{\rightarrow} c_U(A)).$$

Theorem 6.3.6 *For every $A \in \mathcal{U}$ and $\{A_j \mid j \in J\} \subseteq \mathcal{U}$, the mappings $R^{\leftarrow} R^{\rightarrow} : \mathcal{U} \to \mathcal{U}$ and $r^{\leftarrow} r^{\rightarrow} : \mathcal{U} \to \mathcal{U}$ satisfy the following properties:*

(1) $R^{\leftarrow} R^{\rightarrow} A \subseteq A \subseteq r^{\leftarrow} r^{\rightarrow} A$.

(2) $A_1 \subseteq A_2 \implies R^{\leftarrow} R^{\rightarrow} A_1 \subseteq R^{\leftarrow} R^{\rightarrow} A_2$.

(3) $A_1 \subseteq A_2 \implies r^{\rightarrow} r^{\leftarrow} A_1 \subseteq r^{\rightarrow} r^{\leftarrow} A_2$.

(4) $\bigvee_{j \in J} R^{\leftarrow} R^{\rightarrow} A_j \subseteq R^{\leftarrow} R^{\rightarrow} \bigvee_{j \in J} A_j$.

(5) $r^{\leftarrow} r^{\rightarrow} \bigcap_{j \in J} A_j \subseteq \bigcap_{j \in J} r^{\leftarrow} r^{\rightarrow} A_j$.

(6) $R^{\leftarrow} R^{\rightarrow} A = R^{\leftarrow} R^{\rightarrow} (R^{\leftarrow} R^{\rightarrow} A)$.

(7) $r^{\leftarrow} r^{\rightarrow} A = r^{\leftarrow} r^{\rightarrow} (r^{\leftarrow} r^{\rightarrow} A)$.

Theorem 6.3.7 *The mappings $R^{\leftarrow} R^{\rightarrow} : \mathcal{U} \to \mathcal{U}$ and $r^{\leftarrow} r^{\rightarrow} : \mathcal{U} \to \mathcal{U}$ satisfy the following properties where $A \in \mathcal{U}$ and $\{A_j \mid j \in J\}$.*

(1) *If (r, R) is t-serial, $R^{\leftarrow} R^{\rightarrow} U = U$ and $r^{\leftarrow} r^{\rightarrow} \emptyset = \emptyset$.*

(2) *If (r, R) is t-inverse serial, $r^{\leftarrow} r^{\rightarrow} U = U$ and $R^{\leftarrow} R^{\rightarrow} \emptyset = \emptyset$.*

(3) *If (r, R) is t-inverse serial and injective, then*

  (i) $R^{\leftarrow} R^{\rightarrow} \bigvee_{j \in J} A_j = \bigvee_{j \in J} R^{\leftarrow} R^{\rightarrow} A_j$.

  (ii) $r^{\leftarrow} r^{\rightarrow} \bigcap_{j \in J} A_j = \bigcap_{j \in J} r^{\leftarrow} r^{\rightarrow} A_j$.

(4) *If (r, R) is t-serial and injective, then*

  (i) $r^{\leftarrow} r^{\rightarrow} \bigvee_{j \in J} A_j = \bigvee_{j \in J} r^{\leftarrow} r^{\rightarrow} A_j$.

  (ii) $R^{\leftarrow} R^{\rightarrow} \bigcap_{j \in J} A_j = \bigcap_{j \in J} R^{\leftarrow} R^{\rightarrow} A_j$.

(5) *Let (r, R) be t-serial and injective.*

  (i) $r^{\leftarrow} r^{\rightarrow} (R^{\leftarrow} R^{\rightarrow} A) = A$.

  (ii) $R^{\leftarrow} R^{\rightarrow} (r^{\leftarrow} r^{\rightarrow} A) = A$.

Theorem 6.3.8 (i) *Let* (r, R) *be t-serial and* $(R^{\leftarrow}, r^{\leftarrow})$ *be injective. Then for every* $B \in \mathcal{V}$*, the sets*

$$R^{\rightarrow}R^{\leftarrow}B \text{ and } r^{\rightarrow}r^{\leftarrow}B$$

are presectional definable.

(ii) *Let* (r, R) *be inverse serial and* (r, R) *be injective. Then for every set* $B \in \mathcal{V}$*, the sets*

$$R^{\leftarrow}R^{\rightarrow}B \text{ and } r^{\leftarrow}r^{\rightarrow}B$$

are sectional definable.

Proof (i) Note that by Corollary 5.5.2, (r, R) is a difunction. Since

$$R^{\rightarrow}R^{\leftarrow}B, r^{\rightarrow}r^{\leftarrow}B \in \mathcal{V},$$

by Theorem 5.5.4, we obtain that

$$R^{\leftarrow}R^{\rightarrow}R^{\leftarrow}B = r^{\leftarrow}R^{\rightarrow}R^{\leftarrow}B \text{ and } R^{\leftarrow}r^{\rightarrow}r^{\leftarrow}B = r^{\leftarrow}r^{\rightarrow}r^{\leftarrow}B,$$

that is, the sets

$$R^{\rightarrow}R^{\leftarrow}B, r^{\rightarrow}r^{\leftarrow}B$$

are presectional definable. The proof of (ii) is similar.

6.4 Revised Rough Set Approximations

In a natural way, the rough set approximations for two domains of discourse can be defined using the set-valued functions as the generalization of neighborhood operators of the approximation spaces with a single domain. In this case, the approximations of a subset of one of the domains will be a subset of the other domain. This is a natural consequence of the counterparts of the concept of neighborhood operators in generalized approximation spaces. Note that if (U, V, r) is an approximation space with two domains of discourse, the sets $F(u)$ and $G(v)$ presented in Sect. 5.3 are not neighborhoods in the usual sense since the points u and v are in the opposite sides according to the domains of the sets $F(u)$ and $G(v)$ where $u \in U$ and $v \in V$. In order to overcome this inconsistency problem, Pei and Xu defined revised approximations in [5]. As we shall see in the sequel, the revised textural rough sets also provide us complementary observations for the results of Pei given in this section.

Now, let us take the approximation operators

$$\overline{apr}_G : \mathcal{P}(U) \to \mathcal{P}(V),$$

$$\underline{apr}_G : \mathcal{P}(U) \to \mathcal{P}(V),$$

$$\overline{apr}_F : \mathcal{P}(V) \to \mathcal{P}(U),$$

$$\underline{apr}_F : \mathcal{P}(V) \to \mathcal{P}(U).$$

If we consider the composition of the operators $\underline{apr}_F$ and $\overline{apr}_G$, then we obtain the operator

$\overline{apr}_G \circ \underline{apr}_F : \mathcal{P}(V) \to \mathcal{P}(V)$

defined by

$$\forall B \in \mathcal{P}(V),\ (\overline{apr}_G \circ \underline{apr}_F)(B) = \overline{apr}_G(\underline{apr}_F)(B).$$

Similarly, for the operators $\underline{apr}_G$ and $\overline{apr}_F$, we take the operator

$\underline{apr}_G \circ \overline{apr}_F : \mathcal{P}(V) \to \mathcal{P}(V)$

defined by

$$\forall B \in \mathcal{P}(V),\ (\underline{apr}_G \circ \overline{apr}_F)(B) = \underline{apr}_G(\overline{apr}_F)(B).$$

Now let us say

$\overline{Apr}_F = \underline{apr}_G \circ \overline{apr}_F$ and $\underline{Apr}_F = \overline{apr}_G \circ \underline{apr}_F$.

Note that the following diagrams are commutative:

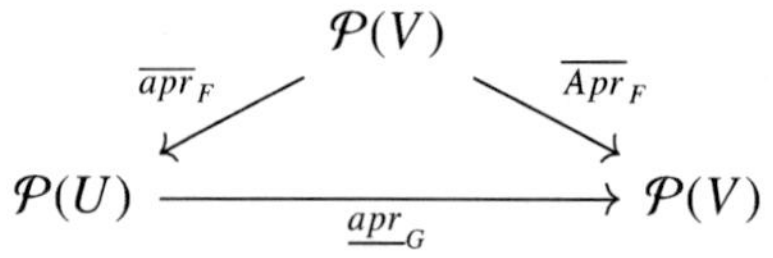

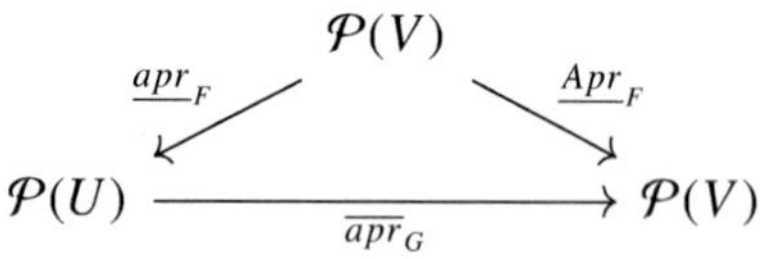

For any subset $B \subseteq V$, the set $\overline{Apr}_F(B)$ is also a subset of the domain V. In any case, the sets $\overline{Apr}_F(B)$ and $\underline{Apr}_F(B)$ consider an information about B taking into account the effects of the elements of U. These sets are called *revised lower approximation and revised upper approximation* set of B with respect to mappings F and G, respectively.

Dually, if we change the roles of the approximations for the subsets of the domain U, then we may also get the revised approximation operators with respect to mappings F and G as follows:

$\overline{Apr}_G = \underline{apr}_F \circ \overline{apr}_G$ and $\underline{Apr}_G = \overline{apr}_F \circ \underline{apr}_G$.

This means that the following diagrams also commute:

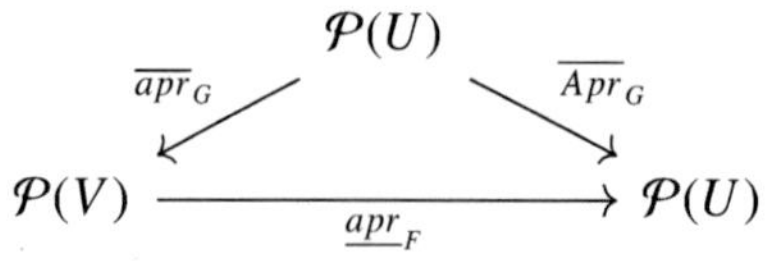

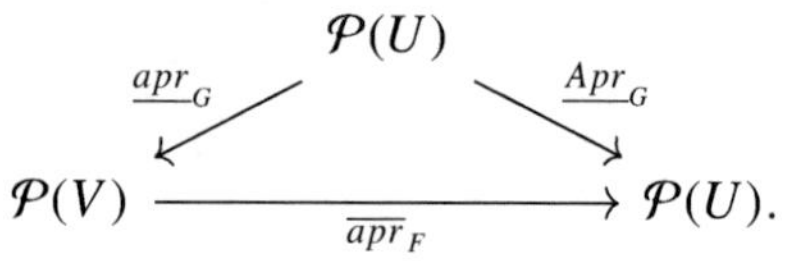

Note that the approximation operators $\overline{Apr}_G$ and $\underline{Apr}_G$ provide us the approximations in the domain U, and for all $A \subseteq U$, we have

$\overline{Apr}_G(A) = \underline{apr}_F(\overline{apr}_G(A))$ and $\underline{Apr}_G(A) = \overline{apr}_F(\underline{apr}_G(A))$.

The following theorem says that the revised approximation operators are dual.

Theorem 6.4.1 *Let (U, V, r) be a generalized approximation space. Then we have the following statements:*

(1) $V \setminus \overline{Apr}_F(B) = \underline{Apr}_F(V \setminus B)$ *and*

$V \setminus \underline{Apr}_F(B) = \overline{Apr}_F(V \setminus B)$.

(2) $U \setminus \overline{Apr}_G(A) = \underline{Apr}_G(U \setminus A)$ *and*

$U \setminus \underline{Apr}_G(A) = \overline{Apr}_G(U \setminus A)$.

Proof (1) By Theorem 6.3.5, for the generalized t-approximation space (U, V, r, R), note that we have the equality

$r^{\to}r^{\leftarrow}B = c_V(R^{\to}R^{\leftarrow}c_V(B))$.
Hence, we conclude that

$$
\begin{aligned}
V \setminus \overline{Apr}_F(B) &= V \setminus (\underline{apr}_G \circ \overline{apr}_F)(B) \\
&= V \setminus (\underline{apr}_G(\overline{apr}_F)(B)) \\
&= c_V(R^{\to}(R^{\leftarrow}B)) = r^{\to}r^{\leftarrow}c_V(B) \\
&= \overline{apr}_G(\underline{apr}_F(V \setminus B)) = (\overline{apr}_G \circ \underline{apr}_F)(V \setminus B) \\
&= \underline{Apr}_F(V \setminus B).
\end{aligned}
$$

For the second equality, again note that by Theorem 6.3.5, we have

$R^{\to}R^{\leftarrow}B = c_V(r^{\to}r^{\leftarrow}c_V(B))$.

Then by Theorem 6.3.2

$$
\begin{aligned}
V \setminus \underline{Apr}_F(B) &= V \setminus (\overline{apr}_G \circ \underline{apr}_F)(B) \\
&= V \setminus (\overline{apr}_G(\underline{apr}_F(B)) \\
&= c_V(r^{\to}r^{\leftarrow}B) = (R^{\to}R^{\leftarrow}c_V(B)) \\
&= \underline{apr}_G(\overline{apr}_F(V \setminus B)) = (\underline{apr}_G \circ \overline{apr}_F)(V \setminus B) \\
&= \overline{Apr}_F(V \setminus B).
\end{aligned}
$$

Theorem 6.4.2 *Let (U, V, r) be a generalized approximation space. Then we have the following statements:*

(i) $\underline{Apr}_F(B) \subseteq B \subseteq \overline{Apr}_F(B)$.

(ii) $\underline{Apr}_G(A) \subseteq A \subseteq \overline{Apr}_G(A)$.

Proof (i) By Lemmas 4.2.15 (1)(ii) and (2)(ii), it is enough to observe that

$$\begin{aligned}\underline{Apr}_F(B) &= (\overline{apr}_G \circ \underline{apr}_F)(B)\\ &= \overline{apr}_G(\underline{apr}_F)(B)\\ &= r^{\rightarrow}r^{\leftarrow}B \subseteq B \subseteq R^{\rightarrow}R^{\leftarrow}B\\ &= \underline{apr}_G(\overline{apr}_F(B)) = (\underline{apr}_G \circ \overline{apr}_F)(B)\\ &= \overline{Apr}_F(B).\end{aligned}$$

(ii) The proof is similar. □

The revised approximations are monotonic and idempotent.

Theorem 6.4.3 *Let (U, V, r) be a generalized approximation space and $A_1, A_2 \subseteq U$ and $B_1, B_2 \subseteq V$. Then we have the following statements:*

(i) $B_1 \subseteq B_2 \implies \underline{Apr}_F(B_1) \subseteq \underline{Apr}_F(B_2)$.

(ii) $B_1 \subseteq B_2 \implies \overline{Apr}_F(B_1) \subseteq \overline{Apr}_F(B_2)$.

(iii) $A_1 \subseteq A_2 \implies \underline{Apr}_G(A_1) \subseteq \underline{Apr}_G(A_2)$.

(iv) $A_1 \subseteq A_2 \implies \overline{Apr}_G(A_1) \subseteq \overline{Apr}_G(A_2)$.

(v) $\underline{Apr}_F(\underline{Apr}_F(B)) = \underline{Apr}_F(B)$.

(vi) $\overline{Apr}_F(\overline{Apr}_F(B)) = \overline{Apr}_F(B)$.

(vii) $\underline{Apr}_G(\underline{Apr}_G(A)) = \underline{Apr}_G(A)$.

(viii) $\overline{Apr}_G(\overline{Apr}_G(A)) = \overline{Apr}_G(A)$.

Proof We give only the proof of (i) and (vi).

(i) Let $B_1 \subseteq B_2$. By Theorem 6.3.3 (ii), we have

$$r^{\rightarrow}r^{\leftarrow}B_1 \subseteq r^{\rightarrow}r^{\leftarrow}B_2.$$

Then

$$\begin{aligned}\underline{Apr}_F(B_1) &= (\overline{apr}_G \circ \underline{apr}_F)(B_1) = \overline{apr}_G(\underline{apr}_F)(B_1)\\ &= r^{\rightarrow}r^{\leftarrow}B_1 \subseteq r^{\rightarrow}r^{\leftarrow}B_2 = \overline{apr}_G(\underline{apr}_F)(B_2)\\ &= (\overline{apr}_G \circ \underline{apr}_F)(B_2) = \underline{Apr}_F(B_2).\end{aligned}$$

(vi) From Theorem 6.3.3 (6), we have

$$r^{\rightarrow}r^{\leftarrow}(r^{\rightarrow}r^{\leftarrow}B) = r^{\rightarrow}r^{\leftarrow}B.$$

Then

$$\begin{aligned}\overline{Apr}_F(\overline{Apr}_F(B)) &= (\overline{apr}_G \circ \underline{apr}_F)((\overline{apr}_G \circ \underline{apr}_F)(B))\\ &= \overline{apr}_G(\underline{apr}_F(\overline{apr}_G(\underline{apr}_F(B))))\\ &= r^{\rightarrow}(r^{\leftarrow}(r^{\rightarrow}r^{\leftarrow}B))\\ &= r^{\rightarrow}r^{\leftarrow}B = (\overline{apr}_G \circ \underline{apr}_F)(B) = \overline{Apr}_F(B).\end{aligned}$$

Theorem 6.4.4 *Let* (U, V, r) *be a generalized approximation space. Then for all family* $\{B_j \mid j \in J\} \subseteq \mathcal{P}(V)$*, we have the following statements:*

(i) $\bigcup_{j\in J} \underline{Apr}_F(B_j) \subseteq \underline{Apr}_F(\bigcup_{j\in J} B_j)$.

(ii) $\underline{Apr}_F(\bigcap_{j\in J} B_j) \subseteq \bigcap_{j\in J} \underline{Apr}_F(B_j)$.

(iii) *If* r *is serial, then* $\underline{Apr}_F(\emptyset) = \emptyset$.

(iv) *If* r *is inverse serial, then* $\underline{Apr}_F(V) = V$.

Proof (i) Let (U, V, r, R) be the corresponding t-generalized approximation space to (U, V, r). By Theorem 6.3.3 (4), for every family $\{B_j \mid j \in J\} \subseteq \mathcal{P}(V)$,

$$\bigvee_{j\in J} r^{\rightarrow}r^{\leftarrow}B_j \subseteq r^{\rightarrow}r^{\leftarrow}\bigvee_{j\in J} B_j.$$

This easily gives that

$$\begin{aligned}\bigcup_{j\in J} \underline{Apr}_F(B_j) &= \bigcup_{j\in J}(\overline{apr}_G \circ \underline{apr}_F)(B_j) = \bigcup_{j\in J} \overline{apr}_G(\underline{apr}_F)(B_j)\\ &= \bigvee_{j\in J} r^{\rightarrow}r^{\leftarrow}B_j \subseteq r^{\rightarrow}r^{\leftarrow}\bigvee_{j\in J} B_j = \overline{apr}_G(\underline{apr}_F(\bigcup_{j\in J} B_j))\\ &= (\overline{apr}_G \circ \underline{apr}_F)(\bigcup_{j\in J} B_j) = \underline{Apr}_F(\bigcup_{j\in J} B_j).\end{aligned}$$

(iii) Let r be serial. Then by Theorem 4.4.6 (i), the corresponding direlation (r, R) is t-serial where $R = (U \times V) \setminus r$. Therefore, by Theorem 6.3.4 (1), the equality $r^{\rightarrow}r^{\leftarrow}\emptyset = \emptyset$ holds.

Hence, we obtain that

$$\underline{Apr}_F(\emptyset) = (\overline{apr}_G \circ \underline{apr}_F)(\emptyset) = \overline{apr}_G(\underline{apr}_F)(\emptyset) = r^{\rightarrow}r^{\leftarrow}\emptyset = \emptyset.$$

Theorem 6.4.5 *Let* (U, V, r) *be a generalized approximation space. Then we have the following statements:*

(1) $\bigcup_{j\in J} \overline{Apr}_F(B_j) \subseteq \overline{Apr}_F(\bigcup_{j\in J} B_j)$.

(2) $\overline{Apr}_F(\bigcap_{j\in J} B_j) \subseteq \bigcap_{j\in J} \overline{Apr}_F(B_j)$.

(3) *If r is serial, then* $\overline{Apr}_F(V) = V$.

(4) *If r is inverse serial, then* $\overline{Apr}_F(\emptyset) = \emptyset$.

Proof (2) From Theorem 6.3.3 (5), note that for every family $\{B_j \mid j \in J\} \subseteq \mathcal{P}(V)$,

$$R^{\rightarrow}R^{\leftarrow}(\bigcap_{j\in J} B_j) \subseteq \bigcap_{j\in J} R^{\rightarrow}R^{\leftarrow}B_j = \bigcap_{j\in J} R^{\rightarrow}R^{\leftarrow}B_j.$$

Hence,

$$\begin{aligned}\overline{Apr}_F(\bigcap_{j\in J} B_j) &= (\overline{apr}_G \circ \underline{apr}_F)(\bigcap_{j\in J} B_j)\\ &= R^{\rightarrow}R^{\leftarrow}(\bigcap_{j\in J} B_j) \subseteq \bigcap_{j\in J} R^{\rightarrow}R^{\leftarrow}B_j\\ &= \bigcap_{j\in J}(\overline{apr}_G \circ \underline{apr}_F)B_j = \bigcap_{j\in J} \overline{Apr}_F(B_j).\end{aligned}$$

(3) Let r be inverse serial. By Corollary 4.4.6 (1), (r, R) is t-inverse serial where $R = (U \times V) \setminus r$. Hence, from Theorem 6.3.4 (1), we obtain that

$$\overline{Apr}_F(V) = (\overline{apr}_G \circ \underline{apr}_F)(V) = R^{\rightarrow}R^{\leftarrow}V = V.$$

Theorem 6.4.6 *If (U, V, r) is a generalized approximation space, then we have the following statements:*

(1) $\bigcup_{j\in J} \underline{Apr}_G(A_j) \subseteq \underline{Apr}_G(\bigcup_{j\in J} A_j)$.

(2) $\underline{Apr}_G(\bigcap_{j\in J} A_j) \subseteq \bigcap_{j\in J} \underline{Apr}_G(A_j)$.

(3) *If r is serial, then* $\underline{Apr}_G(U) = U$.

(4) *If r is inverse serial, then* $\underline{Apr}_G(\emptyset) = \emptyset$.

Proof (1) Considering the corresponding t-generalized approximation space (U, V, r, R), from Theorem 6.3.3 (5), we obtain that

$$\begin{aligned}\bigcup_{j\in J} \underline{Apr}_G(A_j) &= \bigcup_{j\in J} \overline{apr}_F(\underline{apr}_G(A_j))\\ &= \bigcup_{j\in J} R^{\leftarrow}R^{\rightarrow}A_j \subseteq R^{\leftarrow}R^{\rightarrow}\bigcup_{j\in J} A_j\\ &= \overline{apr}_F(\underline{apr}_G \bigcup_{j\in J} A_j) = \underline{Apr}_G(\bigcup_{j\in J} A_j).\end{aligned}$$

(4) Since r is inverse serial, by Corollary 4.4.6 (2), (r, R) is also t-inverse serial where $R = (U \times V) \setminus r$. By Theorem 6.3.4 (2), this gives that

$$\underline{Apr}_G(\emptyset) = \overline{apr}_F(\underline{apr}_G(\emptyset) = R^{\leftarrow} R^{\rightarrow} \emptyset = \emptyset.$$

Theorem 6.4.7 *If (U, V, r) is a generalized approximation space, then we have the following statements:*

(1) $\bigcup_{j \in J} \overline{Apr}_G(A_j) \subseteq \overline{Apr}_G(\bigcup_{j \in J} A_j)$.

(2) $\overline{Apr}_G(\bigcap_{j \in J} A_j) \subseteq \bigcap_{j \in J} \overline{Apr}_G(A_j)$.

(3) *If r is serial, then* $\overline{Apr}_G(\emptyset) = \emptyset$.

(4) *If r is inverse serial, then* $\overline{Apr}_G(U) = U$.

Proof (2) From Theorem 6.3.6 (5), we easily observe that

$$\begin{aligned}
\overline{Apr}_G(\bigcap_{j \in J} A_j) &= (\underline{apr}_F \circ \overline{apr}_G)\big(\bigcap_{j \in J} A_j\big) \\
&= \underline{apr}_F\big(\overline{apr}_G(\bigcap_{j \in J} A_j)\big) \\
&= r^{\leftarrow} r^{\rightarrow}(\bigcap_{j \in J} A_j) \subseteq \bigcap_{j \in J} r^{\leftarrow} r^{\rightarrow} A_j \\
&= \bigcap_{j \in J} \underline{apr}_F\big(\overline{apr}_G(A_j)\big).
\end{aligned}$$

(4) By Corollary 4.4.6 (2), the corresponding direlation is also inverse serial. Then by Theorem 6.3.3 (4), we conclude that

$$\overline{Apr}_G(U) = (\underline{apr}_F \circ \overline{apr}_G)(U) = r^{\leftarrow} r^{\rightarrow} U = U.$$

Theorem 6.4.8 *Let (U, V, r) be a generalized approximation space, $B \subseteq V$ and $\{B_j \mid j \in J\} \subseteq \mathcal{P}(V)$. Then we have the following properties.*

(1) *Let r be inverse serial and injective.*

(i) $\underline{Apr}_F(\bigcap_{j \in J} B_j) = \bigcap_{j \in J} \underline{Apr}_F(B_j)$.

(ii) $\overline{Apr}_F(\bigcup_{j \in J} B_j) = \bigcup_{j \in J} \overline{Apr}_F(B_j)$.

(2) *Let r be serial and injective.*

(i) $\overline{Apr}_F(\bigcap_{j \in J} B_j) = \bigcap_{j \in J} \overline{Apr}_F(B_j)$.

(ii) $\underline{Apr}_F(\bigcup_{j\in J} B_j) = \bigcup_{j\in J} \underline{Apr}_F(B_j)$.

(ii) *If r is inverse serial and injective, then*

$\overline{Apr}_F(B) = \underline{Apr}_F(\overline{Apr}_F(B))$.

(3) *Let r be serial and injective.*

(i) $\overline{Apr}_F(\underline{Apr}_F(B)) \subseteq B$.

(ii) $B \subseteq \underline{Apr}_F(\overline{Apr}_F(B))$.

Proof (1)(i) From Theorem 6.3.4 (3)(i), we easily see that

$$\begin{aligned}
\underline{Apr}_F(\bigcap_{j\in J} B_j) &= (\underline{apr}_G \circ \overline{apr}_F)(\bigcap_{j\in J} B_j) \\
&= \underline{apr}_G(\overline{apr}_F(\bigcap_{j\in J} B_j)) = r^{\leftarrow} r^{\rightarrow} \bigcap_{j\in J} B_j \\
&= \bigcap_{j\in J} r^{\leftarrow} r^{\rightarrow} B_j = \bigcap_{j\in J} \underline{apr}_G(\overline{apr}_F(B_j) \\
&= \bigcap_{j\in J} \underline{Apr}_F(B_j).
\end{aligned}$$

(5)(ii) □

Theorem 6.4.9 *Let (U, V, r) be a generalized approximation space, $A \in \mathcal{U}$ and $\{A_j \mid j \in J\}$. Then we have the following properties.*

(1) *If r is inverse serial and injective, then*

(i) $\underline{Apr}_G(\bigcup_{j\in J} A_j) = \bigcup_{j\in J} \underline{Apr}_G(A_j)$.

(ii) $\overline{Apr}_G(\bigcap_{j\in J} A_j) = \bigcap_{j\in J} \overline{Apr}_G(A_j)$.

(2) *Let r be serial and injective.*

(i) $\overline{Apr}_G(\bigcup_{j\in J} A_j) = \bigcup_{j\in J} \overline{Apr}_G(A_j)$.

(ii) $\underline{Apr}_G(\bigcap_{j\in J} A_j) = \bigcap_{j\in J} \underline{Apr}_G(A_j)$.

(3) *Let r be inverse serial and injective.*

(i) $\overline{Apr}_G(\underline{Apr}_G(A)) = A$.

(ii) $A = \underline{Apr}_G(\overline{Apr}_G A))$.

Proof (3) (i) By Theorem 6.3.7 (3)(i), we conclude that

$$\begin{aligned}\underline{Apr}_G\Big(\bigcup_{j\in J} A_j\Big) &= \underline{apr}_F\big(\overline{apr}_G\big(\bigcup_{j\in J} A_j\big)\big)\\ &= R^{\leftarrow}R^{\rightarrow}\bigvee_{j\in J} A_j = \bigvee_{j\in J} R^{\leftarrow}R^{\rightarrow}A_j\\ &= \bigcup_{j\in J}\underline{Apr}_G(\overline{apr}_G(A_j)) = \bigcup_{j\in J}\underline{Apr}_G(A_j).\end{aligned}$$

(4) (i) By Theorem 6.3.7 (5)(i), we immediately obtain that

$$\begin{aligned}\overline{Apr}_G(\underline{Apr}_G(A)) &= \underline{apr}_F(\overline{apr}_G(\overline{apr}_F(\underline{apr}_G(A))))\\ &= r^{\leftarrow}r^{\rightarrow}(R^{\leftarrow}R^{\rightarrow}A) = A.\end{aligned}$$

Remark 6.4.10 Let (U, V, r) be a generalized approximation space. Note that as it is proved in Theorem 6.4.8, if r is inverse serial and injective, then we have the equalities

$$\bigcup_{j\in J}\underline{Apr}_F(B_j) \subseteq \underline{Apr}_F\Big(\bigcup_{j\in J} B_j\Big),$$

$$\underline{Apr}_F\Big(\bigcap_{j\in J} B_j\Big) = \bigcap_{j\in J}\underline{Apr}_F(B_j), \text{ and}$$

$$\underline{Apr}_F(V) = V.$$

Furthermore, we also have that

$$\overline{Apr}_F\Big(\bigcup_{j\in J} B_j\Big) = \bigcup_{j\in J}\overline{Apr}_F(B_j),$$

$$\overline{Apr}_F\Big(\bigcap_{j\in J} B_j\Big) \subseteq \bigcap_{j\in J}\overline{Apr}_F(B_j), \text{ and}$$

$$\overline{Apr}_F(\emptyset) = \emptyset.$$

On the other hand, by Lemma 4.2.15 (1)(iii), we find

$$\underline{Apr}_F(\emptyset) = \overline{apr}_G(\underline{apr}_F(\emptyset)) = r^{\rightarrow}r^{\leftarrow}\emptyset \subseteq \emptyset,$$

that is, $\underline{Apr}_F(\emptyset) = \emptyset$. Similarly, we obtain that $\overline{Apr}_F(V) = V$.

Corollary 6.4.11 *Let (U, V, r) be a generalized approximation space. Then*

(i) *If r is inverse serial and injective, then $(\underline{Apr}_F, \overline{Apr}_F)$ is an interval structure.*

(ii) *If r is serial and injective, then $(\underline{Apr}_G, \overline{Apr}_G)$ is an interval structure.*

Actually, Corollary 6.4.11 is an expected observation since for an injective and inverse serial relation r, the inverse r^{-1} is an injective function, and by the same way, an injective and serial relation r is also an injective function.

Corollary 6.4.12 *Let (U, V, r) be a generalized approximation space. If r is serial and injective, then the following statements hold:*

(i) $\overline{Apr}_F(\underline{Apr}_F(B)) = B$.

(ii) $B = \underline{Apr}_F(\overline{Apr}_F(B))$.

Proof (i) In view of Theorem 6.3.4 (5) (i), we easily observe that

$$\begin{aligned}\overline{Apr}_F(\underline{Apr}_F(B)) &= ((\underline{apr}_G \circ \overline{apr}_F) \circ (\overline{apr}_G \circ \underline{apr}_F))(B)\\ &= \underline{apr}_G(\overline{apr}_F(\overline{apr}_G(\underline{apr}_F(B)))\\ &= R^{\to}R^{\leftarrow}(r^{\to}r^{\leftarrow}B) = B.\end{aligned}$$

(ii) By Theorem 6.3.4 (5) (ii), we see that

$$\begin{aligned}\underline{Apr}_F(\overline{Apr}_F(B)) &= ((\overline{apr}_G \circ \underline{apr}_F) \circ (\underline{apr}_G \circ \overline{apr}_F))(B)\\ &= \overline{apr}_G(\underline{apr}_F(\underline{apr}_G(\overline{apr}_F(B)))\\ &= r^{\to}r^{\leftarrow}(R^{\to}R^{\leftarrow}B) = B.\end{aligned}$$

Corollary 6.4.13 *Let (U, V, r) be a generalized approximation space.*

(i) *If r is inverse serial and injective, then*

$\forall A \subseteq U$, $\underline{Apr}_G(A)$ *and* $\overline{Apr}_G(A)$ *are* G*-definable.*

(ii) *If r is serial and injective, then*

$\forall B \subseteq V$, $\overline{Apr}_F(B)$ *and* $\underline{Apr}_F(B)$ *are* F*-definable.*

Proof (i) We show that $\overline{apr}_G(\underline{Apr}_G(A)) = \underline{apr}_G(\underline{Apr}_G(A))$. In view of Theorem 4.4.15 (3),

$$\begin{aligned}\overline{apr}_G(\underline{Apr}_G(A)) &= \overline{apr}_G(\overline{apr}_F(\underline{apr}_G(A))\\ &= r^{\to}R^{\leftarrow}R^{\to}A\\ &= R^{\to}R^{\leftarrow}R^{\to}A\\ &= \underline{apr}_G(\overline{apr}_F(\underline{apr}_G(A)))\\ &= \underline{apr}_G(\underline{Apr}_G(A)).\end{aligned}$$

The proof of G-definability of $\overline{Apr}_G(A)$ is similar.

(ii) We prove that $\underline{apr}_F(\overline{Apr}_F(B)) = \overline{apr}_F(\overline{Apr}_F(B))$. By Theorem 4.4.16 (3), we see that

$$\begin{aligned}\underline{apr}_F(\overline{Apr}_F(B)) &= \underline{apr}_F(\underline{apr}_G(\overline{apr}_F(B)) \\ &= r^{\leftarrow}R^{\rightarrow}R^{\leftarrow}B \\ &= R^{\leftarrow}R^{\rightarrow}R^{\leftarrow}B \\ &= \overline{apr}_F(\underline{apr}_G(\overline{apr}_F(B))) \\ &= \overline{apr}_F(\overline{Apr}_F(B)).\end{aligned}$$

The proof of F-definability of $\overline{Apr}_B(B)$ can be given using a similar way.

Corollary 6.4.14 *Let* (U, V, r) *be a generalized approximation space.*

(1) *If* r *is injective and t-serial, then for all* $B \subseteq V$*, the sets*

$\overline{apr}_F(B)$ *and* $\underline{apr}_F(B)$

are successor definable with respect to revised approximations on U*.*

(2) *If* r^{-1} *is injective and* r *inverse serial, then for all* $A \subseteq U$*, the sets*

$\overline{apr}_G(A)$ *and* $\underline{apr}_G(A)$

are predecessor definable with respect to revised approximations on V*.*

Proof (1) Let $B \subseteq V$. Note that by Theorem 4.4.15 (4), we have

$$\begin{aligned}\overline{Apr}_G(\overline{apr}_F(B)) &= \underline{apr}_F(\overline{apr}_G(\overline{apr}_F(B)) \\ &= r^{\leftarrow}R^{\rightarrow}R^{\leftarrow}B \\ &= R^{\leftarrow}R^{\rightarrow}R^{\leftarrow}B \\ &= \overline{apr}_F(\underline{apr}_G(\overline{apr}_F(B))) \\ &= \underline{Apr}_G(\overline{apr}_F(B)),\end{aligned}$$

that is, $\overline{apr}_F(B)$ is successor definable with respect to revised approximations on U. The proof of definability of $\underline{apr}_F(B)$ is similar and omitted.

(2) In view of Theorem 4.4.16 (4), we see that

$$\begin{aligned}\overline{Apr}_F(\overline{apr}_G(A)) &= \underline{apr}_G(\overline{apr}_F(\overline{apr}_G(A))) \\ &= R^{\rightarrow}R^{\leftarrow}r^{\rightarrow}A \\ &= r^{\rightarrow}r^{\leftarrow}r^{\rightarrow}A \\ &= \overline{apr}_G(\overline{apr}_F(\overline{apr}_G(A))) \\ &= \underline{Apr}_F(\overline{apr}_F(A)).\end{aligned}$$

The proof of the definability of $\underline{apr}_G(A)$ is omitted.

If we take $U = V$, then the revised approximation operators $\underline{Apr}_F$, $\overline{Apr}_F$, $\overline{Apr}_G$ and $\underline{Apr}_G$ correspond to the subset approximation operators $\underline{apr}'$, $\overline{apr}'$ of Yao [6] given in Sect. 5.2.

Corollary 6.4.15 *Let (U, r) be an approximation space.*

(1) $\forall A \subseteq U$,

$$\underline{Apr}_F(A) = \underline{apr}'_F(A) \text{ and}$$

$$\overline{Apr}_F(A) = \overline{apr}'_F(A).$$

(2) $\forall A \subseteq U$,

$$\underline{Apr}_G(A) = \underline{apr}'_G(A) \text{ and}$$

$$\overline{Apr}_G(A) = \overline{apr}'_G(A).$$

Proof We only give the proofs of the first equalities of the items (1) and (2), respectively leaving the rest of the equalities to the interested reader. Recall that for $u, v \in U = V$,

$$F(u) = r_s(u) \text{ and } G(v) = r_p(v).$$

Further, by Proposition 5.4.6 (i), $r^{\rightarrow} P_u = r_s(u)$ and $r^{\leftarrow} Q_u = U \setminus r_p(v)$. Moreover, by Theorem 4.2.17(i) and (ii), section operator distributes over join and presection operator distributes over intersection. Therefore,

$$\begin{aligned}
\underline{Apr}_F(A) &= \overline{apr}_G(\underline{apr}_F(A)) \\
&= r^{\rightarrow} r^{\leftarrow} A = r^{\rightarrow} \bigvee \{P_u \mid \forall v \in U, r \not\subseteq \overline{Q}_{(u,v)} \implies P_v \subseteq A\} \\
&= \bigvee \{r^{\rightarrow} P_u \mid \forall v \in U, r \not\subseteq \overline{Q}_{(u,v)} \implies P_v \subseteq A\} \\
&= \bigcup \{r_s(u) \mid \forall v \in U, (u, v) \in r \implies v \in A\} \\
&= \bigcup \{r_s(u) \mid r_s(u) \subseteq A\} = \underline{apr}'_{r_s}(A) = \underline{apr}'_F(A).
\end{aligned}$$

$$\begin{aligned}
\overline{Apr}_G(A) &= \underline{apr}_F(\overline{apr}_G(A)) \\
&= r^{\leftarrow} r^{\rightarrow} A = r^{\leftarrow} \bigcap \{Q_v \mid \forall u, r \not\subseteq \overline{Q}_{(u,v)} \implies A \subseteq Q_u\} \\
&= \bigcap \{r^{\leftarrow} Q_v \mid \forall u, (u, v) \in r \implies u \notin A\} \\
&= \bigcap \{U \setminus r_p(v) \mid r_p(v) \cap A = \emptyset\} \\
&= U \setminus \bigcup \{r_p(v) \mid r_p(v) \cap A = \emptyset\} \\
&= U \setminus \{w \mid \exists v \in U, w \in r_p(v),\ r_p(v) \cap A = \emptyset\} \\
&= \{w \mid \forall v \in U, w \in r_p(v) \implies r_p(v) \cap A \neq \emptyset\} \\
&= \overline{apr}'_{r_p}(A) = \overline{apr}'_G(A).
\end{aligned}$$

References

1. J.W. Grzymala-Busse, W. Rzasa, Definability of approximations for a generalization of the indiscernibility relation, in *Proceedings of the 2007 IEEE Symposium on Foundations of Computational Intelligence* (2007), pp. 65–72
2. J.W. Grzymala-Busse, W.A. Sedelow Jr., On rough sets, and information system homomorphism. Bull. Pol. Acad. Sci. Technol., Sci. **36**(3–4), 233–239 (1988)
3. J.W. Grzymala-Busse, Data with missing attribute values: generalization of indiscernibility relation and rule reduction. Trans. Rough Sets I, LNCS **3100**, 7895 (2004)
4. G. Shafer, Belief functions and possibility measures, in *The Analysis of Fuzzy Information*, vol. I, ed. by J.C. Bezdek (CRC Press, 1987), pp. 51–84
5. D. Pei, Z.-Ben Xu, Rough set models on two universes. Int. J. General Syst. **33**(5), 569–581 (2004)
6. M. Diker, A. Altay Uğur, S. Bayhan, Rough sets and textural neighborhood, in *Rough Sets and Knowledge Technology*. RSKT 2015. Lecture Notes in Computer Science, ed. by D. Ciucci, G. Wang, S. Mitra, W.Z. Wu, vol. 9436 (Springer, 2015), pp. 190–202
7. Y.Y. Yao, Relational interpretations of neigborhood operators and rough set approximations operators. Inf. Sci. **111**(1–4), 239–259 (1998)

Chapter 7
Fuzzy Rough Sets

Essentially, the motivation of various combinations of fuzzy sets and rough sets is to find alternative formulations for imperfect data and knowledge. A fuzzy set is defined with the membership degree of an object of a crisp (ordinary) set, while a rough set is subjected to a pair of approximation sets containing the objects which possibly belong or surely belong to a target set. One of the approaches for the combinations of fuzzy sets and rough sets is based on the propositional fuzzy logical connectives.

7.1 Fuzzy Logical Connectives

In two-valued logic, every proposition is decidable, that is, the truth values of a proposition are true or false denoted by 1 and 0, respectively. The basic logical connectives *disjunction* and *conjunction* are given in Table 7.1.

Note that the conjunction and disjunction can be regarded as the functions

$$\vee : \{0, 1\} \times \{0, 1\} \to \{0, 1\} \text{ and } \quad \wedge : \{0, 1\} \times \{0, 1\} \to \{0, 1\}$$

defined by

$$\begin{aligned} &\wedge(1, 1) = 1, && \vee(1, 1) = 1,\\ &\wedge(1, 0) = 0, && \vee(1, 0) = 1,\\ &\wedge(0, 1) = 0, && \vee(0, 1) = 1,\\ &\wedge(0, 0) = 0, && \vee(0, 0) = 0, \end{aligned}$$

respectively, where $\{0, 1\}$ is the set of truth values of the propositions. If we change the set $\{0, 1\}$ with the unit interval $[0, 1]$, then we obtain the counterparts of the conjunction C and disjunction S, respectively, in fuzzy logic as follows:

$$\begin{aligned} &C : [0, 1] \times [0, 1] \to [0, 1], && S : [0, 1] \times [0, 1] \to [0, 1],\\ &C(1, 1) = 1, && D(1, 1) = 1, \end{aligned}$$

M. Diker, *Texture Spaces*, Studies in Fuzziness and Soft Computing 411,
https://doi.org/10.1007/978-3-031-39748-6_7

Table 7.1 Conjunction and disjunction

p	q	$p \wedge q$	$p \vee q$
1	1	1	1
1	0	0	1
0	1	0	1
0	0	0	0

$$C(1,0) = 1, \qquad D(1,0) = 0,$$
$$C(0,1) = 1, \qquad D(0,1) = 0,$$
$$C(0,0) = 0, \qquad D(0,0) = 0.$$

If a conjunction T satisfies the conditions

$$\forall \eta, \mu, \gamma \in [0,1],$$
$$T(\eta, \mu) = T(\mu, \eta),$$
$$\eta \le \mu \implies T(\eta, \gamma) \le T(\mu, \gamma),$$
$$T(1, \mu) = \mu,$$
$$T(\eta, T(\mu, \gamma)) = T(T(\eta, \mu), \gamma),$$

then T is called *a t-norm*. A t-norm T is said to be *strictly monotone* if

$$T(\eta, \mu) < T(\eta, \lambda)$$

whenever $\eta \in (0,1)$ and $\mu < \lambda$. It is easy to see that if a t-norm T is strictly monotone, then

$$T(\eta, \mu) < T(\eta, \lambda) \implies \mu = \lambda.$$

If a disjunctor S satisfies the conditions

$$\forall x, y, z \in [0,1],$$
$$S(x, y) = S(y, x),$$
$$x \le y \implies S(x, z) \le S(y, z),$$
$$S(1, x) = x,$$
$$S(x, S(y, z)) = S(S(x, y), z),$$

then S is called *a t-conorm*. Note that a norm and a t-conorm are increasing in the first and second components.

The implicator can be also regarded as a function (Table 7.2)

$$\rightarrow : \{0,1\} \times \{0,1\} \rightarrow \{0,1\}$$

Table 7.2 Implication in two-valued logic

p	q	$p \to q$
1	1	1
1	0	0
0	1	1
0	0	1

Table 7.3 Negation

p	$\neg p$
1	0
0	1

defined by

$$\to (1, 1) = 1,$$

$$\to (1, 0) = 0,$$

$$\to (0, 1) = 1,$$

$$\to (0, 0) = 1.$$

Then the counterpart of the implicator in fuzzy logic will be the mapping $I : [0, 1] \times [0, 1] \to [0, 1]$, called *a fuzzy implicator* if it satisfies the following conditions:

$I(0, 0) = 1,$

$I(1, 1) = 1,$

$I(1, 0) = 0,$

$I(0, 1) = 1.$

$\forall x, y, z \in [0, 1],$

if $x \leq y$, then $I(x, z) \geq I(y, z)$,

if $z \leq y$, then $I(x, z) \leq I(x, y)$.

Note that a fuzzy implicator also holds the equality $I(0, 1) = 1$. Further, the negation $\neg$ of a proposition p is defined as in Table 7.3. Note that the negation can also be regarded as a mapping $\neg : \{0, 1\} \to \{0, 1\}$ defined by $\neg(0) = 1$ and $\neg(1) = 0$.

In this case, *a fuzzy negator* will be the mapping $N : [0, 1] \to [0, 1]$ satisfying the conditions

$N(0) = 1$ and $N(1) = 0$, and

$\forall x, y \in [0, 1],\ x \leq y \implies \neg(y) \leq \neg(x).$

A fuzzy negator N is *involutive* if $N(N(\eta)) = \eta$ for all $\eta \in [0, 1]$. An involutive fuzzy negator is called *a strong negator*. Throughout the section, we will use the

Table 7.4 The implicator "$\rightarrow$" can be stated in terms of disjunction

p	$\neg p$	q	$\neg p \vee q$
1	0	1	1
1	0	0	0
0	1	1	1
0	1	1	1

mapping $N : [0, 1] \rightarrow [0, 1]$ defined by $N(\eta) = 1 - \eta$ for all $\eta \in [0, 1]$ which is a strong negator.

For any two propositions p and q, the two-valued implicator denoted by $p \rightarrow q$ is the conjunction $\neg p \vee q$ (for truth values, see Table 7.4). The conjunction form of the implicator in two-valued logic gives a hint in obtaining a different type of the implicator in fuzzy logic. For instance, if D is a t-conorm and N is a strong negator, then the mapping $I_{D,N} : [0, 1] \times [0, 1] \rightarrow [0, 1]$ defined by

$$I_{D,N}(\eta, \mu) = D(N(\eta), \mu)$$

for all $\eta, \mu \in [0, 1]$ is a fuzzy implicator which is called *an S-implicator* based on D and N. If N is a strong negator, then $I_{D,N}$ is called an *S-implicator*. The inspiration for a different type implicator comes from the classical set theory. We know that if $A, B \subseteq \mathcal{P}(U)$ where U is a domain of discourse, then we clearly have that

$$A' \cup B = \bigcup \{C \subseteq U \mid C \cap A \subseteq B\}$$

where $A' = U \setminus A$. Hence, using a t-norm T, we define an implicator called *R-implicator* based on a t-norm T as the mapping $I_T : [0, 1] \times [0, 1] \rightarrow [0, 1]$ defined by

$$I_T(\eta, \mu) = \bigvee \{\gamma \in [0, 1] \mid T(\eta, \gamma) \leq \mu\}$$

for all $\eta, \mu \in [0, 1]$.

A fuzzy implicator I is called *a border implicator* if $I(1, \mu) = \mu$ for all $\mu \in [0, 1]$. We say an implicator I satisfies *the consequent boundary condition* if for all $\eta, \mu \in [0, 1]$, $I(\eta, \mu) \geq \mu$. Note that since an implicator I is decreasing in the first component,

$$\mu = I(1, \mu) \leq I(\eta, \mu),$$

that is, every border implicator satisfies the consequent boundary condition. All implicators given in Table 7.5 are border implicators.

Duality is also an important concept for fuzzy logic, because different forms of the conjunction in two-valued logic provide us different fuzzy logical connectives. In two-valued logic, by interchanging the truth values 0 and 1, we obtain dual statements. That is, the negation gives us dual propositions. For instance, the propositions

$$p \vee q \quad \text{and} \quad \neg p \wedge \neg q$$

Table 7.5 Some well-known fuzzy logical connectives

T-norms	T-conorms
Minimum	Maximum
$T_M(\eta, \mu) = \min\{\eta, \mu\}$	$D_M(\eta, \mu) = \max\{\eta, \mu\}$
Product	Probabilistic sum
$T_P(\eta, \mu) = \eta\mu$	$D_P(\eta, \mu) = \eta + \mu - \eta\mu$
Łukasiewicz	Łukasiewicz
$T_W(\eta, \mu) = \max\{\eta + \mu - 1, 0\}$	$D_W(\eta, \mu) = \min\{1, \eta + \mu\}$
R-implicators	S-implicators
Gödel	Kleene-Dienes
$I_{T_M}(\eta, \mu) = \begin{cases} 1, & \text{if } \eta \leq \mu \\ \mu, & \text{otherwise} \end{cases}$	$I_{D_M,N}(\eta, \mu) = \max\{1 - \eta, \mu\}$
Goguen	Reichenbach
$I_{T_P}(\eta, \mu) = \begin{cases} 1, & \text{if } \eta \leq \mu \\ \dfrac{\mu}{\eta}, & \text{otherwise} \end{cases}$	$I_{D_P,N}(\eta, \mu) = 1 - \eta + \eta\mu$
Łukasiewicz	Łukasiewicz
$I_{T_W}(\eta, \mu) = \min\{1 - \eta + \mu, 1\}$	$I_{D_W,N}(\eta, \mu) = \min\{1 - \eta + \mu, 1\}$

are also dual. In other words, we have

$$\neg(p \vee q) = \neg p \wedge \neg q.$$

In view of this argument, we obtain the mapping $D' : [0, 1] \times [01] \to [0, 1]$ defined by

$$D'(\eta, \mu) = 1 - D(1 - \eta, 1 - \mu)$$

for all $\eta, \mu \in [0, 1]$ where D is a t-conorm. It is easy to see that D' satisfies the conditions due to a t-norm. That is, $D' = T$ is a t-norm.

More generally, let $F, G : [0, 1] \times [01] \to [0, 1]$ be any two mappings. The mapping $G' : [0, 1] \times [01] \to [0, 1]$ defined by

$$G'(\eta, \mu) = 1 - G(1 - \eta, 1 - \mu) \text{ for all } \eta, \mu \in [0, 1],$$

is called *the complement of* G. If $F = G'$, then we say that F and G are *dual*. In this case, if D is a t-conorm, then $T = D'$ is dual of D and it is a t-norm and vice versa. Just as the argument on t-norm and t-conorm, if I_T is an increasing based on a t-norm T, then we easily observe that

$$I'_T(\eta, \mu) = 1 - I_T(1 - \eta, 1 - \mu) = \bigwedge\{\gamma \in [0, 1] \mid D(\eta, \gamma) \geq \mu\}.$$

Then the above equality gives us the mapping $I_D : [0, 1] \times [0, 1] \to [0, 1]$ defined by

$$I_D(\eta, \mu) = \bigwedge\{\gamma \in [0, 1] \mid D(\eta, \gamma) \geq \mu\}$$

for all $\eta, \mu \in [0, 1]$. The mapping I_D is called *a complicator* [1].

In two-valued logic, a conjunction can be stated as

$p \wedge q = \neg(\neg p \vee \neg q).$

The corresponding connective in fuzzy logic may be given as the mapping $C_{I,N} : [0, 1] \times [0, 1] \to [0, 1]$ defined by

$$\forall \eta, \mu \in [0, 1],\ C_{I,N}(\eta, \mu) = N(I(\eta, N(\mu)))$$

where I is an implicator and N is the strong negation. The mapping $C_{I,N}$ is called *the induced conjunction* of I and N where N is the strong negator. Although the inspiration comes from a conjunction in two-valued logic, $C_{I,N}$ may not be a t-norm in fuzzy logic. Another inspiration also comes from the classical set theory. Now let I be an implicator. In view of the equality

$$A' \cap B = \bigcap \{C \subseteq U \mid B \subseteq A \cup C\}$$

where $A, B \subseteq U$ and $A' = U \setminus A$, we may consider the mapping $T_I : [0, 1] \times [0, 1] \to [0, 1]$ defined by

$$T_I(\eta, \mu) = \bigwedge \{\gamma \in [0, 1] \mid \mu \leq I(\eta, \gamma)\}$$

for all $\eta, \mu \in [0, 1]$. The mapping T_I plays an important role for the representations of R-implicators for some special classes of t-norms [2]. However, it may not be a t-norm. If I is increasing in the second component and satisfies the exchange principle and ordering property, then T_I is a t-norm (Fodor and Roubens [3], Theorem 1.14; Gottwald [4], Theorem 5.4.1).

If I is a right-continuous implicator in the second component, then the mappings T_I and $C_{I,N}$ are left-continuous and increasing mappings in the second component. Note that if $T' = D$, then we have $C_{I_T,N}(\eta, \mu) = T_{I_D}(\eta, \mu)$. Indeed, we easily observe that

$$\begin{aligned}
T_{I_D}(\eta, \mu) &= \bigwedge \{\gamma \in [0, 1] \mid \mu \leq I_D(\eta, \gamma)\} = \bigwedge \{\gamma \in [0, 1] \mid \mu \leq D(1 - \eta, \gamma)\} \\
&= \bigwedge \{\gamma \in [0, 1] \mid \mu \leq T'(1 - \eta, \gamma)\} = \bigwedge \{\gamma \in [0, 1] \mid \mu \leq 1 - T(\eta, 1 - \gamma)\} \\
&= \bigwedge \{\gamma \in [0, 1] \mid T(\eta, 1 - \gamma) \leq 1 - \mu\} \\
&= 1 - \bigvee \{1 - \gamma \in [0, 1] \mid T(\eta, 1 - \gamma)\} \leq 1 - \mu \\
&= 1 - \bigvee \{\gamma \in [0, 1] \mid T(\eta, \gamma) \leq 1 - \mu\} = 1 - I_T(\eta, 1 - \mu) \\
&= C_{I_T,N}(\eta, \mu).
\end{aligned}$$

7.2 Continuity of Fuzzy Logical Connectives

The notion of continuity is also an important concept in the theory of propositional fuzzy logical connectives. For instance, a left-continuous t-norm is already a basic tool in probabilistic metric spaces (see, e.g., [5, 6]). Further, a left-continuous t-norm also provides a residuated lattice on the unit interval [0, 1]. Although the continuity can be essentially seen as a notion providing some algebraic properties of the connectives in fuzzy logic, for the sake of the reader, it will be formally useful to consider some point-set topological arguments about the continuity of functions defined on $[0, 1] \times [0, 1] = [0, 1]^2$ in detail.

A topology on a domain of discourse U is a family of subsets of U satisfying the conditions

$U, \emptyset \in \tau,$

$\forall G, H \in \tau,\ G \cap H \in \tau,$ and

$\forall \{G_j \mid j \in J\} \subseteq \tau,\ \bigcup G_j \mid j \in J\} \in \tau.$

Then the pair (U, τ) is called *a topological space* and elements of τ are called *open sets*. An important example to topology is *the usual topology* on the set of real numbers $\mathbb{R}$ defined by

$$\tau_{\mathcal{U}} = \{G \subseteq \mathbb{R} \mid \forall x \in G, \exists (a, b) \subseteq \mathbb{R},\ x \in (a, b) \subseteq G\}.$$

A base for a topology τ is a subfamily $\mathcal{B}$ of τ having the property

$$\forall G \in \tau, \exists \{B_j \in j \in J\} \subseteq \mathcal{B},\ G = \cup\{B_j \mid j \in J\}.$$

In other words, $\mathcal{B}$ is a base for τ if every open set G can be written as a union of some elements of $\mathcal{B}$. It is easy to see that the family $\mathcal{B}_{\mathcal{U}} = \{(a, b) \mid a, b \in \mathbb{R}\}$ is a base for the usual topology $\tau_{\mathcal{U}}$ on $\mathbb{R}$.

The following result gives a way to determine when a family $\mathcal{B}$ is a base for a topology τ on a domain of discourse U.

Let $\mathcal{B} = \{B_j \mid j \in J\} \subseteq \mathcal{P}(U)$. The following conditions are equivalent:

(i) $\mathcal{B}$ is a base for a topology τ on U.

(ii) (a) $\bigcup\{B_j \mid j \in J\} = U$.

(b) $\forall B_1, B_2 \in \mathcal{B},\ x \in B_1 \cap B_2 \implies \exists B \in \mathcal{B},\ x \in B \subseteq B_1 \cap B_2.$

It is easy to see that the family $\mathcal{B}_u = \{]a, b] \mid a, b \in \mathbb{R}\}$ is a base for a topology on $\mathbb{R}$ denoted by τ_u. We call τ_u *the upper limit topology* on $\mathbb{R}$. Similarly, the family $\mathcal{B}_l = \{[a, b) \mid a, b \in \mathbb{R}\}$ is also a base for a topology τ_l on $\mathbb{R}$ which is called *the lower limit topology* on $\mathbb{R}$.

Another important concept in topology is a subspace of a topological space (U, τ). Let $\emptyset \neq A \subseteq U$. Note that the family

$$\tau_A = \{A \cap G \mid G \in \tau\}$$

is a topology on A. The family τ_A is called *an induced topology* or *subspace topology* on A. Consider the usual topological space $(\mathbb{R}, \tau_{\mathcal{U}})$. Let us look at the subspace topology on the unit interval $[0, 1] \subseteq \mathbb{R}$. According to the definition of subspace topology, the open sets of the subspace topology $\tau_{\mathcal{U}}$ in $[0, 1]$ will be as follows:

$(b, 1],\ [0, a),$

$(a, b),\ 0 < a < b < 1$

$[0, 1], \emptyset.$

Note that if $\mathcal{B}$ is a base for a topological space (U, τ), then for any subset $A \subseteq U$, the family

$$\mathcal{B}_A = \{A \cap B \mid B \in \tau\}$$

is also a base for the subspace topology τ_A. In other words, $\mathcal{B}_A$ generates the subspace topology τ_A. As an example, let us take the upper limit topological space $(\mathbb{R}, \tau_u)$. The subspace topology on $[0, 1]$ of $(\mathbb{R}, \tau_u)$ is generated by

$$\begin{aligned}\mathcal{B}^u_{[0,1]} &= \{[0, 1] \cap B \mid B \in \tau_u\}\\ &= \{[0, 1] \cap (a, b] \mid (a, b] \in \mathcal{B}_{\mathcal{U}}\}\\ &= \{(a, b] \mid 0 < a < b < 1\}.\end{aligned}$$

This means that the subspace topology on $[0, 1]$ is also the upper limit topology. Similarly, if we let us take the lower limit topological space $(\mathbb{R}, \tau_l)$, then the subspace topology on $[0, 1]$ is generated by the subintervals $[a, b[$ of $[0, 1]$ on $[0, 1]$ where

$$\mathcal{B}^l_{[0,1]} = \{[a, b) \mid a < b \text{ and } 0 < a < b < 1]\}.$$

Continuity

In Mathematics, continuity of a function is a local concept. Therefore, we define the continuity of a function first at a point of the topological space.

Let (U_1, τ_1) and (U_2, τ_2) be topological spaces, $f : U_1 \to U_2$ be a function and $x_0 \in U$. Then f is said to be *continuous* at x_0 if

for all $H \in \tau_2$ with $f(x_0) \in H$, there exists an open set $G \in \tau_1$ where $x_0 \in G$ such that $f(G) \subseteq H$.

If f is continuous at every point in U_1, then we say that f is τ_1-τ_2 continuous on the whole space (U_1, τ_1).

Theorem 7.2.1 *The following propositions are equivalent:*

(i) *f is τ_1-τ_2 continuous on U.*

(ii) *For all $H \in \tau_2$, $f^{-1}(H) \in \tau_1$.*

(iii) *For all $x \in U$ and for all $B' \in \mathcal{B}_2$ with $f(x) \in B'$, there exists a $B \in \mathcal{B}_1$ where $x \in B$ such that $f(B) \subseteq B'$.*

In Theorem 7.2.1(iii), $\mathcal{B}_1$ and $\mathcal{B}_2$ are the bases for the topologies τ_1 and τ_2, respectively.

Now, let us give the following.

Theorem 7.2.2 *Let $([0, 1], \tau_u)$ be the upper limit topological space and $f : [0, 1] \to [0, 1]$ be an increasing function. Then the following propositions are equivalent.*

(i) *f is τ_u-τ_u continuous at $x_0 \in [0, 1]$.*

(ii) *f is upper semicontinuous at $x_0 \in [0, 1]$, that is, for all $\epsilon > 0$, there exists a $\delta > 0$ such that for all $x \in [0, 1]$,*

$$x_0 - \delta < x < x_0 \implies f(x_0) - \epsilon < f(x).$$

Proof (i) $\implies$ (ii): For the proof, we use Theorem 7.2.1(iii). Let $\epsilon > 0$ be given a sufficiently small number and let $f(x_0) \in (a, b] \in \mathcal{B}$. Since $\epsilon > 0$ is a sufficiently small number, we also may have

$$a < f(x_0) - \epsilon < f(x_0)) \leq b.$$

Note that we also have $(f(x_0) - \epsilon, f(x_0)] \in \mathcal{B}$ and so by the assumption, there exists a $]c, d] \in \mathcal{B}$ with $x_0 \in]c, d]$ such that

$$f((c, d]) \subseteq (f(x_0) - \epsilon, f(x_0)].$$

Since $x_0 \in (c, d]$, there exists a $\delta > 0$ such that $c < x_0 - \delta < x_0 \leq d$. Therefore, we also have the inclusion

$$f((x_0 - \delta, x_0]) \subseteq (f(x_0) - \epsilon, f(x_0)].$$

That is,

$$\forall x,\ x \in (x_0 - \delta, x_0] \implies f(x) \in (f(x_0) - \epsilon, f(x_0)].$$

In other words, we obtain that

$$x_0 - \delta < x < x_0 \implies f(x_0) - \epsilon < f(x).$$

(ii) $\implies$ (i): Let $B' = (a, b] \in \mathcal{B}$ and $f(x_0) \in (a, b]$. Then there exists an $\epsilon > 0$ such that $a < f(x_0) - \epsilon < f(x_0) \leq b$. By the assumption, for some $\delta > 0$

$$x_0 - \delta < x < x_0 \implies f(x_0) - \epsilon < f(x).$$

Since f is increasing, the implication

$$x \in (x_0 - \delta, x_0] \implies f(x) \in (f(x_0) - \epsilon, f(x_0)]$$

holds and so we conclude that

$$f((x_0 - \delta, x_0]) \subseteq (f(x_0) - \epsilon, f(x_0)] \subseteq (a, b] = B'.$$

If we say that $B = (x_0 - \delta, x_0]$, the inclusion $f(B) \subseteq B'$ completes the proof.

Theorem 7.2.3 *Let* $([0, 1], \tau_l)$ *be the lower limit topological space and* $f : [0, 1] \to [0, 1]$ *be a decreasing function. Then the following propositions are equivalent:*

(i) f *is* τ_l-τ_l *continuous at* $x_0 \in [0, 1]$.

(ii) f *is lower semicontinuous, that is, for all* $\epsilon > 0$, *there exists a* $\delta > 0$ *such that for all* $x \in [0, 1]$,

$$x < x_0 + \delta \implies f(x) < f(x_0) + \epsilon.$$

Proof It is similar to the proof of Theorem 7.2.2. □

A function $F : [0, 1] \times [0, 1] \to [0, 1]$ is called

(i) *increasing* if

$\forall x_0, y_0, x_1, y_1 \in [0, 1]$,

$$x_0 \leq x_1 \text{ and, } y_0 \leq y_1 \implies F(x_0, y_0) \leq F(x_1, y_1),$$

(ii) *increasing in the first component* if

for every fixed $y_0 \in [0, 1]$ and for all $x, x' \in [0, 1]$,

$$x \leq x' \implies F(x, y_0) \leq F(x', y_0),$$

(iii) *decreasing in the first component* if

for every fixed $y_0 \in [0, 1]$ and for all $x, x' \in [0, 1]$,

$$x \leq x' \implies F(x', y_0) \leq F(x, y_0).$$

Definitions (ii) and (iii) given above can be similarly stated for the second component.
Product Topology

Let (U_1, τ_1) and (U_2, τ_2) be topological spaces. In a natural way, we may consider a topology on the Cartesian product $U_1 \times U_2$. Indeed, it is easy to see that the family

$$\{G_1 \times G_2 \mid G_1 \in \tau_1 \text{ and } G_2 \in \tau_2\}$$

is a base for a topology denoted by $\tau_1 \times \tau_2$ on $U_1 \times U_2$. The pair $(U_1 \times U_2, \tau_1 \times \tau_2)$ is called *a product topological space* and $\tau_1 \times \tau_2$ is called *product topology* or *box topology* on $U_1 \times U_2$.

Let us consider the product topology $\tau_u \times \tau_u = \tau_u^p$ on $[0, 1]^2$ where τ_u is the upper limit topology on $[0, 1]$.

Theorem 7.2.4 *Let* $F : [0, 1] \times [0, 1] \to [0, 1]$ *be an increasing function and* $(x_0, y_0) \in [0, 1] \times [0, 1]$. *Then the following statements are equivalent:*

(i) F *is* τ_u^p-τ_u *continuous at* (x_0, y_0).

(ii) F *is upper semicontinuous at* $(x_0, y_0) \in [0, 1]^2$, *that is, for each* $\epsilon > 0$ *there exists a* $\delta > 0$ *such that*

$$(x, y) \in (x_0 - \delta, x_0] \times (y_0 - \delta, y_0] \implies F(x_0, y_0) - \epsilon < F(x, y).$$

Proof (i) $\implies$ (ii): Let $\epsilon > 0$ be given and let $F(x_0, y_0) \in (a, b] \in \mathcal{B}$. Since ϵ is a sufficiently small number, we have

$$a < F(x_0, y_0) - \epsilon < F(x_0, y_0) \leq b.$$

Note that we also have

$$F(x_0, y_0) \in (F(x_0, y_0) - \epsilon, F(x_0, y_0)] \in \mathcal{B}.$$

Hence, by the assumption, there exists a $(c, d]\times]e, f] \in \mathcal{B}_p^u$ with $(x_0, y_0) \in (c, d]\times]e, f]$ such that

$$F((c, d] \times (e, f]) \subseteq (F(x_0, y_0) - \epsilon, F(x_0, y_0)].$$

Since $x_0 \in (c, d]$, there exists a $\delta_1 > 0$ such that $c < x_0 - \delta_1 < x_0 < d$. Further, Since $y_0 \in (e, f]$, there exists a $\delta_2 > 0$ such that $e < y_0 - \delta_2 < y_0 < f$. Let $\delta = \min\{\delta_1, \delta_2\}$. Therefore, we also have that

$$F((x_0 - \delta, x_0]\times]y_0 - \delta, y_0]) \subseteq (F(x_0, y_0)) - \epsilon, F(x_0, y_0)].$$

This means that $(x, y) \in (x_0 - \delta, x_0] \times (y_0 - \delta, y_0]$, and this follows that

$$F(x, y) \in (F(x_0, y_0) - \epsilon, F(x_0, y_0)].$$

As a result, we conclude the implication

$$(x, y) \in (x_0 - \delta, x_0] \times (y_0 - \delta, y_0] \implies F(x_0, y_0) - \epsilon < F(x, y)$$

as desired.

(ii) $\implies$ (i): Let $B' = (a, b] \in \mathcal{B}_u$ and $F(x_0, y_0) \in (a, b]$. Then there exists an $\epsilon > 0$ such that

$$a < F(x_0, y_0) - \epsilon < F(x_0, y_0) \leq b.$$

By the assumption, for some $\delta > 0$

$$(x, y) \in (x_0 - \delta, x_0] \times (y_0 - \delta, y_0] \implies F(x_0, y_0) - \epsilon < F(x, y).$$

Since F is increasing,

if $(x, y) \in (x_0 - \delta, x_0] \times (y_0 - \delta, y_0]$, then $F(x, y) \in (F(x_0, y_0) - \epsilon, F(x_0, y_0)]$

and so

$$F((x_0 - \delta, x_0] \times (y_0 - \delta, y_0]) \subseteq (F(x_0, y_0) - \epsilon, F(x_0, y_0)] \subseteq (a, b] = B'.$$

If we say that $B = (x_0 - \delta, x_0] \times (y_0 - \delta, y_0]$, then $F(B) \subseteq B'$ completes the proof.

Theorem 7.2.5 *Let* $F : [0, 1] \times [0, 1] \to [0, 1]$ *be a decreasing function and* $(x_0, y_0) \in [0, 1] \times [0, 1]$. *The following statements are equivalent:*

(i) F *is* τ_{p_l}-τ_l *continuous at* (x_0, y_0).

(ii) F *is lower semicontinuous at* $(x_0, y_0) \in [0, 1]^2$, *that is, for each* $\epsilon > 0$ *there exists a* $\delta > 0$ *such that*

$$(x, y) \in [x_0, x_0 + \delta) \times [y_0, y_0 + \delta) \implies F(x, y) < F(x_0, y_0) + \epsilon.$$

Sequential Continuity and Left Continuity

Note that in a topological space (U, τ), a sequence $(u_n)_{n \in \mathbb{N}}$ converges to an element $u \in U$ if for all $G \in \tau$ with $u \in G$, there exists a number $N(\epsilon) \in \mathbb{N}$ such that

$$\forall n, \; n \geq N(\epsilon) \implies x_n \in G.$$

Then we write $\lim_{n \to \infty} \{u_n \mid n \in \mathbb{N}\} = u$ or, in brief, $\lim_{n \to \infty} u_n = u$. For example, let $(u_n)_{n \in \mathbb{N}}$ be a sequence and suppose that it converges to a point u in the upper limit topological space $([0, 1], \tau_u)$. Then for all $(a, b] \in \tau_u$ with $u \in (a, b]$, there exists a number $N \in \mathbb{N}$ such that the implication $n \geq N \implies u_n \in]a, b]$ holds. Note that for all sufficiently small number $\epsilon > 0$, we also have that $u_n \in (u - \epsilon, b]$, that is, $u - \epsilon < u_n$. For the lower limit topological space $([0, 1], \tau_l)$, if $(u_n)_{n \in \mathbb{N}}$ is a sequence and it converges to a point u in U, then we obtain that for all $\epsilon > 0$, $u_n < u + \epsilon$.

Now, let (U_1, τ_1) and (U_2, τ_2) be topological spaces, $f : U_1 \to U_2$ be a function and $u \in U_1$. Then f is said to be *sequentially continuous* at $u \in U_1$ if for all sequence $(u_n)_{n \in \mathbb{N}}$ in U, the implication

$$\lim_{n \to \infty} \{u_n \mid n \in \mathbb{N}\} = u \implies \lim_{n \to \infty} \{f(u_n) \mid n \in \mathbb{N}\}) = f(u)$$

holds. Sequentially continuous functions preserve the convergence of sequences.

In general, sequential continuity is weaker than continuity.

Theorem 7.2.6 *If* $f : U_1 \to U_2$ *is* τ_1-τ_2 *continuous, it is also sequentially continuous.*

Let (U, τ) be a topological space and $\mathcal{B}$ be a base for τ. A subfamily $\mathcal{B}_u \subseteq \mathcal{B}$ is called *a local base* at the point u if for all $B \in \mathcal{B}$ with $u \in B$, there exists a $B_u \in \mathcal{B}_u$ such that $u \in B_u \subseteq B$.

Then (U, τ) is called *first countable* if for all $u \in U$, there exists a countable local base. Note that if $\mathcal{B}$ is a base for τ, then the family

$$\mathcal{B}_u = \{B \in \mathcal{B} \mid u \in B\}$$

is a local base at $u \in U$. For instance, the topological space $([0, 1], \tau_l)$ is first countable. Indeed, for a point $u \in U$, the family $\mathcal{B}_u = \{[u, u - 1/n[\mid n \in \mathbb{N}\}$ is a countable local base at u for the lower limit topology τ_l. Similarly, the upper limit topology τ_u is also first countable.

Theorem 7.2.7 *Let (U_1, τ_1) and (U_2, τ_2) be any two topological spaces and $f : U_1 \to U_2$ be a function. If (U_1, τ_1) is first countable, then the following statements are equivalent:*

(i) $f : U_1 \to U_2$ is τ_1-τ_2 continuous at $u \in U_1$.

(ii) f is sequentially continuous at $u \in U_1$.

Now, let $F : [0, 1] \times [0, 1] \to [0, 1]$ be a mapping. For any $y_0 \in [0, 1]$, let us consider the function $f_{y_0} : [0, 1] \to [0, 1]$ defined by

$$\forall x \in [0, 1], \ f_{y_0}(x) = F(x, y_0).$$

By the same way, for any $x_0 \in [0, 1]$, let us take the function $f_{x_0} : [0, 1] \to [0, 1]$ defined by

$$\forall y \in [0, 1], \ f_{x_0}(x) = F(x_0, y).$$

Since $([0, 1], \tau_u)$ and $([0, 1], \tau_l)$ are first countable, we may give the following two corollaries which are immediate results of Theorem 7.2.7.

Corollary 7.2.8 *Let f_{x_0} and f_{y_0} be increasing. Then*

(i) $f_{y_0} : ([0, 1], \tau_u) \to ([0, 1], \tau_u)$ *is sequentially continuous if and only if it is upper semicontinuous.*

(ii) $f_{x_0} : ([0, 1], \tau_u) \to ([0, 1], \tau_u)$ *is sequentially continuous if and only if it is upper semicontinuous.*

Corollary 7.2.9 *Let f_{x_0} and f_{y_0} be decreasing.*

(i) *The function* $f_{x_0} : ([0, 1], \tau_l) \to ([0, 1], \tau_l)$ *is sequentially continuous if and only if it is lower semicontinuous.*

(ii) $f_{y_0} : ([0, 1], \tau_l) \to ([0, 1], \tau_l)$ *is sequentially continuous if and only if it is lower semicontinuous.*

Left Continuity

A function $F : [0, 1] \times [0, 1] \to [0, 1]$ is called

(i) *left-continuous (right-continuous)* [7] if

$$\lim_{n\to\infty} \{F(\eta_n, \mu_n) \mid n \in \mathbb{N}\} = F(\lim_{n\to\infty} \{\eta_n \mid n \in \mathbb{N}\}, \lim_{n\to\infty} \{\mu_n \mid n \in \mathbb{N}\})$$

for all non-decreasing (resp., non-increasing) sequences $(\eta_n)_{n\in\mathbb{N}}, (\mu_n)_{n\in\mathbb{N}}$ in $[0, 1]$.

(ii) *left-continuous (right-continuous)* [7] *in the first component* if

$$\lim_{n\to\infty} \{F(\eta_n, \mu) \mid n \in \mathbb{N}\} = F(\lim_{n\to\infty} \{\eta_n \mid n \in \mathbb{N}\}, \mu),$$

for all $\eta, \mu \in [0, 1]$ and non-decreasing (resp., non-increasing) sequences $(\eta_n)_{n\in\mathbb{N}}$, $(\mu_n)_{n\in\mathbb{N}}$ in $[0, 1]$.

(iii) *left-continuous (right-continuous)* [7] *in the second component* if

$$\lim_{n\to\infty} \{F(\eta, \mu_n) \mid n \in \mathbb{N}\} = F(\eta, \lim_{n\to\infty} \{\mu_n \mid n \in \mathbb{N}\})$$

for all $\eta, \mu \in [0, 1]$ and non-decreasing (resp., non-increasing) sequences $(\eta_n)_{n\in\mathbb{N}}$, $(\mu_n)_{n\in\mathbb{N}}$ in $[0, 1]$.

Note that if $f_{y_0} : ([0, 1], \tau_u) \to ([0, 1], \tau_u)$ is sequentially continuous, then for every sequence $(x_n)_{n\in\mathbb{N}} \subseteq [0, 1]$,

$$\begin{aligned}
\lim_{n\to\infty} \{F(x_n, y_0) \mid n \in \mathbb{N}\}) &= \lim_{n\to\infty} f_{y_0}(\{x_n \mid n \in \mathbb{N}\}) \\
&= \lim_{n\to\infty} \{f_{y_0}(x_n) \mid n \in \mathbb{N}\} \\
&= f_{y_0}(\lim_{n\to\infty} \{x_n \mid n \in \mathbb{N}\}) \\
&= F(\lim_{n\to\infty} \{x_n \mid n \in \mathbb{N}\}, y_0).
\end{aligned}$$

Therefore, the sequential continuity of f_{y_0} in fact is the left continuity of the mapping $F : [0, 1] \times [0, 1] \to [0, 1]$ in the first component where the sequences are increasing. The increasing condition of f_{y_0} guarantees the existence of the limit of a sequence. Note that if $(x_n)_{n\in\mathbb{N}}$ is an increasing sequence in $([0, 1], \tau_u)$, then it is convergent. Indeed, if we say $\bigvee x_n = a \leq 1$, then there exists a number N such that for $n \geq N$, $x_n \in]c, a] \in \tau_u$ for some $c \in [0, 1]$. That is, $\lim_{n\to\infty} \{x_n \mid n \in \mathbb{N}\} = a$. This argument can be easily extended to the right continuity.

The first countability is productive. Therefore, the product of any two first countable topological spaces is also first countable. This implies that the product space $([0, 1] \times [0, 1], \tau_u \times \tau_u)$ is also first countable. Thus, by Theorems 7.2.4, and 7.2.5 we may give the following.

Corollary 7.2.10 *Let $F : [0, 1] \times [0, 1] \to ([0, 1]$ be increasing mapping.*

(i) *$F : ([0, 1] \times [0, 1], \tau_u^P) \to ([0, 1], \tau_u)$ is left-continuous if and only if F is upper semicontinuous.*

(ii) *$F : ([0, 1] \times [0, 1], \tau_l^P) \to ([0, 1], \tau_l)$ is right-continuous if and only if F is lower semicontinuous.*

Theorem 7.2.11 *Let $F : [0, 1] \times [0, 1] \to [0, 1]$ be an increasing mapping. Then the following statements are equivalent.*

(i) *F is left-continuous (τ_u^P-τ_u continuous) at (x_0, y_0).*

(ii) *F is left-continuous in each component. In other words, the functions $f_{x_0}, f_{y_0} : [0, 1] \to [0, 1]$ are left-continuous (τ_u-τ_u continuous).*

Proof Necessity: If F is left-continuous at (x_0, y_0), taking the constant sequences $(\mu_n)_{n\in\mathbb{N}} = (y_0)_{n\in\mathbb{N}}$ and $(\eta_n)_{n\in\mathbb{N}} = (x_0)_{n\in\mathbb{N}}$, we clearly see that the functions f_{x_0}, f_{y_0} are left-continuous.

Sufficiency. Let F be left-continuous in each component and $\epsilon > 0$ be given. Take a sequence $(\eta_n, \mu_n)_{n\in\mathbb{N}}$ converging (x_0, y_0) in $([0, 1] \times [0, 1], \tau_{p_u})$. Note that the sequences $(\eta_n)_{n\in\mathbb{N}}$ and $(\mu_n)_{n\in\mathbb{N}}$ also converge to x_0 and y_0, respectively, in $([0, 1], \tau_u)$. Let us consider the increasing sequences $(\beta_n)_{n\in\mathbb{N}}$ and $(\gamma_n)_{n\in\mathbb{N}}$ such that

$$\beta_n \leq \eta_n \text{ and } \gamma_n \leq \mu_n$$

where $\lim_{n\to\infty} \beta_n = x_0$ and $\lim_{n\to\infty} \gamma_n = y_0$ in $([0, 1], \tau_u)$. Since F is left-continuous in the second component, we have

$$\lim_{n\to\infty} \{F(x_0, \gamma_n) \mid n \in \mathbb{N}\} = F(x_0, \lim_{n\to\infty} \{\gamma_n \mid n \in \mathbb{N}\}) = F(x_0, y_0).$$

Then there exists a number N such that for all $n \geq N$,

$$F(x_0, \gamma_n) \in]F(x_0, y_0) - \epsilon, 1] \in \tau_u,$$

that is, for all $n \geq N$,

$$F(x_0, y_0) - \epsilon < F(x_0, \gamma_n).$$

Further, since F is also left-continuous in the first component, for every fixed $n \subset \mathbb{N}$,

$$\lim_{m\to\infty} \{F(\beta_m, \gamma_n) \mid m \in \mathbb{N}\} = F(\lim_{m\to\infty} \{\beta_m \mid m \in \mathbb{N}\}, \gamma_n) = F(x_0, \gamma_n).$$

Then there exists a number M such that for all $m \geq M$,

$$F(\beta_m, \gamma_n) \in (F(x_0, \gamma_n) - \epsilon, 1] \in \tau_u$$

and hence,

$$F(x_0, \gamma_n) - \epsilon < F(\beta_m, \gamma_n).$$

The function F is increasing, and this implies that

$$F(x_0, y_0) - 2\epsilon < F(x_0, \gamma_n) - \epsilon < F(\beta_m, \gamma_n) < F(\eta_m, \mu_n).$$

Finally, for $K = \max\{M, N\}$ and for all $k \geq K$, we conclude that

$$F(x_0, y_0) - 2\epsilon < F(\eta_k, \mu_k).$$

That is, for the arbitrarily chosen ϵ, there exists a number K in $\mathbb{N}$ such that if $k \geq K$, then $F(\eta_k, \mu_k) \in (F(x_0, y_0) - 2\epsilon, 1]$. As a result, F is left-continuous at (x_0, y_0).

The following results give the connection between continuity and distributivity [8].

Lemma 7.2.12 *Let* $F : [0, 1] \times [0, 1] \to [0, 1]$ *be a binary operation.*

(i) *Let F be increasing in the first component. Then*

(a) *F is left-continuous in the first component iff F is infinitely join-distributive in the first component, that is, for all $\eta, \mu_k \in [0, 1]$ and index set $K \neq \emptyset$,*

$$\bigvee\{F(\mu_k, \eta) \mid k \in K\} = F(\bigvee\{\mu_k \mid k \in K\}, \eta).$$

(b) *F is right-continuous in the first component iff F is infinitely meet-distributive in the first component, that is, for all $\eta, \mu_k \in [0, 1]$ and index set $K \neq \emptyset$,*

$$\bigwedge\{F(\mu_k, \eta) \mid k \in K\} = F(\bigwedge\{\mu_k \mid k \in k\}, \eta).$$

(ii) *Let F be decreasing in the first component. Then*

(a) *F is left-continuous in the first component iff F is infinitely meet-distributive in the first component, that is, for all $\eta, \mu_k \in [0, 1]$ and index set $K \neq \emptyset$,*

$$\bigwedge\{F(\eta_k, \mu) \mid k \in K\} = F(\bigwedge\{\eta_k \mid k \in k\}, \mu).$$

(b) *F is right-continuous in the first component iff F is infinitely join-distributive in the first component, that is, for all $\eta, \mu_k \in [0, 1]$ and index set $K \neq \emptyset$,*

$$\bigvee\{F(\eta_k, \mu) \mid k \in K\} = F(\bigvee\{\eta_k \mid k \in K\}, \mu).$$

Further, we obtain similar equivalences if one replaces "in the first component" by "in the second component".

Proof We give only the proof of (i)(a). For the necessity, let F be left-continuous in the first component. Let us take $\{\mu_k \mid k \in K\} \subseteq [0, 1]$ and $\eta \in [0, 1]$. Since $\{\mu_k \mid k \in K\}$ is bounded, it has a least upper bound, say $\bigvee_{k \in K} \mu_k = \mu$. Suppose that for each $n \in \mathbb{N}$, there exists a $\mu_{k(n)} \in \{\mu_k \mid k \in K\} \setminus \{\mu\}$ such that $\mu - \mu_{k(n)} < 1/n$. Note that for $n + 1$, $\mu - \mu_{k(n+1)} < 1/(n + 1)$ and so $\mu_{k(n)} - \mu_{k(n+1)} < -1/(n + 1) < 0$. This means that for all $n \in \mathbb{N}$, $\mu_{k(n)} \le \mu_{k(n+1)}$, that is, the sequence $(\mu_{k(n)})_{n\in\mathbb{N}}$ is non-decreasing. Therefore, we get

$$\lim_{n\to\infty} \mu_{k(n)} = \bigvee_{k\in K} \mu_{k(n)} = \mu$$

and by the choice of the terms of the sequence $(\mu_{k(n)})_{n\in\mathbb{N}}$, for all $n \in \mathbb{N}$, $\mu_{k(n)} < \mu$ (note that $(\mu_{k(n)})_{n\in\mathbb{N}}$ is convergent with respect to upper limit topology τ_u on $[0, 1]$). Since F is left-continuous in the first component, we also conclude that

$$\begin{aligned} F(\bigvee_{k\in K} \mu_k, \eta) &= F(\lim_{n\to\infty} \mu_{k(n)}, \eta) = \lim_{n\to\infty} F(\mu_{k(n)}, \eta) \\ &= \bigvee_{n\in\mathbb{N}} F(\mu_{k(n)}, \eta) = \bigvee_{k\in K} F(\mu_k, \eta), \end{aligned}$$

as desired.

Now, suppose that for some $n \in \mathbb{N}$,

$$\forall k \in K,\ \mu - \mu_k > 1/n.$$

This implies that for all k, $\mu_k < \mu_k + 1/n < \mu$, and therefore, $\bigvee \mu_k = \mu_{k_0}$ for some $k_0 \in K$. Thus, we conclude that

$$F(\bigvee_{k\in K} \mu_k, \eta) = F(\mu_{k_0}, \eta) \le \bigvee_{k\in K} F(\mu_k, \eta).$$

On the other hand, since F is increasing, $F(\mu_k, \eta) \le F(\bigvee_{k\in K} \mu_k, \eta)$ whence $\bigvee_{k\in K} F(\mu_k, \eta) \le F(\bigvee_{k\in K} \mu_k, \eta)$.

For the sufficiency, let $(\mu_k)_{k\in\mathbb{N}}$ be an increasing sequence in $[0, 1]$. Then $(\mu_k)_{k\in\mathbb{N}}$ is convergent with respect to upper limit topology τ_u. If $\lim_{k\to\infty} \mu_k = \mu$, then $\bigvee_{k\in\mathbb{N}} \mu_k = \mu$. This follows that

$$\begin{aligned} F(\lim_{k\to\infty} \mu_k, \eta) &= F(\mu, \eta) = F(\bigvee\{\mu_k \mid k \in K\}, \eta) \\ &= \bigvee\{F(\mu_k, \eta) \mid k \in K\} = \lim_{k\to\infty} \{F(\mu_k, \eta) \mid k \in \mathbb{N}\}. \end{aligned}$$

7.3 Textural Fuzzy Direlations

Through Brown's Representation Theorem, we obtain a natural counterpart of a direlation between fuzzy textures $(W_U, \mathcal{W}_U)$ and $(W_V, \mathcal{W}_V)$ called *textural fuzzy direlation* which are in fact L-fuzzy subsets of the set $U \times M_L \times V$ where U and V are any two domains of discourse and M is a set of all molecules of L [9]. In this section, we take $L = [0, 1]$ and we focus on the fuzzy lattice $\mathcal{F}(U)$ whose elements are mappings $\alpha : U \to [0, 1]$. Fuzzy direlations can be also regarded as the mappings between fuzzy lattices $\mathcal{F}(U)$ and $\mathcal{F}(V)$. The arguments presented in Chap. 4 will be easily translated to the remarkable results for the fuzzy approximation spaces.

A fuzzy relation φ is a mapping from $U \times V$ to $[0, 1]$ and the family of all fuzzy relations is denoted by $\mathcal{F}(U \times V)$. *A t-fuzzy direlation* (ϕ, Φ) is a pair of mappings from $U \times (0, 1] \times V$ to $[0, 1]$ satisfying certain conditions. Essentially, this concept arosed from the textural arguments and is formally obtained via Brown's Representation Theorem (see Theorem 3.1.4). The mappings ϕ and Φ are the fuzzy subsets of $U \times (0, 1] \times V$, that is, $\phi, \Phi \in \mathcal{F}(U \times (0, 1] \times V)$. A t-fuzzy direlation can be regarded as a fuzzy version of direlations between textures. Besides that, it provides reasonable operators between the lattices of fuzzy subsets; t-fuzzy direlations establish a reasonable connection to fuzzy rough set theory. As in the study of algebraic properties of fuzzy logical connectives, the notion of continuity is an indispensable tool for a natural link between the fuzzy relations and t-fuzzy direlations.

Proposition 7.3.1 *Let $(U, \mathcal{U})$, $(V, \mathcal{V})$ and $(W, \mathcal{W})$ be texture spaces. Then*

$$((U \times V) \times W,\ (\mathcal{U} \otimes \mathcal{V}) \otimes \mathcal{W}) \cong (U \times (V \times W),\ (\mathcal{U} \otimes (\mathcal{V} \otimes \mathcal{W})).$$

Proof For the sake of shortness, we denote the product textures

$$((U \times V) \times W,\ (\mathcal{U} \otimes \mathcal{V}) \otimes \mathcal{W}) \text{ and } (U \times (V \times W),\ \mathcal{U} \otimes (\mathcal{V} \otimes \mathcal{W}))$$

by $(S, \mathcal{S})$ and $(T, \mathcal{T})$, respectively. Then the function $\psi : S \to T$ defined by

$$\forall((u, v), w) \in S,\ \psi(((u, v), w)) = (u, (v, w)) \in T$$

is one-to-one and onto. Further, the mapping

$$\psi : \mathcal{S} \to \mathcal{T},\ A \mapsto \psi(A),\ A \in \mathcal{S}$$

is also one-to-one and onto. Hence, ψ is a textural isomorphism.

Let us consider the fuzzy textures

$(W_U, \mathcal{W}_U) = (U \times (0, 1],\ \mathcal{P}(U) \otimes \mathcal{M})$

and

$$(W_V, \mathcal{W}_V) = (V \times (0, 1], \mathcal{P}(V) \otimes \mathcal{M})$$

where $\mathcal{M} = \{(0, r] \mid r \in [0, 1]\}$ and $M = (0, 1]$. Let (r, R) be a direlation from $(W_U, \mathcal{W}_U)$ to $(W_V, \mathcal{W}_V)$. By definition of a direlation, we clearly have that

$$r, R \in \mathcal{P}(W_U) \otimes \mathcal{W}_V = (\mathcal{P}(U \times (0, 1])) \otimes (\mathcal{P}(V) \otimes \mathcal{M}).$$

Note that by Proposition 7.3.1,

$$\begin{aligned} \mathcal{P}(U \times (0, 1]) \otimes (\mathcal{P}(V) \otimes \mathcal{M}) &\cong (\mathcal{P}(U \times (0, 1]) \otimes \mathcal{P}(V)) \otimes \mathcal{M} \\ &= \mathcal{P}(U \times (0, 1] \times V) \otimes \mathcal{M}. \end{aligned}$$

This means that we have $r, R \in \mathcal{P}(U \times (0, 1] \times V) \otimes \mathcal{M}$ up to isomorphism. In view of Brown's Representation Theorem, $\mathcal{F}(U \times (0, 1] \times V)$ is the fuzzy lattice corresponding to texture $\mathcal{P}(U \times (0, 1] \times V) \otimes \mathcal{M}$:

$$\begin{array}{ccc} \mathcal{F}(U) & \qquad & \mathcal{F}(U \times (0, 1] \times V) \\ \big\downarrow \mathfrak{I} & & \big\downarrow \mathfrak{I} \\ \mathcal{P}(U) \otimes \mathcal{M} & & \mathcal{P}(U \times (0, 1] \times V) \otimes \mathcal{M} \end{array}$$

Hence, for the direlation (r, R), there exist $\phi_r, \Phi_R \in \mathcal{F}(U \times (0, 1] \times V)$ such that

$$\mathfrak{I}(\phi_r) = r \text{ and } \mathfrak{I}(\Phi_R) = R.$$

Note that ϕ_r and Φ_R are the mappings from $U \times (0, 1] \times V$ to $[0, 1]$. A fuzzy point $(u, \lambda, v)_\mu$ and fuzzy co-point $(u, \lambda, v)^\mu$ in $\mathcal{F}(U \times (0, 1] \times V)$ may be given as

$$(u, \lambda, v)_\mu(u', \lambda', v') = \begin{cases} \mu, & \text{if } (u, \lambda, v) = (u', \lambda', v') \\ 0, & \text{if } (u, \lambda, v) \neq (u', \lambda', v') \end{cases}$$

and

$$(u, \lambda, v)^\mu(u', \lambda', v') = \begin{cases} \mu, & \text{if } (u, \lambda, v) = (u', \lambda', v') \\ 1, & \text{if } (u, \lambda, v) \neq (u', \lambda', v'), \end{cases}$$

respectively. The elements of the texture $\mathcal{P}(U \times (0, 1] \times V) \otimes \mathcal{M}$ have the form

$$\begin{aligned} \mathfrak{I}(\phi_r) &= \{(u, \lambda, v)_\mu \mid (u, \lambda, v)_\mu \leq \phi_r\} \\ &= \{((u, \lambda, v), \mu) \mid \mu \leq \phi_r(u, \lambda, v)\} \\ &= \{((u, \lambda), (v, \mu)) \mid \mu \leq \phi_r(u, \lambda, v)\} \end{aligned}$$

for every $\phi_r \in \mathcal{F}(U \times (0, 1] \times V)$. Similarly, for the mapping Φ_R, we also have that

$$\mathfrak{I}(\Phi_R) = \{((u, \lambda), (v, \mu)) \mid \mu \leq \Phi_R(u, \lambda, v)\}.$$

Let us denote the p-sets and q-sets of the texture $\mathcal{P}(U \times (0,1] \times V) \otimes \mathcal{M}$ by

$$\overline{P}_{((u,\lambda),(v,\mu))} \text{ and } \overline{Q}_{((u,\lambda),(v,\mu))}.$$

It is easy to see that

$$\overline{P}_{((u,\lambda),(v,\mu))} = \{(u,\lambda)\} \times (\{v\} \times (0,\mu])$$

and

$$\overline{Q}_{((u,\lambda),(v,\mu))} = \big((W_U \setminus \{(u,\lambda)\}) \times W_V\big) \cup (W_U \times (0,\mu]\big).$$

Lemma 7.3.2 $\overline{P}_{((u,\lambda),(v,\mu))} = \mathfrak{I}((u,\lambda,v)_\mu)$ *and* $\overline{Q}_{((u,\lambda),(v,\mu))} = \mathfrak{I}((u,\lambda,v)^\mu)$.

Proof By Lemma 3.3.2, the proof is immediate. □

Corollary 7.3.3 *Let* (r, R) *be a direlation from* $(W_U, \mathcal{W}_U)$ *to* $(W_V, \mathcal{W}_V)$.

(i) $r \not\subseteq \overline{Q}_{((u,\lambda),(v,\mu))}$ *if and only if* $\mu < \phi_r(u,\lambda,v)$.

(ii) $\overline{P}_{((u,\lambda),(v,\mu))} \not\subseteq R$ *if and only if* $\mu > \Phi_R(u,\lambda,v)$.

Proof (i) Let $r \not\subseteq \overline{Q}_{((u,\lambda),(v,\mu))}$. Since $\mathfrak{I}$ is a lattice isomorphism, by Lemma 7.3.2, we have $\mathfrak{I}(\phi_r) \not\subseteq \mathfrak{I}((u,\lambda,v)^\mu)$ where

$$\mathfrak{I}(\phi_r) = r \text{ and } \mathfrak{I}((u,\lambda,v)^\mu) = \overline{Q}_{((u,\lambda),(v,\mu))}.$$

Since I is a lattice isomorphism, we also have $\phi_r \not\leq (u,\lambda,v)^\mu$, that is, $(u,\lambda,v)^\mu < \phi_r$. Then we obtain that $\mu < \phi_r(u,\lambda,v)$. For the sufficiency, if $\mu < \phi_r(u,\lambda,v)$, then we easily find that $r \not\subseteq \overline{Q}_{((u,\lambda),(v,\mu))}$.

(ii) Let $\overline{P}_{((u,\lambda),(v,\mu))} \not\subseteq R$. Again by Lemma 7.3.2, $\mathfrak{I}((u,\lambda,v)_\mu) \not\subseteq \mathfrak{I}(\Phi_R)$ where

$$\mathfrak{I}(\Phi_R) = R \text{ and } \mathfrak{I}((u,\lambda,v)_\mu) = \overline{P}_{((u,\lambda),(v,\mu))}.$$

This implies that $(u,\lambda,v)_\mu) \not\leq \Phi_R$ and so $\Phi_R < (u,\lambda,v)_\mu$. This gives that $\Phi_R(u,\lambda,v) < \mu$. The sufficiency is similar. □

Now, we need the following characterizations.

Theorem 7.3.4 *Let* $r, R \in \mathcal{P}(W_U) \otimes \mathcal{W}_V$. *Then we have the following equivalences:*

(i) r *is a relation if and only if* $\phi_r(u,\lambda,v) = \bigvee\{\phi_r(u,\lambda',v) \mid 0 < \lambda' < \lambda\}$.

(ii) R *is a corelation if and only if* $\Phi_R(u,\lambda,v) = \bigwedge\{\Phi_R(u,\lambda',v) \mid \lambda < \lambda' \leq 1\}$.

Proof (i) Let r be a relation. Suppose that

$$\phi_r(u, \lambda, v) \neq \bigvee\{\phi_r(u, \lambda', v) \mid 0 < \lambda' < \lambda\}.$$

If $\phi_r(u, \lambda, v) < \bigvee\{\phi_r(u, \lambda', v) \mid 0 < \lambda' < \lambda\}$, then for some $\lambda' \in (0, 1]$ with $0 < \lambda' < \lambda$, we have $\phi_r(u, \lambda, v) < \phi(u, \lambda', v)$. Let $\mu = \phi_r(u, \lambda, v)$. Then by Corollary 7.3.3 (i), for $\mu < \phi_r(u, \lambda', v)$, we get $r \not\subseteq \overline{Q}_{((u,\lambda'),(v,\mu))}$. Further, $\lambda' < \lambda$ implies that $P_\lambda = (0, \lambda] \not\subseteq Q_{\lambda'} = (0, \lambda']$. Hence, by Example 4.1.3, we have that $P_{(u,\lambda)} \not\subseteq Q_{(u,\lambda')}$. Then by R1, we obtain $r \not\subseteq \overline{Q}_{((u,\lambda),(v,\mu))}$. Again, by Corollary 7.3.3 (i), we find that $\mu < \phi_r(u, \lambda, v)$ which is an immediate contradiction. Hence, the inequality

$$\bigvee\{\phi_r(u, \lambda', v) \mid 0 < \lambda' < \lambda\} \leq \phi(u, \lambda, v)$$

holds. Now, suppose that

$$\bigvee\{\phi_r(u, \lambda', v) \mid 0 < \lambda' < \lambda\} < \phi_r(u, \lambda, v).$$

Let us choose a $\mu \in (0, 1]$ such that

$$\bigvee\{\phi_r(u, \lambda', v) \mid 0 < \lambda' < \lambda\} < \mu < \phi_r(u, \lambda, v).$$

By Corollary 7.3.3 (i), $\mu < \phi_r(u, \lambda, v)$ implies that $r \not\subseteq \overline{Q}_{((u,\lambda),(v,\mu))}$. Then by R2, there exists a $(u', \lambda') \in U \times (0, 1]$ such that $P_{(u,\lambda)} \not\subseteq Q_{(u',\lambda')}$ and $r \not\subseteq \overline{Q}_{((u',\lambda'),(v,\mu))}$. By Example 4.1.2, $0 < \lambda' < \lambda$ and $u = u'$. Then we also have $r \not\subseteq \overline{Q}_{((u,\lambda'),(v,\mu))}$ whence $\mu < \phi_r(u, \lambda', v)$ by Lemma 7.3.3 (i). This gives the contradiction

$$\mu < \bigvee\{\phi_r(u, \lambda', v) \mid 0 < \lambda' < \lambda\}.$$

For the sufficiency, let $r \in \mathcal{P}(W_U) \otimes \mathcal{W}_V$ and

$$\phi(u, \lambda, v) = \bigvee\{\phi(u, \lambda', v) \mid 0 < \lambda' < \lambda\}.$$

We show that r is a relation. Let $r \not\subseteq \overline{Q}_{((u,\lambda),(v,\mu))}$. By Corollary 7.3.3 (i), $\mu < \phi_r(u, \lambda, v)$. If $P_{(u,\lambda')} \not\subseteq Q_{(u,\lambda)}$, then $0 < \lambda < \lambda'$ and so by the assumption, we conclude that $\phi_r(u, \lambda, v) \leq \phi(u, \lambda', v)$ and so by Corollary 7.3.3 (i), we obtain that $r \not\subseteq \overline{Q}_{((u,\lambda'),(v,\mu))}$. That is r holds the condition R1. For the condition R2, let $r \not\subseteq \overline{Q}_{((u,\lambda),(v,\mu))}$. By Corollary 7.3.3 (i), the equality

$$\mu < \phi_r(u, \lambda, v) = \bigvee\{\phi_r(u, \lambda', v) \mid 0 < \lambda' < \lambda\}$$

holds. Then there exists a $\lambda' \in (0, 1]$ with $0 < \lambda' < \lambda$ such that $\mu < \phi_r(u, \lambda', v)$ Clearly $P_{(u,\lambda)} \not\subseteq Q_{(u,\lambda')}$. By Corollary 7.3.3 (i), we obtain that $r \not\subseteq \overline{Q}_{((u,\lambda'),(v,\mu))}$ as desired, that is, the condition R2 holds.

The proof of the item (ii) is dual and it is left to the interested reader. □

Theorem 7.3.5 (i) *The mappings ϕ_r and Φ_R are increasing for λ for a fixed u and v.*

(ii) *For all $(u, v) \in U \times V$, $\Phi_R(u, 1, v) = 1$.*

Proof By definitions of ϕ_r and Φ_R, the proof is straightforward. □

$$\begin{array}{ccc} \mathcal{F}(U) & \xrightarrow{(\phi_r,\,\Phi_R)} & \mathcal{F}(V) \\ \Big\downarrow \mathfrak{I} & & \Big\downarrow \mathfrak{I} \\ \mathcal{P}(U)\otimes\mathcal{M} & \xrightarrow{(r,\,R)} & \mathcal{P}(V)\otimes\mathcal{M} \end{array}$$

Remark 7.3.6 As is seen in the above diagram, the mappings ϕ_r and Φ_R are the fuzzy subsets of $U \times (0, 1] \times V$ and they correspond to the direlation (r, R) from the fuzzy texture $\mathcal{P}(U) \otimes \mathcal{M}$ to $\mathcal{P}(V) \otimes \mathcal{M}$ under the lattice isomorphism $\mathfrak{I}$. Therefore, the pair (ϕ_r, Φ_R) can also be regarded as a morphism from $\mathcal{F}(U)$ to $\mathcal{F}(V)$.

More generally, Theorem 7.3.4 leads us to the following definition.

Definition 7.3.7 (i) A mapping $\phi \in \mathcal{F}(U \times (0, 1] \times U)$ defined by

$$\forall(u, \lambda, v) \in U \times (0, 1] \times V,\ \phi(u, \lambda, v) = \bigvee\{\phi(u, \lambda', v) \mid 0 < \lambda' < \lambda\}$$

is called *a textural fuzzy relation* or in brief, *t-fuzzy relation* from $\mathcal{F}(U)$ to $\mathcal{F}(V)$.

(i) A mapping $\Phi \in \mathcal{F}(U \times (0, 1] \times U)$ defined by

$$\forall(u, \lambda, v) \in U \times (0, 1] \times V,\ \Phi(u, \lambda, v) = \bigwedge\{\Phi(u, \lambda', v) \mid \lambda < \lambda' \leq 1\}$$

is called *a textural fuzzy corelation* or in brief, *t-fuzzy corelation* from $\mathcal{F}(U)$ to $\mathcal{F}(V)$.

The pair (ϕ, Φ) is called *a t-fuzzy direlation* from $\mathcal{F}(U)$ to $\mathcal{F}(V)$.

Theorem 7.3.8 *If (i, I) is the identity direlation on $\mathcal{P}(U) \otimes \mathcal{W}$, then the corresponding t-fuzzy direlation on $\mathcal{F}(U)$ is the pair (ϕ_i, Φ_I) where*

$$\phi_i(u, \lambda, v) = \begin{cases} \lambda, & \text{if } u = v \\ 0, & \text{if } u \neq v, \end{cases} \qquad \Phi_I(u, \lambda, v) = \begin{cases} \lambda, & \text{if } u = v \\ 1, & \text{if } u \neq v \end{cases}$$

for all $u, v \in U$ and $\lambda \in (0, 1]$.

Proof Note that by Corollary 7.3.3 (i),

$$i \not\subseteq \overline{Q}_{((u,\lambda),(v,\mu))} \text{ if and only if } \mu < \phi_i(u, \lambda, v).$$

That is, by Corollary 4.1.11 (1) and Theorem 3.2.11 (ii), the equivalence

$$u = v \text{ and } \mu < \lambda \text{ if and only if } \mu < \phi_i(u, \lambda, v)$$

holds. Then $u = v$ and $\phi_i(u, \lambda, v) \neq \lambda$. If $\lambda < \phi_i(u, \lambda, v)$, then we obtain an immediate contradiction $\lambda < \lambda$. For the case $\phi_i(u, \lambda, v) < \lambda$, we may choose a $\mu \in (0, 1]$ with $\phi_i(u, \lambda, v) < \mu < \lambda$. However, again by the above equivalence, we obtain

$\mu < \phi_i(u, \lambda, v)$ which is also an immediate contradiction. Now let $u \neq v$. Then again by the same equivalence, for all $\mu \in (0, 1]$, the inequality $\phi_i(u, \lambda, v) \leq \mu$ holds. This means that $\phi_i(u, \lambda, v) = 0$. The proof of the second equality is similar. □

Definition 7.3.9 The pair (ϕ_i, Φ_I) given in Theorem 7.3.8 is called *the identity t-fuzzy direlation* on $\mathcal{F}(U)$.

Definition 7.3.10 Let (ϕ, Φ) be a t-fuzzy direlation on $\mathcal{F}(U)$. If

$$\phi_i \leq \phi \text{ and } \Phi \leq \Phi_I,$$

then (ϕ, Φ) is called *a reflexive t-fuzzy direlation*.

Theorem 7.3.11 *Let r, R and (h, H) be direlations from $(W_U, \mathcal{W}_U)$ to $(W_V, \mathcal{W}_V)$. Then we have the following inclusions:*

(i) $r \subseteq h$ if and only if $\phi_r \leq \phi_h$.

(ii) $R \subseteq H$ if and only if $\Phi_R \leq \Phi_H$.

Proof (i) Let $h \subseteq r$ and $\phi_r \not\leq \phi_h$. Then there exists a $(\lambda, v, \mu) \in U \times (0, 1] \times V$ such that

$$\phi_h((u, \lambda), v) < \mu < \phi_r(u, \lambda, v).$$

Then by Corollary 7.3.3(i), we get

$$r \subseteq \overline{Q}_{((u,\lambda),(v,\mu))} \text{ and } h \not\subseteq \overline{Q}_{((u,\lambda),(v,\mu))}.$$

However, this gives the contradiction $r \not\subseteq h$. The proof of the necessity and the item (ii) is similar.

Theorem 7.3.12 *Let (ϕ, Φ) be a t-fuzzy direlation on $\mathcal{F}(U)$. Then*

(ϕ, Φ) is reflexive if and only if (r, R) is reflexive

where (r, R) is the direlation on $(\mathcal{W}_U, W_U)$ where $\phi = \phi_r$ and $\Phi = \Phi_R$.

Proof The proof is immediate by Theorem 7.3.11.

Now, let (r, R) be a direlation from $(W_U, \mathcal{W}_U)$ to $(W_V, \mathcal{W}_V)$. We know that by Lemma 4.1.16, the inverse $(r, R)^{\leftarrow} = (R^{\leftarrow}, r^{\leftarrow})$ is also a direlation from $(W_V, \mathcal{W}_V)$ to $(W_U, \mathcal{W}_U)$. The following result gives a way to determine the inverse t-fuzzy direlation.

Theorem 7.3.13 *For the fuzzy relation $(\phi_{R^{\leftarrow}}, \Phi_{r^{\leftarrow}})$ from $\mathcal{F}(V)$ to $\mathcal{F}(U)$, $\forall(v, \mu, u) \in V \times (0, 1] \times U$, the following equalities hold:*

$$\phi_{R^{\leftarrow}}(v, \mu, u) = \bigwedge\{\lambda \in (0, 1] \mid \mu \leq \Phi_R(u, \lambda, v)\}, \text{ and}$$

$$\Phi_{r^{\leftarrow}}(v, \mu, u) = \bigvee\{\lambda \in (0, 1] \mid \phi_r(u, \lambda, v) \leq \mu\}.$$

Proof Let us prove the first equality, leaving the second to the interested reader. By Lemma 4.1.23 (ii),

$$\overline{P}_{((u,\lambda),(v,\mu))} \not\subseteq R \text{ if and only if } R^{\leftarrow} \not\subseteq \overline{Q}_{(((v,\mu),(u,\lambda))}.$$

By Corollary 4.1.18, $R^{\leftarrow}$ is a relation and so by Corollary 7.3.3,

$$\mu > \Phi_R(u, \lambda, v) \text{ if and only if } \lambda < \phi_{R^{\leftarrow}}(v, \mu, u), \quad (1)$$

or equivalently,

$$\mu \leq \Phi_R(u, \lambda, v) \text{ if and only if } \phi_{R^{\leftarrow}}(v, \mu, u) \leq \lambda. \quad (2)$$

Since $\phi_{R^{\leftarrow}}$ is a fuzzy relation, we already know that

$$\phi_{R^{\leftarrow}}(v, \mu, u) = \bigvee \{\phi_{R^{\leftarrow}}(v, \mu', u) \mid 0 < \mu' < \mu\}.$$

By the equivalence (2), we clearly have that

$$\phi_{R^{\leftarrow}}(v, \mu, u) \leq \bigwedge \{\lambda \in (0, 1] \mid \mu \leq \Phi_R(u, \lambda, v)\}.$$

For the reverse inequality, suppose that there exists a λ' such that

$$\phi_{R^{\leftarrow}}(v, \mu, u) < \lambda' < \bigwedge \{\lambda \in (0, 1] \mid \mu \leq \Phi_R(u, \lambda, v)\}.$$

Then by the right side of the above inequality, we have $\mu > \Phi_R(u, \lambda', v)$ and so by (1), we obtain that $\lambda' < \phi_{R^{\leftarrow}}(v, \mu, u)$ and this implies a contradiction. □

In view of Theorem 7.3.13, we obtain the following concept.

Definition 7.3.14 Let (ϕ, Φ) be *a fuzzy direlation* from $\mathcal{F}(U)$ to $\mathcal{F}(V)$. Then *the inverse* of $(\phi, \Phi)^{\leftarrow} = (\Phi^{\leftarrow}, \phi^{\leftarrow})$ is defined by

$$\Phi^{\leftarrow}(v, \mu, u) = \bigwedge \{\lambda \in (0, 1] \mid \mu \leq \Phi(u, \lambda, v)\},$$

$$\phi^{\leftarrow}(v, \mu, u) = \bigvee \{\lambda \in (0, 1] \mid \phi(u, \lambda, v) \leq \mu\}.$$

Remark 7.3.15 Note that by Theorem 7.3.13, the pair $(\phi, \Phi)^{\leftarrow} = (\Phi^{\leftarrow}, \phi^{\leftarrow})$ is a fuzzy direlation from $\mathcal{F}(V)$ to $\mathcal{F}(U)$. □

Definition 7.3.16 Let (ϕ, Φ) be a t-fuzzy direlation on $\mathcal{F}(U)$. If

$$\phi = \Phi^{\leftarrow} \text{ and } \Phi = \phi^{\leftarrow},$$

then (ϕ, Φ) is called *a symmetric t-fuzzy direlation.*

Theorem 7.3.17 *Let (ϕ, Φ) be a t-fuzzy direlation on $\mathcal{F}(U)$. Then*

$$(\phi, \Phi) \textit{ is symmetric if and only if } (r, R) \textit{ is symmetric}$$

where (r, R) is the direlation on $(\mathcal{W}_U, W_U)$ where $\phi = \phi_r$ and $\Phi = \Phi_R$.

Proof The proof is immediate by Theorem 7.3.11.

Theorem 7.3.18 *Let (r, R) and (d, D) be direlations from $(W_U, \mathcal{W}_U)$ to $(W_V, \mathcal{W}_V)$ and from $(W_V, \mathcal{W}_V)$ to $(W_Z, \mathcal{W}_Z)$, respectively. Then the pair $(\phi_{d\circ r}, \Phi_{D\circ R})$ is a t-fuzzy direlation from $\mathcal{F}(U)$ to $\mathcal{F}(Z)$ where*

$$\phi_{d\circ r}(u, \lambda, z) = \bigvee\{\phi_d(v, \mu, z) \mid v \in V, 0 < \mu < \phi_r(u, \lambda, v)\}$$

and

$$\Phi_{D\circ R}(u, \lambda, z) = \bigwedge\{\Phi_D(v, \mu, z) \mid v \in V, \Phi_R(u, \lambda, v) < \mu \leq 1\}.$$

Proof We prove the first equality. Let $\gamma < \phi_{d\circ r}(u, \lambda, z)$. Then by Corollary 7.3.3 (i), $d \circ r \not\subseteq \overline{Q}_{((u,\lambda),(z,\gamma))}$. By Definition 4.3.1, for all $\gamma \in (0, 1]$, we clearly have the equivalence

$$\gamma < \phi_{d\circ r}(u, \lambda, z) \Longleftrightarrow \exists (v, \mu) \in W_V \text{ with } \mu < \phi_r(u, \lambda, v) \text{ and } \gamma < \phi_d(v, \mu, z).$$

Note that $\phi_{d\circ r}$ is a t-fuzzy relation and so

$$\phi_{d\circ r}(u, \lambda, z) = \bigvee\{\phi_{d\circ r}(u, \lambda', z) \mid 0 < \lambda' < \lambda\}.$$

Now, for $0 < \lambda' < \lambda$, let $\gamma = \phi_{d\circ r}(u, \lambda', z)$. By the above equivalence, for γ, we get

$$\mu < \phi_r(u, \lambda, v) \text{ and } \phi_{d\circ r}(u, \lambda', z) < \phi_d(v, \mu, z).$$

As a result, we obtain

$$\phi_{d\circ r}(u, \lambda, z) \leq \bigvee\{\phi_d(v, \mu, z) \mid v \in V, 0 < \mu < \phi_r(u, \lambda, v)\}.$$

Suppose that

$$\phi_{d\circ r}(u, \lambda, z) < \bigvee\{\phi_d(v, \mu, z) \mid v \in V, 0 < \mu < \phi_r(u, \lambda, v)\}.$$

Take a γ such that

$$\phi_{d\circ r}(u, \lambda, z) < \gamma < \bigvee\{\phi_d(v, \mu, z) \mid v \in V, 0 < \mu < \phi_r(u, \lambda, v)\}.$$

From the right side of the inequality, there exists a $v \in V$ such that

$$0 < \mu < \phi_r(u, \lambda, v) \text{ and } \gamma < \phi_d(v, \mu, z).$$

Again, by the above equivalence, we get $\gamma < \phi_{d\circ r}(u, \lambda, z)$ and this gives a contradiction.

For the t-fuzzy direlations, the composition can be given as follows.

Definition 7.3.19 Let (ϕ, Φ) and (ψ, Ψ) be t-fuzzy direlations from $\mathcal{F}(U)$ to $\mathcal{F}(V)$ and from $\mathcal{F}(V)$ to $\mathcal{F}(Z)$. Then the pair $(\phi \circ \psi, \Phi \circ \Psi)$ is called *a composition* from $\mathcal{F}(U)$ to $\mathcal{F}(Z)$ where

$$(\psi \circ \phi)(u, \lambda, z) = \bigvee\{\psi(v, \mu, z) \mid v \in V,\ 0 < \mu < \phi(u, \lambda, v)\}$$

and

$$(\Psi \circ \Phi)(u, \lambda, z) = \bigwedge\{\Psi(v, \mu, z) \mid v \in V,\ \Phi(u, \lambda, v) < \mu \leq 1\}.$$

Remark 7.3.20 Note that by Theorem 7.3.18, $(\phi \circ \psi, \Phi \circ \Psi)$ is a t-fuzzy direlation from $\mathcal{F}(U)$ to $\mathcal{F}(Z)$.

Definition 7.3.21 Let (ϕ, Φ) be a t-fuzzy direlation on $\mathcal{F}(U)$. Then

(i) ϕ is *transitive* if $\phi \circ \phi \leq \phi$.

(ii) Φ is *transitive* if $\Phi \leq \Phi \circ \Phi$.

(iii) (ϕ, Φ) is called *transitive* if ϕ and Φ are transitive.

(iv) If (ϕ, Φ) is reflexive, symmetric and transitive, then it is called *an equivalence t-fuzzy direlation*.

Theorem 7.3.22 *Let (ϕ, Φ) be a t-fuzzy direlation from $\mathcal{F}(U)$ to $\mathcal{F}(V)$, and (ϕ_{i_U}, Φ_{I_U}) and (ϕ_{i_V}, Φ_{I_V}) be the identity direlations on $\mathcal{F}(U)$ and $\mathcal{F}(V)$, respectively. Then*

(i) $(\phi, \Phi) \circ (\phi_{i_U}, \Phi_{I_U}) = (\phi, \Phi)$.

(ii) $(\phi_{i_V}, \Phi_{I_V}) \circ (\phi, \Phi) = (\phi, \Phi)$.

(iii) *Let (ψ, Ψ) be a t-fuzzy direlation from $\mathcal{F}(V)$ to $\mathcal{F}(Z)$ and (θ, Θ) be a t-fuzzy direlation from $\mathcal{F}(Z)$ to $\mathcal{F}(T)$. Then the composition of t-fuzzy direlations is associative, that is,*

$$((\theta, \Theta) \circ (\psi, \Psi)) \circ (\phi, \Phi) = (\theta, \Theta) \circ ((\psi, \Psi) \circ (\phi, \Phi)).$$

Proof We give the proof of (i), leaving proofs of (ii) and (iii) to the interested reader. First, we prove that

$$\phi \circ \phi_{i_U} = \phi.$$

By definition of the composition of t-fuzzy relations and t-fuzzy identity relation, for all $u, u' \in U$ and $\lambda \in (0, 1]$, we observe that

$$\begin{aligned}(\phi \circ \phi_{i_U})(u, \lambda, u') &= \bigvee \{\phi(u'', \mu, u') \mid u'' \in U,\ 0 < \mu < \phi_{i_U}(u, \lambda, u'')\} \\ &= \bigvee \{\phi(u, \mu, u') \mid 0 < \mu < \lambda\} \\ &= \phi(u, \lambda, u').\end{aligned}$$

Similarly by definition of the composition of t-fuzzy corelations and t-fuzzy identity corelation, for all $u, u' \in U$ and $\lambda \in (0, 1]$, we conclude that

$$\begin{aligned}(\Phi \circ \Phi_{I_U})(u, \lambda, u') &= \bigwedge \{\Phi(u'', \mu, u') \mid u'' \in U,\ \Phi_{I_U}(u, \lambda, u'') < \mu \leq 1\} \\ &= \bigwedge \{\Phi(u, \mu, u') \mid \lambda < \mu \leq 1\} \\ &= \Phi(u, \lambda, u').\end{aligned}$$

As we have discussed in Sect. 4, the section and presection of a direlation play a similar role as the image and inverse image notions due to ordinary functions. Using the lattice isomorphism $\mathfrak{I}$, we may determine the section and presection of a t-fuzzy direlation as follows.

Theorem 7.3.23 *Let (r, R) be a direlation from $(W_U, \mathcal{W}_U)$ to $(W_V, \mathcal{W}_V)$ and $\alpha \in \mathcal{F}(U)$. Then for all $\beta \in \mathcal{F}(U)$,*

(i) $r^{\to}(\mathfrak{I}(\alpha)) = \mathfrak{I}(\beta)$ *if and only if*

$$\forall v \in V,\ \beta(v) = \bigwedge\{\mu \in (0,1] \mid \lambda < \alpha(u) \implies \phi_r(u,\lambda,v) \le \mu\}.$$

(ii) $R^{\to}(\mathfrak{I}(\alpha)) = \mathfrak{I}(\beta)$ *if and only if*

$$\forall v \in V,\ \beta(v) = \bigvee\{\mu \in (0,1] \mid \alpha(u) < \lambda \implies \mu \le \Phi_r(u,\lambda,v)\}.$$

Proof (i) By Definition 4.2.1 and Lemma 7.3.3, we easily observe that

$$\begin{aligned} r^{\to}(\mathfrak{I}(\alpha)) &= \bigcap\{\overline{Q}_{(v,\mu)} \mid \forall (u,\lambda) \in W_U,\ r \not\subseteq \overline{Q}_{((u,\lambda),(v,\mu))} \implies \mathfrak{I}(\alpha) \subseteq \overline{Q}_{(u,\lambda)}\} \\ &= \bigcap\{\overline{Q}_{(v,\mu)} \mid \forall (u,\lambda) \in W_U,\ \mu < \phi_r(u,\lambda,v) \implies \alpha(u) \le \lambda\} \\ &= \bigcap\{\overline{Q}_{(v,\mu)} \mid \forall (u,\lambda) \in W_U,\ \lambda < \alpha(u) \implies \phi_r(u,\lambda,v) \le \mu\}. \end{aligned}$$

Now, we prove that the following statements are equivalent:

(1) $r^{\to}(\mathfrak{I}(\alpha)) = \mathfrak{I}(\beta)$.

(2) $\beta(v) \le \mu \iff (\lambda < \alpha(u) \implies \phi_r(u,\lambda,v) \le \mu)$.

(1) $\implies$ (2): Let $\lambda < \alpha(u) \implies \phi_r(u,\lambda,v) \le \mu$. Then we have $r^{\to}(\mathfrak{I}(\alpha)) \subseteq \overline{Q}_{(v,\mu)}$. By the assumption, $\mathfrak{I}(\beta) \subseteq \overline{Q}_{(v,\mu)}$ and this gives that $\beta(v) \le \mu$. For the necessity, let $\beta(v) \le \mu$ and $\lambda < \alpha(u)$. Then we get $\beta \le v^{\mu}$, that is, $\mathfrak{I}(\beta) \subseteq \overline{Q}_{(v,\mu)}$. Since $\lambda < \alpha(u)$, $u_\lambda \le \alpha$, that is, $\overline{P}_{(u,\lambda)} \subseteq \mathfrak{I}(\alpha)$. This gives that

$$r^{\to}\overline{P}_{(u,\lambda)} \subseteq r^{\to}\mathfrak{I}(\alpha) = \mathfrak{I}(\beta) \subseteq \overline{Q}_{(v,\mu)}.$$

Therefore, we conclude that $r \subseteq \overline{Q}_{((u,\lambda),(v,\mu))}$, that is, $\phi_r(u,\lambda,v) \le \mu$.

(2) $\implies$ (1): Let $\mathfrak{I}(\beta) \not\subseteq \overline{Q}_{(v,\mu)}$. Then $\beta \not\le v^{\mu}$, that is, $\mu < \beta(v)$. By the assumption, there exists a $\lambda \in (0,1]$ such that $\lambda < \alpha(u)$ and $\phi_r(u,\lambda,v) > \mu$. Then $u_\lambda \le \alpha$ and $r \not\subseteq \overline{Q}_{(u,\lambda),(v,\mu)}$. Then we have

$$\overline{P}_{(u,\lambda)} \subseteq \mathfrak{I}(\alpha) \text{ and } r^{\to}\overline{P}_{(u,\lambda)} \not\subseteq \overline{Q}_{(v,\mu)}.$$

Therefore, $r^{\to}\overline{P}_{(u,\lambda)} \subseteq r^{\to}\mathfrak{I}(\alpha)$ and so $r^{\to}\mathfrak{I}(\alpha) \not\subseteq \overline{Q}_{(v,\mu)}$. As a result, we obtain that $r^{\to}(\mathfrak{I}(\alpha)) \subseteq \mathfrak{I}(\beta)$. Now let $r^{\to}(\mathfrak{I}(\alpha)) \not\subseteq \overline{Q}_{(v,\mu)}$. Then there exists a $\lambda \in (0,1]$ such that $\phi_r(u,\lambda,\mu) > \mu$. By the assumption, we have $\mu < \beta(v)$ and hence, $\beta \not\le v^{\mu}$. This implies that $\mathfrak{I}(\beta) \not\subseteq \overline{Q}_{(v,\mu)}$, and we obtain that $\mathfrak{I}(\beta) \subseteq r^{\to}(\mathfrak{I}(\alpha))$.

Now let us show that the equivalence given in (i) holds.

Necessity: Let $\lambda < \alpha(u) \implies \phi_r(u, \lambda, v) \leq \mu$. Then by (2), we obtain $\beta(v) \leq \mu$. This gives that

$$\beta(v) \leq \bigwedge\{\mu \in (0, 1] \mid \lambda < \alpha(u) \implies \phi_r(u, \lambda, v) \leq \mu\}.$$

Now let

$$\beta(v) < \bigwedge\{\mu \in (0, 1] \mid \lambda < \alpha(u) \implies \phi_r(u, \lambda, v) \leq \mu\}.$$

Let us take a $\mu' \in (0, 1]$ such that

$$\beta(v) < \mu' < \bigwedge\{\mu \in (0, 1] \mid \lambda < \alpha(u) \implies \phi_r(u, \lambda, v) \leq \mu\}.$$

By (2), the left side of the inequality implies that

$$\lambda < \alpha(u) \implies \phi_r(u, \lambda, v) \leq \mu'.$$

However, this gives an immediate contradiction that

$$\bigwedge\{\mu \in (0, 1] \mid \lambda < \alpha(u) \implies \phi_r(u, \lambda, v) \leq \mu\} \leq \mu'.$$

Sufficiency is similar.

(ii) The proof is left to the interested reader. □

Definition 7.3.24 (i) Let ϕ be a t-fuzzy relation from $\mathcal{F}(U)$ to $\mathcal{F}(V)$ and $\alpha \in \mathcal{F}(U)$. Let us consider the mapping $\phi^{\rightarrow} : \mathcal{F}(U) \rightarrow \mathcal{F}(V)$ defined by

$$\forall v \in V, \ (\phi^{\rightarrow}\alpha)(v) = \bigwedge\{\mu \in (0, 1] \mid \lambda < \alpha(u) \implies \phi(u, \lambda, v) \leq \mu\}.$$

The fuzzy subset $\phi^{\rightarrow}\alpha \in \mathcal{F}(V)$ is called *the α-section* of the t-fuzzy relation ϕ.

(ii) Let Φ be a t-fuzzy corelation from $\mathcal{F}(U)$ to $\mathcal{F}(V)$ and $\alpha \in \mathcal{F}(U)$. Let us consider the mapping $\Phi^{\rightarrow} : \mathcal{F}(U) \rightarrow \mathcal{F}(V)$ defined by

$$\forall v \in V, \ (\Phi^{\rightarrow}\alpha)(v) = \bigvee\{\mu \in (0, 1] \mid \alpha(u) < \lambda \implies \mu \leq \Phi(u, \lambda, v)\}.$$

The fuzzy subset $\Phi^{\rightarrow}\alpha \in \mathcal{F}(V)$ is called *the α-section* of the t-fuzzy corelation Φ.

Presections

Just as the presections of direlations, we can also define the presections for t-fuzzy direlations. Let (ϕ, Φ) be a t-fuzzy direlation from $\mathcal{F}(U)$ to $\mathcal{F}(V)$. By Theorem 7.3.13, we know that $(\Phi^{\leftarrow}, \phi^{\leftarrow})$ is a t-fuzzy direlation from $\mathcal{F}(V)$ to $\mathcal{F}(U)$. Since $\Phi^{\leftarrow}$ is a t-fuzzy relation and $\phi^{\leftarrow}$ is a t-fuzzy corelation, for every $\beta \in \mathcal{F}(V)$, we can consider the β-sections as

$$(\Phi^{\leftarrow})^{\rightarrow}\beta \text{ and } (\phi^{\leftarrow})^{\rightarrow}\beta.$$

In other words, a presection is defined as a section of $\Phi^{\leftarrow}$ and $\phi^{\leftarrow}$, respectively. For the β-presections, we use the brief notations

$$\Phi^{\leftarrow}\beta = (\Phi^{\leftarrow})^{\rightarrow}\beta \text{ and } \phi^{\leftarrow}\beta = (\phi^{\leftarrow})^{\rightarrow}\beta.$$

The following result gives an alternative formulation for the β-presections.

Theorem 7.3.25 *Let (ϕ, Φ) be a t-fuzzy direlation from $\mathcal{F}(U)$ to $\mathcal{F}(V)$ and $\beta \in \mathcal{F}(V)$. Then for all $(v, \mu, u) \in V \times (0, 1] \times U$, we have the following equalities:*

(i) $(\phi^{\leftarrow}\beta)(u) = \bigvee\{\lambda \in (0, 1] \mid \phi(u, \lambda, v) \leq \beta(v), \forall v \in V\}$,

(ii) $(\Phi^{\leftarrow}\beta)(u) = \bigwedge\{\lambda \in (0, 1] \mid \beta(v) \leq \Phi(u, \lambda, v), \forall v \in V\}$.

Proof (i) By Definition 7.3.24 (i), we have

$$(\phi^{\leftarrow}\beta)(u) = \bigvee\{\lambda \in (0, 1] \mid \beta(v) < \mu \implies \lambda \leq \phi^{\leftarrow}(v, \mu, u)\}.$$

Suppose that $\beta(v) < \mu \implies \lambda \leq \phi^{\leftarrow}(v, \mu, u)$ and $\phi(u, \lambda, v) \nleq \beta(v)$. Let us choose a $\mu \in (0, 1]$ such that $\beta(v) < \mu < \phi(u, \lambda, v)$. Since ϕ is a t-fuzzy relation, there exists a number $\lambda' \in (0, 1]$ such that $0 < \lambda' < \lambda$ and $\beta(v) < \mu < \phi(u, \lambda', v)$. By the assumption, we have $\lambda \leq \phi^{\leftarrow}(v, \mu, u)$ and so by Definition 7.3.14, there exists a λ_1 with $\lambda' < \lambda_1 < \lambda$ such that $\phi(u, \lambda_1, v) \leq \mu$. By Theorem 7.3.5 (i), we also have that $\phi(u, \lambda', v) \leq \mu$ and this gives a contradiction. Hence, we obtain that $\phi(u, \lambda, v) \leq \beta(v)$. Then we clearly have

$$(\phi^{\leftarrow}\beta)(u) \leq \bigvee\{\lambda \in (0, 1] \mid \phi(u, \lambda, v) \leq \beta(v), \forall v \in V\}.$$

For the reverse inequality, let $\phi(u, \lambda, v) \leq \beta(v)$ and $\beta(v) < \mu$. Then $\phi(u, \lambda, v) < \mu$ and so $\lambda \leq \phi^{\leftarrow}(v, \mu, u)$, that is,

$$\bigvee\{\lambda \in (0, 1] \mid \phi(u, \lambda, v) \leq \beta(v), \forall v \in V\} \leq (\phi^{\leftarrow}\beta)(u)$$

as desired.

The mappings $\phi^{\rightarrow}$, $\Phi^{\rightarrow}$, $\phi^{\leftarrow}$ and $\Phi^{\leftarrow}$ are monotonic.

Theorem 7.3.26 (i) *For all $\alpha_1, \alpha_2 \in \mathcal{F}(U)$*

$$\alpha_1 \leq \alpha_2 \implies \phi^{\rightarrow}\alpha_1 \leq \phi^{\rightarrow}\alpha_2,$$

$$\alpha_1 \leq \alpha_2 \implies \Phi^{\rightarrow}\alpha_1 \leq \Phi^{\rightarrow}\alpha_2.$$

(ii) *For all $\beta_1, \beta_2 \in \mathcal{F}(U)$*

$$\beta_1 \leq \beta_2 \implies \phi^{\leftarrow}\beta_1 \leq \phi^{\leftarrow}\beta_2,$$

$$\beta_1 \leq \beta_2 \implies \Phi^{\leftarrow}\beta_1 \leq \Phi^{\leftarrow}\beta_2.$$

Proof Let us prove the first equality in (i), leaving the rest of the proof to the interested reader. Let $\alpha_1 \leq \alpha_2$. Then for all $u \in U$, $\alpha_1(u) \leq \alpha_2(u)$. Then clearly, for every $v \in \mathcal{F}(V)$,

$$\begin{aligned}\mathcal{A}_2 = \{\mu \in (0, 1] \mid \lambda < \alpha_2(u) \implies \phi(u, \lambda, v) \leq \mu\}\\ \subseteq \{\mu \in (0, 1] \mid \lambda < \alpha_1(u) \implies \phi(u, \lambda, v) \leq \mu\} = \mathcal{A}_1.\end{aligned}$$

Therefore, for all $\mu \in \mathcal{A}_2$,

$$(\phi^{\rightarrow}\alpha_1)(v) = \bigwedge \mathcal{A}_1 \leq \mu.$$

This gives that

$$(\phi^{\rightarrow}\alpha_1)(v) \leq \bigwedge \mathcal{A}_2 = (\phi^{\rightarrow}\alpha_2)(v).$$

That is, $\phi^{\rightarrow}\alpha_1 \leq \phi^{\rightarrow}\alpha_2$.

Theorem 7.3.27 *Let (ϕ_1, Φ_1) and (ϕ_2, Φ_2) be any two t-fuzzy direlations from $\mathcal{F}(U)$ to $\mathcal{F}(V)$ and $\alpha \in \mathcal{F}(U)$. Then*

(i) $\phi_1 \leq \phi_2$ *if and only if for all $\alpha \in \mathcal{F}(U)$, $\phi_1^{\rightarrow}\alpha \leq \phi_2^{\rightarrow}\alpha$.*

(ii) $\Phi_1 \leq \Phi_2$ *if and only if for all $\alpha \in \mathcal{F}(U)$, $\Phi_1^{\rightarrow}\alpha \leq \Phi_2^{\rightarrow}\alpha$.*

Proof (i) Let $\phi_1 \leq \phi_2$. Then for all $(u, \lambda, v) \in U \times (0, 1] \times V$, we have

$$\phi_1(u, \lambda, v) \leq \phi_2(u, \lambda, v).$$

Note that

$$\begin{aligned}\mathcal{A}_2 = \{\mu \in (0, 1] \mid \lambda < \alpha(u) &\implies \phi_2(u, \lambda, v) \leq \mu\} \\ \subseteq \{\mu \in (0, 1] \mid \lambda < \alpha(u) &\implies \phi_1(u, \lambda, v) \leq \mu\} = \mathcal{A}_1.\end{aligned}$$

Therefore,

$$\begin{aligned}(\phi^{\rightarrow}\alpha_1)(v) &= \bigwedge\{\mu \in (0, 1] \mid \lambda < \alpha(u) \implies \phi_1(u, \lambda, v) \leq \mu\} \\ &\leq \bigwedge\{\mu \in (0, 1] \mid \lambda < \alpha(u) \implies \phi_2(u, \lambda, v) \leq \mu\} \\ &= (\phi^{\rightarrow}\alpha_2)(v).\end{aligned}$$

Finally, we obtain $\phi^{\rightarrow}\alpha_1 \leq \phi^{\rightarrow}\alpha_2$ as desired.

(ii) The proof is similar.

Corollary 7.3.28 *Let (ϕ, Φ) be a t-fuzzy direlation from $\mathcal{F}(U)$ to $\mathcal{F}(V)$. Then for all $\alpha \in \mathcal{F}(U)$ and $\beta \in \mathcal{F}(V)$, we have the following equalities:*

(i) $\phi^{\rightarrow}\alpha = \bigvee_{\lambda < \alpha(u)} \phi^{\rightarrow}u_{\lambda}, \ \Phi^{\rightarrow}\alpha = \bigwedge_{\alpha(u) < \lambda} \Phi^{\rightarrow}u^{\lambda}.$

(ii) $\phi^{\leftarrow}\beta = \bigwedge_{\beta(v) < \mu} \phi^{\leftarrow}v^{\mu}, \ \Phi^{\leftarrow}\beta = \bigvee_{\mu < \beta(v)} \Phi^{\leftarrow}v_{\mu}.$

Proof We know that by Lemma 1.2.4, for all $\alpha \in \mathcal{F}(U)$ we have

$$\alpha = \bigvee_{\lambda < \alpha(u)} u_{\lambda} \text{ and } \alpha = \bigwedge_{\alpha(u) < \lambda} u^{\lambda}.$$

Now, the proof is immediate from Theorem 7.3.39.

The following lemma formally gives us the connection between the presection and section of a t-fuzzy direlation.

Lemma 7.3.29 $\phi^{\leftarrow} = (\phi^{\leftarrow})^{\rightarrow}$ *and* $\Phi^{\leftarrow} = (\Phi^{\leftarrow})^{\rightarrow}$.

Proof First, for any fuzzy co-point $v^{\mu} \in \mathcal{F}(V)$, we prove that for all $u \in U$,

$$\phi^{\leftarrow}(v^{\mu})(u) = (\phi^{\leftarrow})^{\rightarrow}(v^{\mu})(u) \text{ and } \Phi^{\leftarrow}(v_{\mu})(u) = (\Phi^{\leftarrow})^{\rightarrow}(v_{\mu})(u).$$

By definition of presection, we have

$$\begin{aligned}\phi^{\leftarrow}(v^{\mu})(u) &= \bigvee\{t \in (0,1] \mid \phi(u,t,v') \leq v^{\mu}(v')\}\\ &= \bigvee\{t \in (0,1] \mid \phi(u,t,v) \leq \mu\}.\end{aligned}$$

By definition of section, we also get

$$\begin{aligned}(\phi^{\leftarrow})^{\rightarrow}(v^{\mu})(u) &= \bigvee\{t \in (0,1]) \mid v^{\mu}(v') < s \implies t \leq \phi^{\leftarrow}(v', s, u)\}\\ &= \bigvee\{t \in (0,1] \mid \mu < s \implies \phi(u,t,v) \leq s\}.\end{aligned}$$

Let $\phi(u,t,v) \leq \mu$ and $\mu < s$. Then $\phi(u,t,v) \leq s$ and so $\phi^{\leftarrow}(v^{\mu})(u) \leq (\phi^{\leftarrow})^{\rightarrow}(v^{\mu})(u)$. Now, let $\mu < s \implies \phi(u,t,v) \leq s$. Suppose that $\mu < \phi(u,t,v)$. Take a $z \in (0,1]$ such that $\mu < z < \phi(u,t,v)$. Then $\mu < z$ and $\phi(u,t,v) \nleq z$ which is a contradiction. Hence, we obtain $(\phi^{\leftarrow})^{\rightarrow}(v^{\mu})(u) \leq \phi^{\leftarrow}(v^{\mu})(u)$.

Now let $\beta \in \mathcal{F}(V)$. Then by Corollary 7.3.28 (i), we conclude that

$$\begin{aligned}\phi^{\leftarrow}\beta = \phi^{\leftarrow} \bigwedge_{\beta(v)<\mu} v^{\mu} &= \bigwedge_{\beta(v)<\mu} \phi^{\leftarrow} v^{\mu}\\ &= \bigwedge_{\beta(v)<\mu} (\phi^{\leftarrow})^{\rightarrow} v^{\mu}\\ &= (\phi^{\leftarrow})^{\rightarrow} \bigwedge_{\beta(v)<\mu} v^{\mu} = (\phi^{\leftarrow})^{\rightarrow}\beta.\end{aligned}$$

The proof of the second equality is similar.

Lemma 7.3.30 *Let* (ϕ, Φ) *be a t-fuzzy direlation from* $\mathcal{F}(U)$ *to* $\mathcal{F}(V)$. *Then for all* $u \in U$ *and* $v \in V$, *we have the following equalities:*

(i) $\phi^{\rightarrow}(u_{\lambda})(v) = \phi(u, \lambda, v)$, $\Phi^{\rightarrow}(u^{\lambda})(v) = \Phi(u, \lambda, v)$.

(ii) $\phi^{\leftarrow}(v^{\mu})(u) = \phi^{\leftarrow}(v, \mu, u)$, $\Phi^{\leftarrow}(v_{\mu})(u) = \Phi^{\leftarrow}(v, \mu, u)$.

Proof (i) Let u_{λ} be any fuzzy point in $\mathcal{F}(U)$. By definition of a section of a t-fuzzy relation, we have

$$\phi^{\rightarrow}(u_\lambda)(v) = \bigwedge\{\mu \in (0,1] \mid \lambda' < \lambda \implies \phi(u,\lambda',v) \le \mu\}.$$

Note that $\phi(u,\lambda,v) = \bigvee\{\phi(u,\lambda',v) \mid 0 < \lambda' < \lambda\}$. Let $0 < \lambda' < \lambda$. Then $\phi(u,\lambda',v) \le \phi(u,\lambda,v)$. Therefore, if we say $\mu = \phi(u,\lambda,v)$, then we find $\phi^{\rightarrow}(u_\lambda)(v) \le \phi(u,\lambda,v)$.

Take a $\mu \in (0,1]$ where $\lambda' < \lambda \implies \phi(u,\lambda',v) \le \mu$. Then clearly, we get $\phi(u,\lambda,v) \le \mu$ and so $\phi(u,\lambda,v) \le \phi^{\rightarrow}(u_\lambda)(v)$.

Now let u^λ be a fuzzy co-point in $\mathcal{F}(U)$. We show that $\Phi^{\rightarrow}(u^\lambda)(v) = \Phi(u,\lambda,v)$ where $\Phi(u,\lambda,v) = \bigwedge\{\Phi(u,\lambda',v) \mid \lambda < \lambda' \le 1\}$. By definition of a section of a t-fuzzy corelation, we have

$$\begin{aligned}\Phi^{\rightarrow}(u^\lambda)(v) &= \bigvee\{\mu \in (0,1] \mid u^\lambda(u') < s \implies \mu < \Phi(u',s,v)\}\\ &= \bigvee\{\mu \in (0,1] \mid \lambda < s \implies \mu < \Phi(u,s,v)\}.\end{aligned}$$

Let $\lambda < \lambda' \le 1$. Then $\Phi(u,\lambda,v) \le \Phi(u,\lambda',v)$ and so $\Phi(u,\lambda,v) \le \Phi^{\rightarrow}(u^\lambda)(v)$ for $\mu = \Phi(u,\lambda,v)$. Let $\mu \in (0,1]$ and $\lambda < s \implies \mu < \Phi(u,s,v)$. For $\lambda < \lambda' \le 1$, we get $\mu < \Phi(u,\lambda',v)$ and so $\Phi^{\rightarrow}(u^\lambda)(v) \le \Phi(u,\lambda,v)$.

(ii) By Theorem 4.2.12, we have $\phi^{\leftarrow}v^\mu = (\phi^{\leftarrow})^{\rightarrow}v^\mu$ and $\Phi^{\leftarrow}v_\mu = (\Phi^{\leftarrow})^{\rightarrow}v_\mu$. Then by (i), we obtain the equalities.

The following result will be useful in the sequel.

Theorem 7.3.31 *Let (ϕ,Φ) be a t-fuzzy direlation from $\mathcal{F}(U)$ to $\mathcal{F}(V)$ and $\alpha \in \mathcal{F}(U)$. Then we have the following equalities:*

(i)

$$\bigvee_{\lambda<\alpha(u)} \phi(u,\lambda,v) = \bigvee_{u\in U} \phi(u,\alpha(u),v).$$

(ii)

$$\bigwedge_{\alpha(u)<\lambda} \Phi(u,\lambda,v) = \bigwedge_{u\in U} \Phi(u,\alpha(u),v).$$

Proof (i) Let $\lambda < \alpha(u)$ and $u \in U$. Since ϕ is a t-fuzzy relation, we clearly have $\phi(u,\lambda,v) \le \phi(u,\alpha(u),v)$ and then we get $\bigvee_{\lambda<\alpha(u)} \phi(u,\lambda,v) \le \bigvee_{u\in U} \phi(u,\alpha(u),v)$.

Now, assume that

$$\bigvee_{\lambda<\alpha(u)} \phi(u,\lambda,v) < \bigvee_{u\in U} \phi(u,\alpha(u),v).$$

Then for some $\mu' \in (0,1]$,

$$\bigvee_{\lambda<\alpha(u)} \phi(u,\lambda,v) < \mu' < \bigvee_{u\in U} \phi(u,\alpha(u),v).$$

Then by the first inequality, we conclude that for all $u \in U$ and $\lambda \in (0, 1]$,

$$\lambda < \alpha(u) \implies \phi(u, \lambda, v) < \mu'. \qquad (*)$$

By the second, for some $u' \in U$

$$\mu' < \phi(u', \alpha(u'), v).$$

However, by definition of t-fuzzy relation,

$$\phi(u', \alpha(u'), v) = \bigwedge\{\phi(u', \lambda', v) \mid 0 < \lambda' < \alpha(u')\}$$

and therefore, for some $\lambda' < \alpha(u')$, we find $\mu' < \phi(u', \lambda', v)$. Then by the implication (*) above, we also get that $\phi(u', \lambda', v) < \mu'$. This gives an immediate contradiction.

(ii) Let $u \in U$ and $\alpha(u) < \lambda$. Since Φ is a t-fuzzy corelation, we have $\Phi(u, \alpha(u), v))$ $\leq \Phi(u, \lambda, v)$. Then we get $\bigwedge_{u \in U} \Phi(u, \alpha(u), v) \leq \bigwedge_{\alpha(u) < \lambda} \Phi(u, \lambda, v)$. Now, assume that

$$\bigwedge_{u \in U} \Phi(u, \alpha(u), v) < \lambda' < \bigwedge_{\alpha(u) < \lambda} \Phi(u, \lambda, v)$$

for some $\lambda' \in (0, 1]$. Then, following a similar way as in the first item we obtain a contradiction. □

The following results present alternative equalities for the sections and presections due to t-fuzzy direlations.

Corollary 7.3.32 *Let (ϕ, Φ) be a t-fuzzy direlation from $\mathcal{F}(U)$ to $\mathcal{F}(V)$. Then for all $\alpha \in \mathcal{F}(U)$ and $v \in V$, we have the following equalities:*

(i) $(\phi^{\to}\alpha)(v) = \bigvee_{u \in U} \phi(u, \alpha(u), v)$.

(ii) $(\Phi^{\to}\alpha)(v) = \bigwedge_{u \in U} \Phi(u, \alpha(u), v)$.

Proof In view of Corollary 7.3.28, Lemma 7.3.30 and Theorem 7.3.31, for all $\alpha \in \mathcal{F}(U)$ and $v \in V$, we conclude that

$$\begin{aligned}(\phi^{\to}\alpha)(v) &= (\bigvee_{\lambda < \alpha(u)} \phi^{\to}u_\lambda)(v) = \bigvee_{\lambda < \alpha(u)} \phi^{\to}u_\lambda(v) \\ &= \bigvee_{\lambda < \alpha(u)} \phi(u, \lambda, v) = \bigvee_{u \in U} \phi(u, \alpha(u), v)\end{aligned}$$

and

$$\begin{aligned}(\Phi^{\to}\alpha)(v) &= (\bigwedge_{\alpha(u) < \lambda} \Phi^{\to}u^\lambda)(v) = \bigwedge_{\alpha(u) < \lambda} \Phi^{\to}u^\lambda(v) \\ &= \bigwedge_{\alpha(u) < \lambda} \Phi(u, \lambda, v) = \bigwedge_{u \in U} \Phi(u, \alpha(u), v).\end{aligned}$$

Note that if Φ is a t-fuzzy corelation from $\mathcal{F}(U)$ to $\mathcal{F}(V)$, then $\Phi^{\leftarrow}$ is a t-fuzzy relation from $\mathcal{F}(V)$ to $\mathcal{F}(U)$ where

$$v \in V, \mu \in (0,1],\ \Phi^{\leftarrow}(v,\mu,u) = \bigwedge\{\lambda \in (0,1] \mid \mu \leq \Phi(u,\lambda,v)\}.$$

Further, $\phi^{\leftarrow}$ is a t-fuzzy corelation where

$$\phi^{\leftarrow}(v,\mu,u) = \bigvee\{\lambda \in (0,1] \mid \phi(u,\lambda,v) \leq \mu\}.$$

The proof of the following result is left to the interested reader.

Corollary 7.3.33 *Let (ϕ, Φ) be a t-fuzzy direlation from $\mathcal{F}(U)$ to $\mathcal{F}(V)$. Then for all $\beta \in \mathcal{F}(V)$ and $u \in U$, the following equalities hold:*

(i) $(\Phi^{\leftarrow}\beta)(u) = \bigvee_{v\in V} \Phi^{\leftarrow}(v, \beta(v), u).$

(ii) $(\phi^{\leftarrow}\beta)(u) = \bigwedge_{v\in V} \phi^{\leftarrow}(v, \beta(v), u).$

Complemented t-Fuzzy Direlations
Let (r, R) be a direlation from $(W_U, \mathcal{W}_U)$ to $(W_V, \mathcal{W}_V)$. We know that by Theorem 4.2.20, the pair (R', r') is also a direlation from $(W_U, \mathcal{W}_U)$ to $(W_V, \mathcal{W}_V)$. Then by Theorem 7.3.4, there exists a t-fuzzy direlation $(\phi_{R'}, \Phi_{r'})$ from $\mathcal{F}(U)$ to $\mathcal{F}(V)$ corresponding to (R', r'). The following result provides a fruitful formulation for the t-fuzzy direlation $(\phi_{R'}, \Phi_{r'})$.

Lemma 7.3.34 *For the complemented fuzzy texture*

$$(W_U, \mathcal{W}_U, c_1) = (U \times (0,1], \mathcal{P}(U) \otimes \mathcal{M}, c_1),$$

we have the following equivalences where $c_1 = c_{U\times(0,1]}$:

(i) $P_{(u_1,\lambda_1)} \not\subseteq c_1(P_{(u,\lambda)})$ *if and only if* $u = u_1$ *and* $1 - \lambda < \lambda_1$.

(ii) $c_1(Q_{(u,\lambda)}) \not\subseteq Q_{(u_1,\lambda_1)}$ *if and only if* $u = u_1$ *and* $\lambda_1 < 1 - \lambda$.

Proof (i) Let $P_{(u_1,\lambda_1)} \not\subseteq c(P_{(u,\lambda)})$. Note that

$$P_{(u_1,\lambda_1)} = \{u_1\} \times (0,\lambda_1] \text{ and } c(P_{(u,\lambda)}) = c_1(\{u\} \times (0,\lambda])).$$

Then by Corollary 3.2.10,

$$\begin{aligned}\{u_1\} \times (0,\lambda_1] \not\subseteq c_1(\{u\} \times (0,\lambda] &= c_1((\{u\} \times (0,1]) \cap (U \times (0,\lambda]) \\ &= (U \setminus \{u\} \times (0,1])) \cup (U \times (0, 1-\lambda]).\end{aligned}$$

Therefore, we have

$$\{u_1\} \times (0,\lambda_1] \not\subseteq (U \setminus \{u\} \times (0,1]) \text{ and } \{u_1\} \times (0,\lambda_1] \not\subseteq U \times (0, 1-\lambda].$$

This clearly implies that $u = u_1$ and $1 - \lambda < \lambda_1$ as it is desired. Sufficiency is similar.

(ii) Let $c_1(Q_{(u,\lambda)}) \not\subseteq Q_{(u_1,\lambda_1)}$. Note that

$$Q_{(u,\lambda)} = ((U \setminus \{u\}) \times (0, 1]) \cup (U \times (0, \lambda])$$

and

$$Q_{(u_1,\lambda_1)} = ((U \setminus \{u_1\}) \times (0, 1]) \cup (U \times (0, \lambda_1]).$$

Hence,

$$c_1(((U \setminus \{u\}) \times (0, 1]) \cup (U \times (0, \lambda])) \not\subseteq ((U \setminus \{u_1\}) \times (0, 1]) \cup (U \times (0, \lambda_1])),$$

that is,

$$((\{u\} \times (0, 1]) \cap (U \times (0, 1 - \lambda])) \not\subseteq ((U \setminus \{u_1\}) \times (0, 1]) \cup (U \times (0, \lambda_1]).$$

As a result, we obtain that

$$((\{u\} \times (0, 1]) \not\subseteq ((U \setminus \{u_1\}) \times (0, 1]) \cup (U \times (0, \lambda_1])$$

and

$$(U \times (0, 1 - \lambda])) \not\subseteq ((U \setminus \{u_1\}) \times (0, 1]) \cup (U \times (0, \lambda_1]).$$

This means that

$$u = u_1 \text{ and } \lambda_1 < 1 - \lambda.$$

Sufficiency is left to the reader.

Theorem 7.3.35 *For the t-fuzzy direlation* $(\phi_{R'}, \Phi_{r'})$*, we have*

$$\phi_{R'}(u, \lambda, v) = \begin{cases} 1 - \Phi_R(u, 1 - \lambda, v), & \lambda \neq 1 \\ \bigvee\{1 - \Phi_R(u, \lambda', v) \mid \lambda' \in (0, 1]\}, & \lambda = 1 \end{cases}$$

and

$$\Phi_{r'}(u, \lambda, v) = \begin{cases} 1 - \phi_r(u, 1 - \lambda, v), & \lambda \neq 1 \\ 1, & \lambda = 1 \end{cases}$$

for all $u, v \in U$ *and* $\lambda \in (0, 1]$.

Proof We give the proof of the first equality leaving the second to the interested reader. Let us consider the complementations $c_1 = c_{U\times(0,1]}$ and $c_2 = c_{V\times(0,1]}$ on the fuzzy textures $(W_U, \mathcal{W}_U)$ and $(W_V, \mathcal{W}_V)$, respectively. Since R' is a corelation from $(W_U, \mathcal{W}_U)$ to $(W_V, \mathcal{W}_V)$, by Theorem 7.3.4,

$$R' = \bigvee\{\overline{P}_{((u,\lambda),(v,\mu))} \mid \exists(u_1, \lambda_1), (v_1, \mu_1) \text{ with}$$
$$\overline{P}_{((u_1,\lambda_1),(v_1,\mu_1))} \not\subseteq R,\ P_{(u_1,\lambda_1)} \not\subseteq c_1(P_{(u,\lambda)}) \text{ and } c_2(Q_{(v,\mu)}) \not\subseteq Q_{(v_1,\mu_1)}\}.$$

By Lemma 7.3.34, we also have that

$$R' = \bigvee\{\overline{P}_{((u,\lambda),(v,\mu))} \mid \exists\lambda_1, \mu_1 \in (0, 1], \text{ with}$$
$$P_{((u,\lambda_1),(v,\mu_1))} \not\subseteq R,\ 1 - \lambda < \lambda_1 \text{ and } \mu_1 < 1 - \mu\}.$$

Therefore,

$$R' \nsubseteq \overline{Q}_{((u,\lambda),(v,\mu'))} \text{ if and only if } \exists \lambda_1, \mu_1, \overline{P}_{((u,\lambda_1),(v,\mu_1))} \nsubseteq R,$$
$$\exists \mu,\ \mu' < \mu < 1, 1-\lambda < \lambda_1 \text{ and } \mu_1 < 1-\mu.$$

In other words, by Corollary 7.3.3,

$$\mu' < \phi_{R'}(u, \lambda, v) \text{ if and only if } \exists \lambda_1, \mu_1, \Phi_R(u, \lambda_1, v) < \mu_1,$$
$$\exists \mu,\ \mu' < \mu < 1,\ \ 1-\lambda < \lambda_1 \text{ and } \mu_1 < 1-\mu. \quad (*)$$

Let $\lambda \neq 1$. Then $1-\lambda \in (0, 1]$. Let $\mu' < \phi_{R'}(u, \lambda, v)$ where $\mu' = \phi_{R'}(u, \lambda', v)$ and $0 < \lambda' < \lambda$. Then for some λ_1, μ_1 and μ, by Theorem 7.3.5 (i),

$$\Phi_R(u, 1-\lambda, v) \leq \Phi_R(u, \lambda_1, v) < \mu_1 < 1-\mu < 1-\mu'.$$

This implies that $\mu' < 1 - \Phi_R(u, 1-\lambda, v)$. That is,

$$\bigvee\{\phi_{R'}(u, \lambda', v) \mid 0 < \lambda' < \lambda\} = \phi_{R'}(u, \lambda, v) \leq 1 - \Phi_R(u, 1-\lambda, v).$$

Now suppose that

$$\phi_{R'}(u, \lambda, v) < 1 - \Phi_R(u, 1-\lambda, v).$$

Let us take a $\mu' \in (0, 1]$ such that

$$\phi_{R'}(u, \lambda, v) < \mu' < 1 - \Phi_R(u, 1-\lambda, v), \text{ and}$$

take μ, μ_1 with

$$\Phi_R(u, 1-\lambda, v) < \mu_1 < 1-\mu < 1-\mu'.$$

Note that $\mu' \neq 1$. Further, take a λ_1 where $1-\lambda < \lambda_1$. Then by Theorem 7.3.5 (i), we clearly have $\Phi_R(u, \lambda_1, v) < \mu_1$. Hence, the right side of the equivalence (*) holds and so we conclude that $\mu' < \phi_{R'}(u, \lambda, v)$ which is an immediate contradiction.

If $\lambda = 1$, one can follow the same way for the proof.

Definition 7.3.36 Let (ϕ, Φ) be a t-fuzzy direlation from $\mathcal{F}(U)$ to $\mathcal{F}(V)$. Then the pair $(\phi, \Phi)' = (\Phi', \phi')$ is called the complement of (ϕ, Φ) where

$$\Phi'(u, \lambda, v) = \begin{cases} 1 - \Phi(u, 1-\lambda, v), & \lambda \neq 1 \\ \bigvee\{1 - \Phi(u, \lambda', v) \mid \lambda' \in (0, 1]\}, & \lambda = 1. \end{cases}$$

and

$$\phi'(u, \lambda, v) = \begin{cases} 1 - \phi(u, 1-\lambda, v), & \lambda \neq 1 \\ 1, & \lambda = 1 \end{cases}$$

for all $u, v \in U$ and $\lambda \in (0, 1]$. We say (ϕ, Φ) is *complemented* if $\phi' = \Phi$ and $\Phi' = \phi$. Note that by Theorem 7.3.35, the pair (Φ', ϕ') is also a t-fuzzy direlation from $\mathcal{F}(U)$ to $\mathcal{F}(V)$.

Remark 7.3.37 Note that if (ϕ, Φ) is a fuzzy direlation from $\mathcal{F}(U)$ to $\mathcal{F}(V)$, then there exists a unique direlation (r, R) from $(W_U, \mathcal{W}_U)$ to $(W_V, \mathcal{W}_V)$ where

$$\phi = \phi_r \text{ and } \Phi = \Phi_R.$$

Therefore, the basic properties of fuzzy direlations with respect to sections and presections can be proved using two different ways. The first one is to give the solutions in their own right, that is to follow the basic definitions and interrelations due to fuzzy direlations, and the second is to use only the corresponding direlations.

For the second way, we need the following proposition.

Proposition 7.3.38 *Let (ϕ, Φ) be a fuzzy direlation from $\mathcal{F}(U)$ to $\mathcal{F}(V)$ and (r, R) be the direlation from $(W_U, \mathcal{W}_U)$ to $(W_V, \mathcal{W}_V)$ where $\phi = \phi_r$ and $\Phi = \Phi_R$.*

(i) *The following diagrams are commutative:*

$$\begin{array}{ccc} \mathcal{F}(U) & \xrightarrow{\ \phi^{\to}\ } & \mathcal{F}(V) \\ \Big\downarrow{\scriptstyle \Im} & & \Big\downarrow{\scriptstyle \Im} \\ \mathcal{P}(U)\otimes\mathcal{M} & \xrightarrow{\ r^{\to}\ } & \mathcal{P}(V)\otimes\mathcal{M} \end{array} \qquad \begin{array}{ccc} \mathcal{F}(U) & \xrightarrow{\ \Phi^{\to}\ } & \mathcal{F}(V) \\ \Big\downarrow{\scriptstyle \Im} & & \Big\downarrow{\scriptstyle \Im} \\ \mathcal{P}(U)\otimes\mathcal{M} & \xrightarrow{\ R^{\to}\ } & \mathcal{P}(V)\otimes\mathcal{M}. \end{array}$$

That is, for all $\alpha \in \mathcal{F}(U)$, we have

$$r^{\to}\Im(\alpha) = \Im(\phi^{\to}\alpha) \text{ and } R^{\to}\Im(\alpha) = \Im(\Phi^{\to}\alpha).$$

(ii) *The following diagrams are commutative:*

$$\begin{array}{ccc} \mathcal{F}(V) & \xrightarrow{\ \phi^{\leftarrow}\ } & \mathcal{F}(U) \\ \Big\downarrow{\scriptstyle \Im} & & \Big\downarrow{\scriptstyle \Im} \\ \mathcal{P}(V)\otimes\mathcal{M} & \xrightarrow{\ r^{\leftarrow}\ } & \mathcal{P}(U)\otimes\mathcal{M} \end{array} \qquad \begin{array}{ccc} \mathcal{F}(V) & \xrightarrow{\ \Phi^{\leftarrow}\ } & \mathcal{F}(U) \\ \Big\downarrow{\scriptstyle \Im} & & \Big\downarrow{\scriptstyle \Im} \\ \mathcal{P}(V)\otimes\mathcal{M} & \xrightarrow{\ R^{\leftarrow}\ } & \mathcal{P}(U)\otimes\mathcal{M}. \end{array}$$

That is, for all $\beta \in \mathcal{F}(V)$, we have

$$r^{\leftarrow}\Im(\beta) = \Im(\phi^{\leftarrow}\beta) \text{ and } R^{\leftarrow}\Im(\beta) = \Im(\Phi^{\leftarrow}\beta).$$

Proof As we know, a β-presection is also a section of the direlation $(R^{\leftarrow}, r^{\leftarrow})$ where

$$\Phi^{\leftarrow}\beta = (\Phi^{\leftarrow})^{\to}\beta \text{ and } \phi^{\leftarrow}\beta = (\phi^{\leftarrow})^{\to}\beta$$

for all $\beta \in \mathcal{F}(V)$. Therefore, the proof is immediate from Theorem 7.3.23.

Theorem 7.3.39 *Let* (ϕ, Φ) *be a t-fuzzy direlation from* $\mathcal{F}(U)$ *to* $\mathcal{F}(V)$. *Then we have the following equalities:*

(i) $\phi^{\rightarrow} \bigvee_{j \in J} \alpha_j = \bigvee_{j \in J} \phi^{\rightarrow} \alpha_j$ *and* $\Phi^{\leftarrow} \bigvee_{j \in J} \beta_j = \bigvee_{j \in J} \Phi^{\leftarrow} \beta_j$.

(ii) $\phi^{\leftarrow} \bigwedge_{j \in J} \beta_j = \bigwedge_{j \in J} \phi^{\leftarrow} \beta_j$ *and* $\Phi^{\rightarrow} \bigwedge_{j \in J} \alpha_j = \bigwedge_{j \in J} \Phi^{\rightarrow} \alpha_j$.

Proof We prove the equalities given in (i), leaving (ii) to the interested reader. Note that from Proposition 7.3.38 (i) and Theorem 4.2.17,

$$
\begin{aligned}
\mathfrak{I}(\phi^{\rightarrow} \bigvee_{j \in J} \alpha_j) &= r^{\rightarrow} \mathfrak{I}(\bigvee_{j \in J} \alpha_j) = r^{\rightarrow} \bigvee_{j \in J} \mathfrak{I}(\alpha_j) \\
&= \bigvee_{j \in J} r^{\rightarrow} \mathfrak{I}(\alpha_j) = \bigvee_{j \in J} \mathfrak{I}(\phi^{\rightarrow} \alpha_j) \\
&= \mathfrak{I}(\bigvee_{j \in J} \phi^{\rightarrow} \alpha_j),
\end{aligned}
$$

and since $\mathfrak{I}$ is a lattice isomorphism, we obtain that

$$
\phi^{\rightarrow} \bigvee_{j \in J} \alpha_j = \bigvee_{j \in J} \phi^{\rightarrow} \alpha_j.
$$

The rest of the proof is similar.

Theorem 7.3.40 *Let* (ϕ, Φ) *be a t-fuzzy direlation from* $\mathcal{F}(U)$ *to* $\mathcal{F}(V)$. *Then for all* $\alpha \in \mathcal{F}(U)$ *and* $\beta \in \mathcal{F}(V)$, *we have the following properties:*

(i) $\bigvee_{j \in J} \phi^{\leftarrow} \beta_j \leq \phi^{\leftarrow} \bigvee_{j \in J} \beta_j$ *and* $\Phi^{\leftarrow} \bigwedge_{j \in J} \beta_j \leq \bigwedge_{j \in J} \Phi^{\leftarrow} \beta_j$.

(ii) $\bigvee_{j \in J} \Phi^{\rightarrow} \alpha_j \leq \Phi^{\rightarrow} \bigvee_{j \in J} \alpha_j$ *and* $\phi^{\rightarrow} \bigwedge_{j \in J} \alpha_j \leq \bigwedge_{j \in J} \phi^{\rightarrow} \alpha_j$.

Proof From Proposition 7.3.38 and Theorem 4.2.17, the proof is straightforward.

Theorem 7.3.41 *Let* (ϕ, Φ) *be a t-fuzzy direlation from* $\mathcal{F}(U)$ *to* $\mathcal{F}(V)$. *Then*

(i) $\forall \alpha \in \mathcal{F}(U),\ \phi^{\rightarrow}(\phi^{\leftarrow}(\phi^{\rightarrow}\alpha)) = \phi^{\rightarrow}\alpha$.

(ii) $\forall \alpha \in \mathcal{F}(U),\ \Phi^{\rightarrow}(\phi^{\leftarrow}(\Phi^{\rightarrow}\alpha)) = \Phi^{\rightarrow}\alpha$.

(iii) $\forall \beta \in \mathcal{F}(V),\ \phi^{\leftarrow}(\phi^{\rightarrow}(\phi^{\leftarrow}\beta)) = \phi^{\leftarrow}\beta$.

(iv) $\forall \beta \in \mathcal{F}(V),\ \Phi^{\leftarrow}(\Phi^{\rightarrow}(\Phi^{\leftarrow}\beta)) = \Phi^{\leftarrow}\beta$.

Proof (i) From Theorem 4.2.19 (i) for all $\Im(\alpha) \in \mathcal{W}_U$, we have

$r^{\rightarrow}(r^{\leftarrow}(r^{\rightarrow}\Im(\alpha))) = r^{\rightarrow}\Im(\alpha)$.

Then by Theorem 7.3.38, we get

$$\begin{aligned}\Im(\phi^{\rightarrow}(\phi^{\leftarrow}(\phi^{\rightarrow}\alpha))) &= r^{\rightarrow}\Im((\phi^{\leftarrow}(\phi^{\rightarrow}\alpha)) = r^{\rightarrow}r^{\leftarrow}(\Im(\phi^{\rightarrow}\alpha)) \\ &= r^{\rightarrow}(r^{\leftarrow}(r^{\rightarrow}\Im(\alpha))) = r^{\rightarrow}\Im(\alpha) = \Im(\phi^{\rightarrow}\alpha)\end{aligned}$$

and so since $\Im$ is a lattice isomorphism, we obtain that

$\phi^{\rightarrow}(\phi^{\leftarrow}(\phi^{\rightarrow}\alpha)) = \phi^{\rightarrow}\alpha$.

The rest of the proof is similar.

Theorem 7.3.42 *Let* (ϕ, Φ) *be a t-fuzzy direlation from* $\mathcal{F}(U)$ *to* $\mathcal{F}(V)$

(1) (i) $\phi^{\rightarrow}\mathbf{0} = \mathbf{0}$.

(ii) $\forall\alpha \in \mathcal{F}(U),\ \alpha \leq \phi^{\leftarrow}(\phi^{\rightarrow}\alpha)$.

$\forall\beta \in \mathcal{F}(V),\ \phi^{\rightarrow}(\phi^{\leftarrow}\beta) \leq \beta$.

(2) (i) $\Phi^{\rightarrow}\mathbf{1} = \mathbf{1}$.

(ii) $\forall\beta \in \mathcal{F}(V),\ \beta \leq \Phi^{\rightarrow}(\Phi^{\leftarrow}\beta)$.

$\forall\alpha \in \mathcal{F}(U),\ \Phi^{\leftarrow}(\Phi^{\rightarrow}\alpha) \leq \alpha$.

Proof (1)(i) Using the definition of a section,

$$\begin{aligned}\forall v \in V,\ \ (\phi^{\rightarrow}\mathbf{0})(v) &= \bigwedge\{\mu \in (0,1] \mid \lambda < \mathbf{0}(u) \implies \phi(u,\lambda,v) \leq \mu\} \\ &= \bigwedge\{\mu \in (0,1] \mid \lambda < 0 \implies \phi(u,\lambda,v) \leq \mu\} \\ &= \bigwedge_{\mu\in(0,1]} \mu = 0 = \mathbf{0}(v).\end{aligned}$$

(ii) Let us consider the direlation (r, R) from $(W_U, \mathcal{W}_U)$ to $(W_V, \mathcal{W}_V)$ where

$\phi = \phi_r$ and $\Phi = \Phi_R$.

Then for $\alpha \in \mathcal{F}(U)$, by Theorem 4.2.15(1)(ii), we already have

$\Im(\alpha) \subseteq r^{\leftarrow}r^{\rightarrow}\Im(\alpha) = r^{\leftarrow}\Im(\phi^{\rightarrow}\alpha) = \Im(\phi^{\leftarrow}(\phi^{\rightarrow}\alpha))$.

Since I is lattice isomorphism from $\mathcal{F}(U)$ to $\mathcal{W}_U$, then we obtain

$\alpha \leq \phi^{\leftarrow}(\phi^{\rightarrow}\alpha)$.

For the second inequality, by Theorem 4.2.15(1)(iii), we also easily observe that for all $\beta \in \mathcal{F}(V)$,

$$\Im(\phi^{\rightarrow}(\phi^{\leftarrow}\beta)) = r^{\rightarrow}\Im(\phi^{\leftarrow}\beta) = r^{\rightarrow}r^{\leftarrow}\Im(\beta) \subseteq \Im(\beta).$$

Finally, we obtain that $\phi^{\rightarrow}(\phi^{\leftarrow}\beta) \leq \beta$.

(ii) The proof is similar and it is left to the interested reader.

Theorem 7.3.43 *Let (ϕ, Φ) be a t-fuzzy direlation from $\mathcal{F}(U)$ to $\mathcal{F}(V)$ and $(\phi, \Phi)' = (\Phi', \phi')$ be the complement of it. Then the following equalities hold:*

(i) $\forall\alpha \in \mathcal{F}(U),\ (\phi')^{\rightarrow}\alpha = \mathbf{1} - \phi^{\rightarrow}(\mathbf{1} - \alpha)$,

$\forall\alpha \in \mathcal{F}(U),\ (\Phi')^{\rightarrow}\alpha = \mathbf{1} - \Phi^{\rightarrow}(\mathbf{1} - \alpha)$.

(ii) $\forall\beta \in \mathcal{F}(V),\ (\phi')^{\leftarrow}\beta = \mathbf{1} - \phi^{\leftarrow}(\mathbf{1} - \beta)$,

$\forall\beta \in \mathcal{F}(V),\ (\Phi')^{\leftarrow}\beta = \mathbf{1} - \Phi^{\leftarrow}(\mathbf{1} - \beta)$.

Proof We prove (i). The proof of the rest of the items is similar. Let $v \in V, \alpha \in \mathcal{F}(U)$ and $\alpha(u) \neq 1$ for some $u \in U$. Then by Definition 7.3.36, we have that

$$\phi'(u, \alpha(u), v) = 1 - \phi(u, 1 - \alpha(u), v).$$

Therefore, by Corollary 7.3.32 (ii),

$$\begin{aligned}
((\phi')^{\rightarrow}\alpha)(v) &= \bigwedge_{u\in U} \phi'(u, \alpha(u), v)\\
&= \bigwedge_{u\in U} 1 - \phi(u, 1 - \alpha(u), v) \quad (\text{for } \lambda \neq 1)\\
&= 1 - \bigvee_{u\in U} \phi(u, 1 - \alpha(u), v)\\
&= 1 - \bigvee_{u\in U} (\phi(u, (\mathbf{1} - \alpha)(u), v))\\
&= 1 - (\phi^{\rightarrow}(\mathbf{1} - \alpha)(v)) = (\mathbf{1} - \phi^{\rightarrow}(\mathbf{1} - \alpha))(v).
\end{aligned}$$

If $\alpha(u) = 1$, then by Theorem 7.3.5 (ii), $\phi'(u, \alpha(u), v) = \phi'(u, 1, v) = 1$. Then for all $v \in V$,

$$((\phi')^{\rightarrow}\alpha)(v) = \bigwedge_{u\in U} \phi'(u, 1, v) = 1.$$

Finally, by Theorem 7.3.42 (1)(i), we obtain that

$$\begin{aligned}
(\mathbf{1} - \phi^{\rightarrow}(\mathbf{1} - \alpha(u)))(v) &= (\mathbf{1} - \phi^{\rightarrow}(\mathbf{1} - \mathbf{1}))(v)\\
&= (\mathbf{1} - \phi^{\rightarrow}(\mathbf{0}))(v)\\
&= (\mathbf{1} - \mathbf{0})(v) = \mathbf{1}(v) = 1.
\end{aligned}$$

Corollary 7.3.44 *Let (ϕ, Φ) be a complemented t-fuzzy direlation from $\mathcal{F}(U)$ to $\mathcal{F}(V)$. Then we have*

(i) $\forall\alpha \in \mathcal{F}(U),\ \Phi^{\rightarrow}\alpha = \mathbf{1} - \phi^{\rightarrow}(\mathbf{1} - \alpha)$,

$\forall \alpha \in \mathcal{F}(U),\ \phi^{\rightarrow}\alpha = \mathbf{1} - \Phi^{\rightarrow}(\mathbf{1} - \alpha)$.

(ii) $\forall \beta \in \mathcal{F}(V),\ \Phi^{\leftarrow}\beta = \mathbf{1} - \phi^{\leftarrow}(\mathbf{1} - \beta)$,

$\forall \beta \in \mathcal{F}(V),\ \phi^{\leftarrow}\beta = \mathbf{1} - \Phi^{\leftarrow}(\mathbf{1} - \beta)$.

Proof Since, (ϕ, Φ) is complemented, $\phi' = \Phi$ and $\phi = \Phi'$ and then the proof is immediate by Theorem 7.3.39.

In fact, the following theorem is an immediate result of Theorem 7.3.35 and Definition 7.3.36.

Theorem 7.3.45 *Let (ϕ, Φ) be a t-fuzzy direlation from $\mathcal{F}(U)$ to $\mathcal{F}(V)$ and $(\phi, \Phi)' = (\Phi', \phi')$ be the complement of it.*

(i) $(\phi')' = \phi$ *and* $(\Phi')' = \Phi$.

(ii) $(\phi')^{\leftarrow} = (\phi^{\leftarrow})'$ *and* $(\Phi')^{\leftarrow} = (\Phi^{\leftarrow})'$.

(iii) *The identity fuzzy direlation (ϕ_i, Φ_I) on $\mathcal{F}(U)$ is complemented.*

Proof Let us consider the direlation (r, R) from $(W_U, \mathcal{W}_U)$ to $(W_V, \mathcal{W}_V)$ where

$\phi = \phi_r$ and $\Phi = \Phi_R$.

(i) From Theorem 7.3.35 and Definition 7.3.36, if $\phi = \phi_r$, then $\phi' = \Phi_{r'}$ and so by the same reason $(\phi')' = \phi_{(r')'}$. However, by Proposition 4.2.29(i), we have $(r')' = r$ and then $(\phi')' = \phi_r = \phi$. The proof of the second equality is similar.

(ii) Note that $\phi' = \Phi_{r'}$ and again in view of Theorem 7.3.35, Definition 7.3.36 and Proposition 4.2.29(ii), we conclude that

$$(\phi')^{\leftarrow} = (\Phi_{r'})^{\leftarrow} = (\Phi_{(r')^{\leftarrow}} = \Phi_{(r^{\leftarrow})'} = (\Phi_{r^{\leftarrow}})' = ((\phi_r)^{\leftarrow})' = (\phi^{\leftarrow})'.$$

The second equality is similar.

(iii) Note that by Theorem 4.2.23, (i, I) is complemented, that is $i' = I$ and $i = I'$. Therefore,

$(\phi_i)' = \Phi_{i'} = \Phi_I$ and $(\Phi_I)' = \phi_{I'} = \phi_i$,

that is, (ϕ_i, Φ_I) is complemented as desired.

7.4 Fuzzy Direlations Defined by Fuzzy Logic Connectives

Continuity of binary operations defined on the product $[0, 1] \times [0, 1]$ provide us an important class of fuzzy direlations. For every fuzzy relation $\varphi : U \times V \rightarrow [0, 1]$, there exists a t-fuzzy direlation in terms of a continuous fuzzy logic connective.

Theorem 7.4.1 (i) *If* $F : [0,1] \times [0,1] \to [0,1]$ *is an increasing and left-continuous operation in the second component, then the mapping* $\phi : U \times (0,1] \times V \to [0,1]$ *defined by*

$$\phi(u, \lambda, v) = F(\varphi(u, v), \lambda),$$

for all $u \in U$, $v \in V$ *and* $\lambda \in (0,1]$, *is a t-fuzzy relation.*

(ii) *If* $G : [0,1] \times [0,1] \to [0,1]$ *is an increasing and right-continuous operation in the second component, then the mapping* $\Phi : U \times (0,1] \times V \to [0,1]$ *defined by*

$$\Phi(u, \lambda, v) = G(\varphi(u, v), \lambda),$$

for all $u \in U$, $v \in V$ *and* $\lambda \in (0,1]$, *is a t-fuzzy corelation.*

Proof (i) Since F is left-continuous in the second component, by Lemma 7.2.12, it is also infinitely join-distributive in the second component, and so we immediately have that

$$\begin{aligned}\bigvee_{0<\lambda'<\lambda} \phi(u, \lambda', v) &= \bigvee_{0<\lambda'<\lambda} F(\varphi(u, v), \lambda') \\ &= F(\varphi(u, v), \bigvee_{0<\lambda'<\lambda} \lambda') \\ &= F(\varphi(u, v), \lambda) = \phi(u, \lambda, v),\end{aligned}$$

that is, ϕ is a t-fuzzy relation.

(ii) Since G is right-continuous in the second component, by Lemma 7.2.12, it is also infinitely meet-distributive in the second component and hence, again by Lemma 7.2.12, we observe that

$$\begin{aligned}\bigwedge_{\lambda<\lambda'\leq 1} \Phi(u, \lambda', v) &= \bigwedge_{\lambda<\lambda'\leq 1} G(\varphi(u, v), \lambda') \\ &= G(\varphi(u, v), \bigwedge_{\lambda<\lambda'\leq 1} \lambda') \\ &= G(\varphi(u, v), \lambda) = \Phi(u, \lambda, v).\end{aligned}$$

Therefore, Φ is a t-fuzzy corelation.

Fuzzy logic connectives are also binary operations and in particular, a t-norm and an implicator are increasing binary operations in the second component. This gives the following remarkable result.

Theorem 7.4.2 (i) *If* T *is a left-continuous t-norm, then the mapping* $\phi : U \times (0,1] \times V \to [0,1]$ *defined by*

$$\phi(u, \lambda, v) = T(\varphi(u, v), \lambda)$$

for all $u \in U$, $v \in V$ *and* $\lambda \in (0,1]$ *is a t-fuzzy relation.*

(ii) *Let I be a right-continuous implicator in the second component. Then the mapping $\Phi : U \times (0, 1] \times V \to [0, 1]$ defined by*

$$\Phi(u, \lambda, v) = I(\varphi(u, v), \lambda)$$

for all $u \in U$, $v \in V$ and $\lambda \in (0, 1]$ is a t-fuzzy corelation.

(iii) *The pair (ϕ, Φ) is a t-fuzzy direlation from $\mathcal{F}(U)$ to $\mathcal{F}(V)$.*

Proof Since T and I are increasing in the second component, by Theorem 7.4.1, the pair (ϕ, Φ) is fuzzy direlation. □

Example 7.4.3 Let us consider the Łukasiewicz t-norm

$$T_W(\eta, \mu) = \max\{\eta + \mu - 1, 0\}$$

and the Gödel implicator

$$I_{T_M}(\eta, \mu) = \begin{cases} 1, & \text{if } \eta \leq \mu \\ \mu, & \text{otherwise.} \end{cases}$$

Then for every fuzzy relation $\varphi \in \mathcal{F}(U \times V)$, the pair (ϕ, Φ) is a fuzzy direlation where

$$\phi(u, \lambda, v) = (\varphi(u, v) + \lambda - 1) \vee 0$$

and

$$\Phi(u, \lambda, v) = \begin{cases} 1, & \text{if } \varphi(u, v) \leq \lambda \\ \lambda, & \text{otherwise} \end{cases}$$

for all $(u, \lambda, v) \in U \times (0, 1] \times V$.

If I is a right-continuous implicator, we obtain two different fuzzy direlations as follows.

Theorem 7.4.4 *Let I be a right-continuous implicator in the second component.*

(i) *The pair (T_I, I) constitutes a t-fuzzy direlation (ϕ, Φ) where*

$$\phi(u, \lambda, v) = T_I(\varphi(u, v), \lambda), \;\; \Phi(u, \lambda, u) = I(\varphi(u, v), \lambda)$$

for all $u \in U$, $v \in V$ and $\lambda \in (0, 1]$.

(ii) *The pair $(C_{I,N}, I)$ gives a t-fuzzy direlation (ϕ, Φ) defined by*

$$\phi(u, \lambda, v) = C_{I,N}(\varphi(u, v), \lambda), \;\; \Phi(u, \lambda, v) = I(\varphi(u, v), \lambda),$$

for all $u \in U$, $v \in V$ and $\lambda \in (0, 1]$.

Proof (i) Since T_I and $C_{I,N}$ are left-continuous and increasing mappings in the second component, the proof is immediate by Theorem 7.4.1.

Example 7.4.5 (i) Since the Gödel implicator I_{T_M} is right-continuous in the second component, by Theorem 7.4.4 (i), the pair (ϕ, Φ) is a t-fuzzy direlation where

$$\phi(u, \lambda, v) = \bigwedge\{\mu \in [0, 1] \mid \lambda \leq I_{T_M}(\varphi(u, v), \mu)\} = \lambda$$

and

$$\Phi(u, \lambda, v) = \begin{cases} 1, & \text{if } \varphi(u, v) \leq \lambda \\ \lambda, & \text{otherwise} \end{cases}$$

for all $(u, \lambda, v) \in U \times (0, 1] \times V$.

(ii) By Theorem 7.4.4 (ii), the pair (ϕ, Φ) is also a t-fuzzy direlation where

$$\phi(u, \lambda, v) = \begin{cases} 0, & \text{if } \varphi(u, v) \geq \lambda \\ 1 - \lambda, & \text{otherwise} \end{cases}$$

and

$$\Phi(u, \lambda, v) = \begin{cases} 1, & \text{if } \varphi(u, v) \leq \lambda \\ \lambda, & \text{otherwise} \end{cases}$$

for all $(u, \lambda, v) \in U \times (0, 1] \times V$.

Theorem 7.4.6
(1) *If ϕ is a t-fuzzy relation determined by a t-norm T as in Theorem 7.4.2 (i), then we have the following equalities:*

(i) $\forall u \in U, v \in V$ *and* $\mu \in (0, 1]$, $\phi^{\leftarrow}(v, \mu, u) = I_T(\varphi(u, v), \mu)$.

(ii) $\forall u \in U, v \in V$ *and* $\lambda \in (0, 1]$, $\phi'(u, \lambda, v) = I_{D,N}(\varphi(u, v), \lambda)$ *where* $D = T'$.

(2) *If Φ is a t-fuzzy corelation defined by an implicator I as in Theorem 7.4.2 (ii), then*

(i) $\forall u \in U, v \in V$ *and* $\mu \in (0, 1]$, $\Phi^{\leftarrow}(v, \mu, u) = T_I(\varphi(u, v), \mu)$.

(ii) $\forall u \in U, v \in V$ *and* $\lambda \in (0, 1]$, $\Phi'(u, \lambda, v) = C_{I,N}(\varphi(u, v), \lambda)$.

Proof (1) (i)

$$\begin{aligned} \phi^{\leftarrow}(v, \mu, u) &= \bigvee\{\lambda \in (0, 1] \mid \phi(u, \lambda, v) \leq \mu\} \\ &= \bigvee\{\lambda \in (0, 1] \mid T(\varphi(u, v), \lambda) \leq \mu\}. \\ &= I_T(\varphi(u, v), \mu) \end{aligned}$$

(ii)

$$\begin{aligned}\phi'(u,\lambda,v) &= 1-\phi(\varphi(u,v),1-\lambda)\\ &= 1-T(\varphi(u,v),1-\lambda)\\ &= T'(1-\varphi(u,v),\lambda)\\ &= D(1-\varphi(u,v),\lambda)\\ &= I_{D,N}(\varphi(u,v),\lambda).\end{aligned}$$

(2) (i)

$$\begin{aligned}\Phi^{\leftarrow}(v,\mu,u) &= \bigwedge\{\lambda\in(0,1]\mid \mu\le\Phi(u,\lambda,v)\}\\ &= \bigwedge\{\lambda\in(0,1]\mid \mu\le I(\varphi(u,v),\lambda)\}\\ &= T_I(\varphi(u,v),\mu).\end{aligned}$$

(ii) For $\lambda\neq 1$,

$$\begin{aligned}\Phi'(u,\lambda,v) &= 1-\Phi(u,1-\lambda,v)\\ &= 1-I(\varphi(u,v),1-\lambda)\\ &= C_{I,N}(\varphi(u,v),\lambda).\end{aligned}$$

Further, for $\lambda=1$, we also have

$$\begin{aligned}\Phi'(u,1,v) &= \bigvee\{1-\Phi(u,\mu,v)\mid \mu\in(0,1]\}\\ &= \bigvee\{1-I(\varphi(u,v),\mu)\mid \mu\in(0,1]\}\\ &= 1-\bigwedge\{I(1-\varphi(u,v),\mu)\mid \mu\in(0,1]\}\\ &= 1-I(1-\varphi(u,v),\bigwedge_{\mu\in(0,1]}\mu)\\ &= 1-I(1-\varphi(u,v),0)\\ &= C_{I,N}(\varphi(u,v),1).\end{aligned}$$

Example 7.4.7 (i) By Theorem 7.4.6(1)(i), for

$$\phi(u,v,\lambda)=T_W(\varphi(u,v),\lambda)=(\varphi(u,v)+\lambda-1)\vee 0,$$

the t-fuzzy corelation is

$$\phi^{\leftarrow}(v,\mu,u)=\bigwedge\{\lambda\in(0,1]\mid \varphi(u,v)+\lambda-1\le\mu\},$$

and so $(\phi,\phi^{\leftarrow})$ is a t-fuzzy direlation.

(ii) For all $(u,\lambda,v)\in U\times(0,1)\times V$, let

$$\phi(u,\lambda,v)=T_P(\varphi(u,v),\lambda)=\varphi(u,v)\lambda.$$

Then by Theorem 7.4.6(1)(ii),

$$\phi'(u, \lambda, v) = 1 - \varphi(u, v) + \varphi(u, v)\lambda$$

is the corresponding t-fuzzy corelation. □

Using the duality of fuzzy logic connectives, we may also obtain the complemented t-fuzzy direlations.

Theorem 7.4.8 *Let I be a right-continuous S-implicator based on a t-conorm D and strong negator N, that is, $I = I_{D,N}$. Further, let $T = D'$. If ϕ and Φ are defined as in Theorem 7.4.2 (i) and (ii) respectively, that is, (ϕ, Φ) is determined by $(T, I_{D,N})$, then we have the following propositions:*

(a) *The pair (ϕ, Φ) is a complemented t-fuzzy direlation.*

(b) *The t-fuzzy inverse direlation $(\Phi^{\leftarrow}, \phi^{\leftarrow})$ holds the equalities*

$$\Phi^{\leftarrow}(v, \mu, u) = C_{I_T,N}(\varphi(u, v), \mu), \text{ and}$$

$$\phi^{\leftarrow}(v, \mu, u) = I_T(\varphi(u, v), \mu)$$

for all $u \in U$, $v \in V$ and $\mu \in (0, 1]$ where I_T is an R-implicator based on t-norm T and $C_{I_T,N}$ is the induced conjunctor of I_T and the standard negator N. Further, $(\Phi^{\leftarrow}, \phi^{\leftarrow})$ is also complemented.

Proof (a) Under the given conditions, we have the equations

$$\phi(u, \lambda, v) = T(\varphi(u, v), \lambda) \text{ and } \Phi(u, \lambda, v) = I_{D,N}(\varphi(u, v), \lambda)$$

where $T = D'$. Note that

$$\begin{aligned}\phi'(u, \lambda, v) &= 1 - \phi(u, 1 - \lambda, v)\\ &= 1 - T(\varphi(u, v), 1 - \lambda)\\ &= 1 - T(1 - (1 - \varphi(u, v), 1 - \lambda)\\ &= T'(1 - \varphi(u, v), \lambda) = D(1 - \varphi(u, v), \lambda)\\ &= I_{D,N}(\varphi(u, v), \lambda) = \Phi(u, \lambda, v)\end{aligned}$$

and for $\lambda \neq 1$,

$$\begin{aligned}\Phi'(u, \lambda, v) &= 1 - \Phi(u, 1 - \lambda, v)\\ &= 1 - I_{D,N}(\varphi(u, v), 1 - \lambda)\\ &= 1 - D(1 - \varphi(u, v), 1 - \lambda)\\ &= D'(\varphi(u, v), \lambda) = T(\varphi(u, v), \lambda)\\ &= \phi(u, \lambda, v).\end{aligned}$$

Further, for $\lambda = 1$, we also have

$$\begin{aligned}\Phi'(u,1,v) &= \bigvee\{1-\Phi(u,\mu,v) \mid \mu \in (0,1]\}\\ &= \bigvee\{1-I_{D,N}(\varphi(u,v),\mu) \mid \mu \in (0,1]\}\\ &= \bigvee\{1-D(1-\varphi(u,v),\mu) \mid \mu \in (0,1]\}\\ &= 1-\bigwedge\{D(1-\varphi(u,v),\mu) \mid \mu \in (0,1]\}\\ &= 1-D(1-\varphi(u,v),0)\\ &= 1-D(1-\varphi(u,v),1-1)\\ &= D'(\varphi(u,v),1) = T(\varphi(u,v),1) = \phi(u,1,v).\end{aligned}$$

That is, (ϕ, Φ) is complemented.

(b) Note that by definition of inverse t-fuzzy relation and inverse t-fuzzy corelation, we immediately get that

$$\begin{aligned}&\phi^{\leftarrow}(v,\mu,u) = \bigvee\{\lambda \in (0,1] \mid \phi(u,\lambda,v) \le \mu\}\\ &= \bigvee\{\lambda \in (0,1] \mid T(\varphi(u,v),\lambda) \le \mu\} = I_T(\varphi(u,v),\mu)\end{aligned}$$

and

$$\begin{aligned}\Phi^{\leftarrow}(v,\mu,u) &= \bigwedge\{\lambda \in (0,1] \mid \mu \le \Phi(u,\lambda,v)\}\\ &= \bigwedge\{\lambda \in (0,1] \mid \mu \le I_{D,N}(\varphi(u,v),\lambda)\}\\ &= \bigwedge\{\lambda \in (0,1] \mid \mu \le D(1-\varphi(u,v),\lambda)\}\\ &= I_D(1-\varphi(u,v),\mu) = I_T'(1-\varphi(u,v),\mu)\\ &= 1-I_T(\varphi(u,v),1-\mu)\\ &= C_{I_T,N}(\varphi(u,v),\mu).\end{aligned}$$

Theorem 7.4.9

Let (ϕ, Φ) be a t-fuzzy direlation defined by the pair $(C_{I,N}, I)$ as in Theorem 7.4.4(ii). Then the following propositions hold:

(a) *(ϕ, Φ) is a complemented t-fuzzy direlation.*

(b) *The inverse t-fuzzy direlation $(\Phi^{\leftarrow}, \phi^{\leftarrow})$ satisfies the equalities*

$\Phi^{\leftarrow}(v,\mu,u) = T_I(\varphi(u,v),\mu)$, *and*

$\phi^{\leftarrow}(v,\mu,u) = I_{C_{I,N}}(\varphi(u,v),\mu)$

for all $u \in U$, $v \in V$ and $\mu \in (0,1]$. Further, $(\Phi^{\leftarrow}, \phi^{\leftarrow})$ is also complemented.

Proof (a) Let us prove that (ϕ, Φ) is complemented. Indeed,

$$\begin{aligned}&\phi'(u,\lambda,v) = 1-\phi(u,1-\lambda,v) = 1-C_{I,N}(\varphi(u,v),1-\lambda)\\ &= 1-(1-I(\varphi(u,v),\lambda) = I(\varphi(u,v),\lambda) = \Phi(u,\lambda,v).\end{aligned}$$

Using a similar way, we may also obtain that $\Phi' = \phi$.

(b) Note that

$$\begin{aligned}\Phi^{\leftarrow}(v, \mu, u) &= \bigwedge\{\lambda \in (0,1] \mid \mu \leq \Phi(u, \lambda, v)\} \\ &= \bigwedge\{\lambda \in (0,1] \mid \mu \leq I(\varphi(u, v), \lambda)\} \\ &= T_I(\varphi(u, v), \mu)\end{aligned}$$

and

$$\begin{aligned}\phi^{\leftarrow}(v, \mu, u) &= \bigvee\{\lambda \in (0,1] \mid \phi(u, \lambda, v) \leq \mu\} \\ &= \bigvee\{\lambda \in (0,1] \mid C_{I,N}(\varphi(u, v), \lambda) \leq \mu\} \\ &= I_{C_{I,N}}(\varphi(u, v), \mu).\end{aligned}$$

Note that fuzzy logic connectives corresponding to inverse operations and complementation are given in Table 7.8.

Example 7.4.10 If T is a left-continuous t-norm, then the R-implicator I_T is right-continuous in the second component (Theorem 2.5.7, [2]). Then by Theorem 7.4.2, every pair (ϕ, Φ) given in Table 7.6 constitutes a t-fuzzy direlation. Further, every pair (ϕ, Φ) given in Table 7.7 constitutes a complemented t-fuzzy direlation (Table 7.8).

Table 7.6 Theorem 7.4.2: t-Fuzzy direlations (ϕ, Φ) with respect to t-norms and implicators

t-Fuzzy relations: ϕ	t-Fuzzy corelations: Φ
$T_M(\varphi(u, v), \lambda)$	$I_{T_M}(\varphi(u, v), \lambda)$
$T_P(\varphi(u, v), \lambda)$	$I_{T_P}(\varphi(u, v), \lambda)$
$T_W(\varphi(u, v), \lambda)$	$I_{T_W}(\varphi(u, v), \lambda)$

Table 7.7 Theorems 7.4.8 and 7.4.9: Complemented t-fuzzy direlations

t-Fuzzy relations: ϕ	t-Fuzzy corelations: Φ	
$T_M(\varphi(u, v), \lambda)$	$I_{D_M,N}(\varphi(u, v), \lambda)$	Gödel
$T_P(\varphi(u, v), \lambda)$	$I_{D_P,N}(\varphi(u, v), \lambda)$	Goguen
$T_W(\varphi(u, v), \lambda)$	$I_{D_W,N}(\varphi(u, v), \lambda)$	Łukasiewicz
$T(\varphi(u, v), \lambda)$	$I_{D,N}(\varphi(u, v), \lambda)$	$T = D'$
$C_{I,N}(\varphi(u, v), \lambda)$	$I(\varphi(u, v), \lambda)$	
$T_I(\varphi(u, v), \lambda)$	$I_{C_I,N}(\varphi(u, v), \lambda)$	

Table 7.8 Inverse operation, complementation and corresponding fuzzy logic connectives: if the t-fuzzy relation is complemented, then the corresponding fuzzy rough set systems are same

(ϕ, Φ)	$(\phi, \Phi)^{\leftarrow}$		(Φ', ϕ')	$\Phi = (\Phi^{\leftarrow})^{\leftarrow}$	$\phi = (\phi^{\leftarrow})^{\leftarrow}$	
(T, I)	(T_I, I_T)		$(C_{I,N}, I_{D,N})$	$I = I_{T_I}$	$T = T_{I_T}$	$T' = D$
$(T, I_{D,N})$	$(C_{I_{T,N}}, I_T)$	$T = D'$	$(T, I_{D,N})$			
$(C_{I,N}, I)$	$(T_I, I_{C_I,N})$		$(C_{I,N}, I)$			

7.5 Adjointness and Duality

As we know if (ϕ, Φ) is a t-fuzzy direlation, then the inverse $(\Phi^{\leftarrow}, \phi^{\leftarrow})$ is also a t-fuzzy direlation.

Theorem 7.5.1

(i) *The t-fuzzy corelation $\phi^{\leftarrow}$ is an adjoint of ϕ, that is, $(\phi^{\rightarrow}, \phi^{\leftarrow})$ is a Galois connection.*

(ii) *The t-fuzzy corelation $\Phi^{\leftarrow}$ is an adjoint of Φ, that is, $(\Phi^{\rightarrow}, \Phi^{\leftarrow})$ is a Galois connection.*

Proof (i) We show that for all $\alpha \in \mathcal{F}(U)$ and $\beta \in \mathcal{F}(V)$,

$$\phi^{\rightarrow}\alpha \leq \beta \text{ if and only if } \alpha \leq \phi^{\leftarrow}\beta.$$

If $\phi^{\rightarrow}\alpha \leq \beta$, then by Theorem 7.3.26 (ii), $\phi^{\leftarrow}\phi^{\rightarrow}\alpha \leq \phi^{\leftarrow}\beta$ and so by Theorem 7.3.42 (1)(ii),

$$\alpha \leq \phi^{\leftarrow}\phi^{\rightarrow}\alpha \leq \phi^{\leftarrow}\beta.$$

Conversely, if $\alpha \leq \phi^{\leftarrow}\beta$, then again by Lemma 7.3.26(i), $\phi^{\rightarrow}\alpha \leq \phi^{\rightarrow}\phi^{\leftarrow}\beta$, and therefore, by Theorem 7.3.42 (1)(ii),

$$\phi^{\rightarrow}\alpha \leq \phi^{\rightarrow}\phi^{\leftarrow}\beta \leq \beta$$

as desired.

(ii) The proof is similar. □

Theorem 7.5.2 *Let us consider the t-fuzzy relation $(\phi, \phi^{\leftarrow})$ as*

$$\phi(u, \lambda, v) = T(\varphi(u, v), \lambda) \text{ and } \phi^{\leftarrow}(v, \mu, u) = I_T(\varphi(u, v), \mu)$$

for all $(u, \lambda, v) \in U \times (0, 1) \times V$ where T is left-continuous t-norm. Then the Galois connectivity of $(\phi, \phi^{\leftarrow})$ corresponds to the adjoint property of the t-norm T and R-implicator I_T based on T.

Proof If T is left-continuous, T and I_T form an adjoint pair, that is,

$$T(\varphi(u, v), \lambda) \leq \mu \text{ if and only if } I_T(\varphi(u, v), \mu) \geq \lambda.$$

By definition of ϕ and $\phi^{\leftarrow}$, we have

$$\phi(u, \lambda, v) \leq \mu \text{ if and only if } \phi^{\leftarrow}(v, \mu, u) \geq \lambda.$$

From Lemma 7.3.30, for every $(u, v) \in U \times V$ and $(\lambda, \mu) \in (0, 1] \times (0, 1]$,

$$(\phi^{\rightarrow} u_\lambda)(v) = \phi(u, \lambda, v) \text{ and } (\phi^{\leftarrow} v^\mu)(u) = \phi^{\leftarrow}(v, \mu, u).$$

Thus,

$$\phi^{\rightarrow} u_\lambda(v) \leq \mu \text{ if and only if } \lambda \leq (\phi^{\leftarrow} v^\mu)(u)$$

and so we obtain that

$$\phi^{\rightarrow} u_\lambda \leq v^\mu \text{ if and only if } u_\lambda \leq \phi^{\leftarrow} v^\mu.$$

Now take $\alpha \in \mathcal{F}(U)$ and $\beta \in \mathcal{F}(V)$ where

$$\alpha = \bigvee_{\lambda < \alpha(u)} u_\lambda \text{ and } \beta = \bigvee_{\beta(v) < \mu} v^\mu.$$

Note that by Theorem 7.3.28, we also have that

$$\phi^{\rightarrow}\alpha = \bigvee_{\lambda < \alpha(u)} \phi^{\rightarrow} u_\lambda \text{ and } \phi^{\leftarrow}\beta = \bigwedge_{\beta(v) < \mu} \phi^{\leftarrow} v^\mu.$$

Now let $\phi^{\rightarrow}\alpha \leq \beta$. Clearly, for all $\lambda < \alpha(u)$ we have $\phi^{\rightarrow} u_\lambda \leq \beta$ and hence, for all $\beta(v) < \mu$, this gives that $\phi^{\rightarrow} u_\lambda \leq v^\mu$ and so by Theorem 7.5.1, we get $u_\lambda \leq \phi^{\leftarrow} v^\mu$.

Finally,

$$\alpha = \bigvee_{\lambda < \alpha(u)} u_\lambda \leq \phi^{\leftarrow} v^\mu,$$

that is,

$$\alpha = \bigvee_{\lambda < \alpha(u)} u_\lambda \leq \bigwedge_{\beta(v) < \mu} \phi^{\leftarrow} v^\mu = \phi^{\leftarrow}\beta.$$

Clearly, the reverse implicator also holds. □

As is explained in Sect. 7.1, if I is a right-continuous implicator in the second component, then the mappings T_I defined by

$$T_I(\varphi(u, v), \lambda) = \bigwedge\{\mu \in [0, 1] \mid I(\varphi(u, v), \mu) \geq \lambda\}$$

is a left-continuous and increasing mapping in the second component, so we have the following result.

Theorem 7.5.3 *Let us define the t-fuzzy direlation* $(\Phi^{\leftarrow}, \Phi)$ *as*

$$\Phi^{\leftarrow}(u, \lambda, v) = T_I(\varphi(u, v), \lambda) \text{ and } \Phi(v, \mu, u) = I(\varphi(u, v), \mu)$$

for all $u, \lambda, v) \in U \times (0, 1) \times V$ *where* I *is a right-continuous implicator in the second component. Then the Galois connectivity of* $(\Phi^{\leftarrow}, \Phi)$ *corresponds to the adjoint property*

$$T_I(\varphi(u, v), \lambda) \leq \mu \text{ if and only if } I(\varphi(u, v), \mu) \geq \lambda$$

of T_I *and R-implicator* I.

If a t-fuzzy relation ϕ is defined with a t-norm which is continuous in the second component, then the inverse and the complement of ϕ give us two different fuzzy rough set systems as follows.

Theorem 7.5.4 *Let* (U, φ) *be a fuzzy approximation space with a single domain and* (ϕ, Φ) *be a fuzzy direlation on* $\mathcal{F}(U)$. *Suppose that* T *is a left-continuous t-norm in the second component where*

$$\phi(u, v, \lambda) = T(\varphi(u, v), \lambda)$$

for all $u, v \in U$ *and* $\lambda \in (0, 1]$.

(i) *The pair* $(\phi, \phi^{\leftarrow})$ *generates the fuzzy rough set model* $(\underline{apr}, \overline{apr})$ *where*

$$\underline{apr}(\alpha)(v) = \bigwedge_{u \in U} I_T(\varphi(u, v), \alpha(u)),$$

$$\overline{apr}(\alpha)(v) = \bigvee_{u \in U} T(\varphi(u, v), \alpha(u)).$$

(ii) *The pair* (ϕ, ϕ') *generates the fuzzy rough set model* $(\underline{apr'}, \overline{apr})$ *where*

$$\underline{apr'}(\alpha)(v) = \bigwedge_{u \in U} C_{I_D, N}(\varphi(u, v), \alpha(u)),$$

$$\overline{apr}(\alpha)(v) = \bigvee_{u \in U} T(\varphi(u, v), \alpha(u)).$$

Proof (i) By Theorem 7.4.6 (i), we know that

$$\phi^{\leftarrow}(u, \lambda, v) = I_T(\varphi(u, v), \lambda),$$

and hence,

$$\phi^{\leftarrow}\alpha(v) = \bigwedge_{u \in U} I_T(\varphi(u, v), \alpha(u)) = \underline{apr}(\alpha)(v).$$

Clearly, we also have

$$(\phi^{\rightarrow}\alpha)(v) = \bigvee_{u \in U} T(\varphi(u, v), \alpha(u)) = \overline{apr}(\alpha)(v).$$

(ii) By Theorem 7.4.6 (ii),

$$\phi'(u, \lambda, v) = C_{I_D,N}(\varphi(u, v), \lambda),$$

where $D = T'$. Then we obtain

$$\underline{apr}'(\alpha)(v) = \bigwedge_{u \in U} C_{I_D,N}(\varphi(u, v), \alpha(u)),$$

as desired.

Theorem 7.5.5 *Let (U, φ) be a fuzzy approximation space with a single domain and (ϕ, Φ) be a fuzzy direlation on $\mathcal{F}(U)$. Suppose that I is a right-continuous implicator in the second component where*

$$\Phi(u, v, \lambda) = I(\varphi(u, v), \lambda)$$

for all $u, v \in U$ and $\lambda \in (0, 1]$.

(i) *The pair $(\Phi^{\leftarrow}, \Phi)$ generates the fuzzy rough set model $(\underline{apr}', \overline{apr})$ where*

$$\underline{apr}'(\alpha)(v) = \bigwedge_{u \in U} T_{I,N}(\varphi(u, v), \alpha(u)),$$

$$\overline{apr}(\alpha)(v) = \bigvee_{u \in U} I(\varphi(u, v), \alpha(u)).$$

(ii) *The pair (Φ', Φ) generates the fuzzy rough set model $(\underline{apr}', \overline{apr})$ where*

$$\underline{apr}'(\alpha)(v) = \bigwedge_{u \in U} C_{I,N}(\varphi(u, v), \alpha(u)),$$

$$\overline{apr}(\alpha)(v) = \bigvee_{u \in U} I(\varphi(u, v), \alpha(u)).$$

Proof It is omitted. □

Remark 7.5.6 Theorems 7.5.4 and 7.5.5 show that the inverse of a t-fuzzy relation corresponds to the adjoint property of an implicator I_T based on a t-norm T. Further, the complement of a t-fuzzy relation is related with the duality operation of fuzzy logical connectives.

7.6 The Well-Known Fuzzy Rough Set Models Obtained by t-Fuzzy Direlations

Fuzzy rough set models based on implicator and conjunctor are considered for a single domain of discourse by L. D'eer et al. in [10]. Generally, the special cases are subjected to the types of implicator and conjunctor or fuzzy relations [11–15]. Let U and V be any two domains of discourse where $\mathcal{F}(U)$ and $\mathcal{F}(V)$ are the families of all fuzzy subsets of U and V, respectively. Let φ be a fuzzy relation, that is, $\varphi \in \mathcal{F}(U \times V)$. Then the triple (U, V, φ) is called *a generalized fuzzy approximation space* with two domains of discourse. More generally, the lower and upper approximations of a fuzzy subset α of U in V are defined by

$$\underline{apr}(\alpha)(v) = \bigwedge_{u\in U} I(\varphi(u, v), \alpha(u)),$$

$$\overline{apr}(\alpha)(v) = \bigvee_{u\in U} T(\varphi(u, v), \alpha(u))$$

for all $v \in V$ where I and T are implicator and conjunctor, respectively.

Besides the above approximation operators, we are also interested in the approximations defined by

$$\overline{apr}'(\beta)(u) = \bigvee_{v\in V} T_I(\varphi(u, v), \beta(v)),$$

$$\underline{apr}'(\beta)(u) = \bigwedge_{v\in V} I_T(\varphi(u, v), \beta(v))$$

where $\beta \in \mathcal{F}(V)$, and

$$T_I(\eta, \mu) = \bigwedge\{\gamma \in [0, 1] \mid \mu \leq I(\eta, \gamma)\},$$

$$I_T(\eta, \mu) = \bigvee\{\gamma \in [0, 1] \mid T(\eta, \gamma) \leq \mu\}$$

for all $\eta, \mu \in [0, 1]$.

For all $\beta \in \mathcal{F}(V)$, we also consider the upper approximation as

$$\overline{apr}'(\beta)(u) = \bigvee_{v\in V} C_{I,N}(\varphi(u, v), \beta(v))$$

where $C_{I,N}$ is the induced conjunctor of the residual implicator I and the strong negator N (see [13] for single domains of discourse).

Now, we will observe how the interval structures $(\phi^{\rightarrow}, \Phi^{\rightarrow})$ and $(\Phi^{\leftarrow}, \phi^{\leftarrow})$ generate well-known fuzzy rough set models with two domains of discourse given in Table 7.7. In fact, the following corollaries are immediate results of Corollaries 7.3.32 and 7.3.33.

Corollary 7.6.1 *Let us consider the complemented t-fuzzy direlation (ϕ, Φ) corresponding to minimum t-norm T_M and the Klenee-Dienes S-implicator $I_{D_M,N}$ where*

$$\phi(u, \lambda, v) = T_M(\varphi(u, v), \lambda) \text{ and } \Phi(u, \lambda, v) = I_{D_M,N}(\varphi(u, v)), \lambda).$$

Then we have the following propositions.

(i) *The pair $(\phi^{\rightarrow}, \Phi^{\rightarrow})$ generates the fuzzy rough set model of Dubois and Prade* [16] *on (U, V, φ) corresponding to the fuzzy rough model $(\underline{apr}, \overline{apr})$, that is, for all $\alpha \in \mathcal{F}(U)$,*

$$\phi^{\rightarrow}\alpha = \overline{apr}(\alpha) \text{ and } \Phi^{\rightarrow}\alpha = \underline{apr}(\alpha).$$

Table 7.9 Fuzzy rough set approximations given in Corollaries 7.6.1 and 7.6.2: the approximations given in view of the semantic link with Pawlak's rough sets of a generalized fuzzy approximation space (U, V, φ) are the same as the approximations of generalized t-fuzzy approximation space (U, V, ϕ, Φ). Note that $C_{I_{T_M},N} = T_{I_{D_M}}$ and $C_{I_{T_P},N} = T_{I_{D_P}}$

$(\phi^{\rightarrow}\alpha)(v) = \overline{apr}(\alpha)(v) = \bigvee_{u\in U} T_M(\varphi(u, v), \alpha(u))$	$(\Phi^{\rightarrow}\alpha)(v) = \underline{apr}(\alpha)(v) = \bigwedge_{u\in U} I_{D_M,N}(\varphi(u, v), \alpha(u))$
$(\Phi^{\leftarrow}\beta)(u) = \overline{apr}'(\beta)(u) = \bigvee_{v\in V} C_{I_{T_M},N}(\varphi(u, v), \beta(v))$	$(\phi^{\leftarrow}\beta)(u) = \underline{apr}'(\beta)(u) = \bigwedge_{v\in V} I_{T_M}(\varphi(u, v), \beta(v))$
$(\phi^{\rightarrow}\alpha)(v) = \overline{apr}(\alpha)(v) = \bigvee_{u\in U} T_P(\varphi(u, v), \alpha(u))$	$(\Phi^{\rightarrow}\alpha)(v) = \underline{apr}(v) = \bigwedge_{u\in U} I_{D_P,N}(\varphi(u, v), \alpha(u))$
$(\Phi^{\leftarrow}\beta)(u) = \overline{apr}'(\beta)(u) = \bigvee_{v\in V} C_{I_{T_P},N}(\varphi(u, v), \beta(v))$	$(\phi^{\leftarrow}\beta)(u) = \underline{apr}'(\beta)(u) = \bigwedge_{v\in V} I_{T_P}(\varphi(u, v), \beta(v))$

(ii) *The pair* $(\Phi^{\leftarrow}, \phi^{\leftarrow})$ *also generates the fuzzy rough set model of Mi and Zhang* [13] *on* (U, V, φ) *with respect to Gödel implicator* [17], *and the conjunctor induced by Gödel implicator and the standard negation* N*, that is, for all* $\beta \in \mathcal{F}(V)$ *(Table 7.9),*

$$\Phi^{\leftarrow}\beta = \overline{apr}'(\beta) \text{ and } \phi^{\leftarrow}\beta = \underline{apr}'(\beta).$$

Proof (i) By Corollary 7.3.32 and Theorem 7.4.2, for all $\alpha \in \mathcal{F}(U)$ and $v \in V$, we conclude that

$$(\phi^{\rightarrow}\alpha)(v) = \bigvee_{u\in U} \phi(u, \alpha(u), v) = \bigvee_{u\in U} T_M(\varphi(u, v), \alpha(u)) = \overline{apr}(\alpha)(v).$$

and

$$(\Phi^{\rightarrow}\alpha)(v) = \bigwedge_{u\in U} \Phi(u, \alpha(u), v) = \bigwedge_{u\in U} I_{D_M,N}(\varphi(u, v)), \alpha(u)) = \underline{apr}(\alpha)(v).$$

(ii) By Corollary 7.3.33 and Theorem 7.4.6, for all $\beta \in \mathcal{F}(V)$, we obtain that

$$\Phi^{\leftarrow}\beta(u) = \bigvee_{v\in V} \Phi^{\leftarrow}(v, \beta(v), u) = \bigvee_{v\in V} C_{I_{T_M},N}(\varphi(u, v), \beta(v)) = \overline{apr}'\beta(u)$$

and

$$\phi^{\leftarrow}\beta(u) = \bigwedge_{v\in V} \phi^{\leftarrow}(v, \beta(v), u) = \bigwedge_{v\in V} I_{T_M}(\varphi(u, v), \beta(v)) = \underline{apr}'\beta(u).$$

Corollary 7.6.2 *Let us consider the t-fuzzy direlation* (ϕ, Φ) *corresponding to product t-norm and Reichenbach S-implicator where*

$$\phi(u, \lambda, v) = T_P(\varphi(u, v), \lambda) \text{ and } \Phi(u, \lambda, v) = I_{D_P,N}(\varphi(u, v), \lambda)$$

for all $u \in U$, $v \in V$ *and* $\lambda \in (0, 1]$. *Then*

(i) *The pair* $(\phi^{\rightarrow}, \Phi^{\rightarrow})$ *generates the fuzzy rough set model* $(\underline{apr}, \overline{apr})$ *on* (U, V, φ) *with respect to product t-norm and Reichenbach S-implicator, that is,*

$\phi^{\rightarrow}\alpha = \overline{apr}(\alpha)$ *and* $\Phi^{\rightarrow}\alpha = \underline{apr}(\alpha)$.

(ii) *The pair* $(\Phi^{\leftarrow}, \phi^{\leftarrow})$ *also generates the fuzzy rough set model* $(\underline{apr}', \overline{apr}')$ *of Mi and Zhang* [13] *with respect to the Goguen implicator, and the conjunctor induced by the Goguen implicator and the negator. That is, for all* $\alpha \in \mathcal{F}(U)$ *and* $\beta \in \mathcal{F}(V)$, *we have the following equalities:*

$\Phi^{\leftarrow}\beta = \overline{apr}'(\beta)$ *and* $\phi^{\leftarrow}\beta = \underline{apr}'(\beta)$.

Proof (i) By Corollary 7.3.32 and Theorem 7.4.2, for all $\alpha \in \mathcal{F}(U)$ and $v \in V$, we conclude that

$$(\phi^{\rightarrow}\alpha)(v) = \bigvee_{u\in U} \phi(u, \alpha(u), v) = \bigvee_{u\in U} T_P(\phi(u, v), \alpha(u)) = \overline{apr}(\alpha)(v)$$

and

$$(\Phi^{\rightarrow}\alpha)(v) = \bigwedge_{u\in U} \Phi(u, \alpha(u), v) = \bigwedge_{u\in U} I_{D_P,N}(\varphi(u, v), \alpha(u)) = \underline{apr}(\alpha)(v).$$

(ii) By Corollary 7.3.33 and Theorem 7.4.6, for all $\beta \in \mathcal{F}(V)$ and $v \in V$, it follows that

$$\phi^{\leftarrow}\beta(u) = \bigwedge_{v\in V} \phi^{\leftarrow}(v, \beta(v), u) = \bigwedge_{v\in V} I_{T_P}(\varphi(u, v), \beta(v)) = \underline{apr}'\beta(u)$$

and

$$\Phi^{\leftarrow}\beta(u) = \bigvee_{v\in V} \Phi^{\leftarrow}(v, \beta(v), u) = \bigvee_{v\in V} C_{I_{T_P},N}(\varphi(u, v), \beta(v)) = \overline{apr}'\beta(u).$$

7.7 Basic Properties of Fuzzy Relations

Seriality

Just as the seriality of a relation in rough set theory, the seriality of a fuzzy relation also plays an important role in fuzzy rough set theory. For an ordinary relation, the seriality is equivalent to say that every successor neighborhood is a non-empty set. By the same way, inverse seriality can be stated that every predecessor neighborhood is non-empty. In fuzzy set theory, the seriality is given in some different forms. The most used and natural fuzzy counterparts of the serialities are as follows.

Table 7.10 $\varphi \in \mathcal{F}(U \times V)$

φ	v_1	v_2	v_3	v_4	$\dots$	v_n	$\dots$
u_1	1	$\frac{2}{3}$	$\frac{3}{4}$	$\frac{4}{5}$	$\dots$	$\frac{n}{n+1}$	$\dots$
u_2	0	$1-\frac{1}{2}$	$1-\frac{1}{3}$	$1-\frac{1}{4}$	$\dots$	$1-\frac{1}{n}$	$\dots$

Definition 7.7.1 Let φ be a fuzzy relation, that is, $\varphi \in \mathcal{F}(U \times V)$. Then φ is called

(i) *serial* if for all $u \in U$, $\bigvee_{v \in V} \varphi(u, v) = 1$,

(ii) *inverse serial* if for all $v \in V$, $\bigvee_{u \in U} \varphi(u, v) = 1$,

(iii) *f-serial* if $\forall u \in U, \exists v \in V, \varphi(u, v) = 1$,

(iv) *f-inverse serial* if $\forall v \in V, \exists u \in U, \varphi(u, v) = 1$.

Clearly, if φ is f-serial, then it is also serial and if φ is f-inverse serial, then it is also inverse serial. In particular, for infinite systems, the converse of these implications is not necessarily true.

Example 7.7.2 Let us consider the domains of discourse

$$U = \{u_1, u_2\} \text{ and } V = \{v_n \mid n \in \mathbb{N}\}$$

where $\mathbb{N} = \{1, 2, 3, \dots\}$. Let $\varphi : U \times V \to [0, 1]$ be the fuzzy relation given in Table 7.10. For $u_2 \in U$, since $\varphi(u_2, v_n) \neq 1$ for all $n \in \mathbb{N}$, φ clearly is not f-serial. However, φ is serial, since

$$\bigvee_{n \in \mathbb{N}} \varphi(u_1, v_n) = \varphi(u_1, v_1) = 1$$

and

$$\bigvee_{n \in \mathbb{N}} \varphi(u_2, v_n) = \bigvee_{n \in \mathbb{N}} (1 - 1/n) = 1.$$

Theorem 7.7.3 *Let (r, R) be a direlation from $(W_U, \mathcal{W}_U)$ to $(W_V, \mathcal{W}_V)$. Then the following statements are equivalent:*

(i) *(r, R) is t-serial.*

(ii) $\forall u \in U,\ \lambda' < \lambda \implies \exists v \in V,\ \Phi_R(u, \lambda', v) < \phi_r(u, \lambda, v)$.

Proof (i) $\implies$ (ii): Let (r, R) be t-serial. Then $\forall u, u' \in U$ and $\lambda, \lambda' \in (0, 1]$,

$$\overline{P}_{(u,\lambda)} \not\subseteq \overline{Q}_{(u',\lambda')} \implies \exists\, (v,\mu) \in W_V \text{ with } r \not\subseteq \overline{Q}_{((u,\lambda),(v,\mu))} \text{ and } \overline{P}_{((u',\lambda'),(v,\mu))} \not\subseteq R.$$

By Example 4.1.2 and Lemma 7.3.3, we also have

$$u = u', \lambda' < \lambda \implies \exists\, (v,\mu) \in W_V \text{ with } \mu < \phi_r(u, \lambda, v) \text{ and } \mu > \Phi_R(u', \lambda', v).$$

Note that Φ_R is a t-fuzzy corelation, and so for $\lambda > \lambda'$, we also have that $\mu > \Phi_R(u', \lambda, v)$. Hence, condition (ii) holds.

(ii) $\Longrightarrow$ (i): Let $u, u' \in U, \lambda, \lambda' \in (0, 1]$ and $\overline{P}_{(u,\lambda)} \nsubseteq \overline{Q}_{(u',\lambda')}$. Then by Lemma 7.3.3, $u = u', \lambda' < \lambda$. By the assumption

$$\mu < \phi_r(u, \lambda, v) \text{ and } \mu > \Phi_R(u, \lambda', v)$$

for some $v \in V$ and $\mu \in (0, 1]$, that is, we obtain $\Phi_R(u, \lambda', v) < \phi_r(u, \lambda, v)$.

Definition 7.7.4 Let (ϕ, Φ) be a t-fuzzy direlation from $\mathcal{F}(U)$ to $\mathcal{F}(V)$. Then (ϕ, Φ) is called *serial* if

$$\forall u \in U,\ \lambda' < \lambda \Longrightarrow \exists v \in V,\ \Phi(u, \lambda', v) < \phi(u, \lambda, v).$$

Theorem 7.7.5 *Let (r, R) be a direlation from $(W_U, \mathcal{W}_U)$ to $(W_V, \mathcal{W}_V)$. Then the following statements are equivalent:*

(i) *(r, R) is t-inverse serial.*

(ii) $\forall v \in V, \mu' < \mu \Longrightarrow \exists u \in U,\ \exists\lambda \in (0, 1],\ \mu' < \phi_r(u, \lambda, v)$ *and* $\mu > \Phi_R(u, \lambda, v)$.

Proof It is similar to the proof of Theorem 7.7.3.

Definition 7.7.6 Let (ϕ, Φ) be a t-fuzzy direlation from $\mathcal{F}(U)$ to $\mathcal{F}(V)$. Then (ϕ, Φ) is called *inverse serial* if

$$\forall v \in V,\ \mu' < \mu \Longrightarrow \exists u \in U, \exists\lambda \in (0, 1], \\ \mu' < \phi(u, \lambda, v) \text{ and } \mu > \Phi(u, \lambda, v).$$

We say (ϕ, Φ) is *compatible* if it is serial and inverse serial.

Theorem 7.7.7 *Let (ϕ, Φ) be a t-fuzzy direlation from $\mathcal{F}(U)$ to $\mathcal{F}(V)$ and (r, R) be the direlation from $(W_U, \mathcal{W}_U)$ to $(W_V, \mathcal{W}_V)$ where $\phi = \phi_r$ and $\Phi = \Phi_R$. Then*

(i) *(ϕ, Φ) is serial if and only if (r, R) is t-serial.*

(ii) *(ϕ, Φ) is inverse serial if and only if (r, R) is t-inverse serial.*

Proof The proof immediately follows from definitions. □

Theorem 7.7.8 *Let (ϕ, Φ) be a t-fuzzy direlation from $\mathcal{F}(U)$ to $\mathcal{F}(V)$. Then*

(i) *(ϕ, Φ) is serial if and only if $(\Phi^{\leftarrow}, \phi^{\leftarrow})$ is inverse serial.*

(ii) *(ϕ, Φ) is inverse serial if and only if $(\Phi^{\leftarrow}, \phi^{\leftarrow})$ is serial.*

Proof We prove (i) leaving the proof of (ii) to the interested reader. Let (ϕ, Φ) be serial. Take $u \in U,\ \lambda', \lambda \in (0, 1]$ where $\lambda' < \lambda$. We show that

$$\lambda' < \Phi^{\leftarrow}(v, \mu, u) \text{ and } \lambda > \phi^{\leftarrow}(v, \mu, u)$$

for some $v \in V$ and $\mu \in (0, 1]$. By the assumption, we have $\Phi(u, \lambda', v) < \phi(u, \lambda, v)$ for some $v \in V$. Choose a $\mu \in (0, 1]$ such that $\Phi(u, \lambda', v) < \mu < \phi(u, \lambda, v)$. Then by Definition 7.3.14, we get $\lambda' < \Phi^{\leftarrow}(v, \mu, u)$ and $\lambda > \phi^{\leftarrow}(v, \mu, u)$, that is, $(\Phi^{\leftarrow}, \phi^{\leftarrow})$ is inverse serial. Necessity is similar. □

The seriality and inverse seriality of a fuzzy relation can be given in terms of a t-norm and an implicator.

Theorem 7.7.9 *Let φ be a fuzzy relation, T be a left-continuous t-norm and I be a right-continuous border implicator in the first component. Then we have the following equivalences:*

(i) *φ is serial if and only if*

$$\forall u \in U,\ \lambda' < \lambda \implies \exists v \in V,\ I(\varphi(u,v),\lambda') < T(\varphi(u,v),\lambda),$$

(ii) *φ is inverse serial if and only if*

$$\forall v \in V,\ \mu' < \mu \implies \exists u \in U, \exists \lambda \in (0,1], \quad \mu' < T(\varphi(u,v),\lambda) \text{ and } \mu > I(\varphi(u,v),\lambda).$$

Proof (i) $\implies$: Let $u \in U$ and $\lambda' < \lambda$. Then we may easily observe that

$$\begin{aligned}
\lambda' < \lambda &\implies I(1,\lambda') < T(1,\lambda) \\
&\implies I(\bigvee_{v\in V} \varphi(u,v),\lambda') < T(\bigvee_{v\in V} \varphi(u,v),\lambda) \\
&\implies \bigvee_{v\in V} I(\varphi(u,v),\lambda') < \bigvee_{v\in V} T(\varphi(u,v),\lambda) \\
&\implies \forall v \in V,\ I(\varphi(u,v),\lambda') < \bigvee_{v\in V} T(\varphi(u,v),\lambda) \\
&\implies \exists v \in V,\ I(\varphi(u,v),\lambda') < T(\varphi(u,v),\lambda).
\end{aligned}$$

$\Longleftarrow$: Suppose that φ is not serial. Then for some $u \in U$, we have $\bigvee_{v\in V} \varphi(u,v) < 1$. Let us choose $\lambda, \lambda' \in (0,1]$ such that $\bigvee_{v\in V} \varphi(u,v) < \lambda' < \lambda < 1$. By the assumption, we have

$$\exists v \in V,\ I(\varphi(u,v),\lambda') < T(\varphi(u,v),\lambda).$$

Since I satisfies the border condition, it also satisfies the consequent boundary condition. Then we easily get that

$$\lambda' \le I(\varphi(u,v),\lambda').$$

Further, by the assumption, we have $\varphi(u,v) < \lambda'$ and so

$$T(\varphi(u,v),\lambda) \le T(\lambda',1) = \lambda'.$$

This gives the immediate contradiction $\lambda' < \lambda'$.

(ii) $\implies$: Let $v \in V$ and $\mu' < \mu$. Take a λ where $\mu' < \lambda < \mu$. Then

$$\mu' < \lambda \Longrightarrow \mu' < T(1, \lambda) = T(\bigvee_{u \in U} \varphi(u, v), \lambda) = \bigvee_{u \in U} T(\varphi(u, v), \lambda)$$
$$\Longrightarrow \exists u \in U,\ \mu' < T(\varphi(u, v), \lambda).$$

Now

$$\lambda < \mu \Longrightarrow I(1, \lambda) < \mu \Longrightarrow I(\bigvee_{u \in U} \varphi(u, v), \lambda) < \mu$$
$$\Longrightarrow \bigvee_{u \in U} I(\varphi(u, v), \lambda) < \mu$$
$$\Longrightarrow \forall u \in U,\ I(\varphi(u, v), \lambda) < \mu.$$

Consequently, we obtain that for some u, $\mu' < T(\varphi(u, v), \lambda)$ and $\mu > I(\varphi(u, v), \lambda)$.

$\Longleftarrow$: Suppose that φ is not inverse serial. Then for some $v \in V$, we have $\bigvee_{u \in U} \varphi(u, v) < 1$. Let us choose $\mu, \mu' \in (0, 1]$ such that $\bigvee_{u \in U} \varphi(u, v) < \mu' < \mu < 1$. By the assumption, for some $u \in U$ and for some $\lambda \in (0, 1]$, we have

$$\mu' < T(\varphi(u, v), \lambda), \text{ and } \mu > I(\varphi(u, v), \lambda).$$

Since T is increasing in both components, we easily get that

$$\mu' < T(\varphi(u, v), \lambda) \leq T(\mu', 1) = \mu'.$$

This gives the immediate contradiction $\mu' < \mu'$.

Corollary 7.7.10 *Let T be a left-continuous t-norm in the first component and I be a border and right-continuous implicator in the second component. Let us consider the t-fuzzy direlation (ϕ, Φ) given in Theorem 7.4.2. Then we have the following equivalences:*

(i) *φ is serial if and only if (ϕ, Φ) is serial.*

(ii) *φ is inverse serial if and only if (ϕ, Φ) is inverse serial.*

Proof (i) Note that the t-fuzzy direlation (ϕ, Φ) is defined by

$$\phi(u, \lambda, v) = T(\varphi(u, v), \lambda) \text{ and } \Phi(u, \lambda, v) = I(\varphi(u, v), \lambda)$$

for all $u \in U$, $v \in V$ and $\lambda \in (0, 1]$. If φ is serial, then by Theorem 7.7.9,

$$\forall u \in U,\ \lambda' < \lambda \implies \exists v \in V,\ I(\varphi(u, v), \lambda') < T(\varphi(u, v), \lambda).$$

That is,

$$\forall u \in U,\ \lambda' < \lambda \implies \exists v \in V,\ \Phi(u, \lambda', v) < \phi(u, \lambda, v).$$

Hence, (ϕ, Φ) is serial. Necessity is similar.

Note that the following fuzzy logic systems satisfy the conditions of Theorem 7.7.9. Then for each fuzzy logic system, we may give a different formulation due to serialities:

Product t-norm and Reichenbach S-implicator:

A fuzzy relation φ is $(T_P, I_{D_P,N})$-serial if

$$\forall u \in U,\ \lambda' < \lambda \implies \exists v \in V, \varphi(u, v) > \frac{1}{\lambda - \lambda' + 1},\ and$$

$(T_P, I_{D_P,N})$-inverse serial if

$$\forall v \in V,\ \mu' < \mu \implies \exists u \in U, \exists \lambda \in (0, 1],$$
$$\mu' < \varphi(u, v)\lambda \text{ and } \mu > 1 - \varphi(u, v) + \varphi(u, v)\lambda.$$

Minimum t-norm and Kleene-Dienes S-implicator:

A fuzzy relation φ is $(T_M, I_{D_M,N})$-serial if

$$\forall u \in U,\ \lambda' < \lambda \implies \exists v \in V,\ (1 - \varphi(u, v)) \vee \lambda' < \varphi(u, v) \wedge \lambda,\ and$$

$(T_M, I_{D_M,N})$-inverse serial if

$\forall v \in V,\ \mu' < \mu \implies \exists u \in U, \exists \lambda \in (0, 1],\ \mu' < \varphi(u, v) \wedge \lambda$ and $\mu > (1 - \varphi(u, v)) \vee \lambda$.

Łukasiewicz t-norm and Łukasiewicz S-implicator:

A fuzzy relation φ is $(T_W, I_{D_W,N})$-serial if

$$\forall u \in U,\ \lambda' < \lambda \implies \exists v \in V,\ (1 - \varphi(u, v) + \lambda') \wedge 1 < (\varphi(u, v) + \lambda - 1) \vee 0,\ and$$

$(T_W, I_{D_W,N})$-inverse serial if

$$\forall v \in V,\ \mu' < \mu \implies \exists u \in U,\ \exists \lambda \in (0, 1],$$
$$\mu' < (\varphi(u, v) + \lambda - 1) \vee 0 \text{ and } \mu > (1 - \varphi(u, v) + \lambda) \wedge 1.$$

Injectivity

Theorem 7.7.11 *Let (r, R) be a direlation from $(W_U, \mathcal{W}_U)$ to $(W_V, \mathcal{W}_V)$. Then the following statements are equivalent:*

(i) *(r, R) is injective.*

(ii) *$\forall u, u' \in U,\ v \in V$ and $\lambda, \lambda' \in (0, 1]$,*

$$\Phi_R(u', v, \lambda') < \phi_r(u, v, \lambda) \implies u = u' \text{ and } \lambda' < \lambda.$$

Proof (i) $\Longrightarrow$ (ii): Let (r, R) be injective. Then by Definition 4.4.8,

$$\forall (u, \lambda), (u', \lambda') \in U \times (0, 1] \text{ and } (v, \mu) \in V \times (0, 1],$$
$$r \nsubseteq \overline{Q}_{((u,\lambda),(v,\mu))} \text{ and } \overline{P}_{((u',\lambda'),(v,\mu))} \nsubseteq R \Longrightarrow \overline{P}(u, \lambda) \nsubseteq \overline{Q}_{(u',\lambda')}.$$

By Example 4.1.2 and Lemma 7.3.3,

$$\mu < \phi_r(u, \lambda, v) \text{ and } \mu > \Phi_R(u', \lambda', v) \Longrightarrow u = u' \text{ and } \lambda' < \lambda,$$

that is,

$$\phi_r(u, \lambda, v) > \Phi_R(u', \lambda', v) \Longrightarrow u = u' \text{ and } \lambda' < \lambda.$$

The necessity is similar.

Definition 7.7.12 Let (ϕ, Φ) be a t-fuzzy direlation from $\mathcal{F}(U)$ to $\mathcal{F}(V)$. Then (ϕ, Φ) is called *injective* if

$$\forall u, u' \in U,\ v \in V \text{ and } \lambda, \lambda' \in (0, 1],$$
$$\Phi(u', v, \lambda') < \phi(u, v, \lambda) \Longrightarrow u = u' \text{ and } \lambda' < \lambda.$$

Then we may give the following definition.

Definition 7.7.13 We call φ is

(i) (T, I)- *injective* if

$$\forall u, u' \in U,\ v \in V \text{ and } \lambda, \lambda' \in (0, 1],$$
$$I(\varphi(u', v), \lambda') < T(\varphi(u, v), \lambda) \Longrightarrow u = u' \text{ and } \lambda' < \lambda.$$

(ii) (T_I, I_T)- *injective* if

$$\forall v, v' \in V,\ u \in U \text{ and } \lambda, \lambda' \in (0, 1],$$
$$I_T(\varphi(u, v'), \lambda') < T_I(\varphi(u, v), \lambda) \Longrightarrow v = v' \text{ and } \lambda' < \lambda.$$

Now let φ be a fuzzy relation, T be a left-continuous t-norm and I be a right-continuous border implicator in the first component.

Theorem 7.7.14 *Let (ϕ, Φ) be the t-fuzzy direlation from $\mathcal{F}(U)$ to $\mathcal{F}(V)$ defined by*

$$\phi(u, \lambda, v) = T(\varphi(u, v), \lambda) \text{ and } \Phi(u, \lambda, v) = I(\varphi(u, v), \lambda)$$

for all $u \in U$, $v \in V$ and $\lambda \in (0, 1]$.

Then

(i) *φ is (T, I)-injective if and only if (ϕ, Φ) is injective.*

(ii) *φ is (T_I, I_T)-injective if and only if $(\Phi^{\leftarrow}, \phi^{\leftarrow})$ is injective.*

Proof Immediate. □

Definition 7.7.15 A fuzzy relation $\varphi \in \mathcal{F}(U \times U)$ is called

(i) T*-identity* if for all $u, v \in U$ and $\lambda \in (0, 1]$,

$T(\varphi(u, v), \lambda) = u_\lambda(v)$, and

(ii) I*-identity* if for all $u, v \in U$ and $\lambda \in (0, 1]$,

$I(\varphi(u, v), \lambda) = u^\lambda(v)$,

(iii) (T, I)*-identity* if φ is T-identity and I-identity.

Theorem 7.7.16 *Let* (ϕ, Φ) *be the t-fuzzy direlation from* $\mathcal{F}(U)$ *to* $\mathcal{F}(U)$ *defined by*

$$\phi(u, \lambda, v) = T(\varphi(u, v), \lambda) \text{ and } \Phi(u, \lambda, v) = I(\varphi(u, v), \lambda)$$

for all $u \in U$, $v \in V$ *and* $\lambda \in (0, 1]$. *Then*

φ *is* (T, I)*-identity if and only if* (ϕ, Φ) *is the identity on* $\mathcal{F}(U)$,

that is,

$$\phi = \phi_i \text{ and } \Phi = \Phi_I$$

where (i, I) *is the identity direlation on* $(W_U, \mathcal{W}_U)$.

Definition 7.7.17 $\varphi \in \mathcal{F}(U \times U)$ is called (T, I)*-reflexive* if

$\forall u, v \in U$ and $\lambda \in (0, 1]$,

$$u_\lambda(v) \leq T(\varphi(u, v), \lambda)) \text{ and } I(\varphi(u, v), \lambda) \leq u^\lambda(v).$$

Definition 7.7.18 $\varphi \in \mathcal{F}(U \times U)$ is called (T, I)*-symmetric* if

$\forall u, v \in U$ and $\lambda \in (0, 1]$,

$$I(\varphi(u, v), \lambda) = I_T(\varphi(u, v), \lambda) \text{ and } T_I(\varphi(u, v), \lambda) = T(\varphi(u, v), \lambda).$$

Definition 7.7.19 $\varphi \in \mathcal{F}(U \times U)$ is called (T, I)*-transitive* if

$\forall u, z \in U$ and $\lambda \in (0, 1]$,

$$\bigvee\{T(\varphi(v, z), \mu) \mid v \in V,\ 0 < \mu < T(\varphi(u, v), \lambda)\} \leq T(\varphi(u, z), \lambda)$$

and

$$I(\varphi(u, z), \lambda) \leq \bigwedge\{I(\varphi(v, z), \mu) \mid v \in V,\ I(\varphi(u, v), \lambda) < \mu \leq 1\}.$$

Definition 7.7.20 A mapping $\varphi \in \mathcal{F}(U \times U)$ is called *an* (T, I)*-equivalence fuzzy relation* if it is (T, I)-reflexive, (T, I)-symmetric and (T, I)-transitive.

Theorem 7.7.21 *Let (ϕ, Φ) be the t-fuzzy direlation from $\mathcal{F}(U)$ to $\mathcal{F}(U)$ defined by*

$$\phi(u, \lambda, v) = T(\varphi(u, v), \lambda) \text{ and } \Phi(u, \lambda, v) = I(\varphi(u, v), \lambda)$$

for all $u \in U$, $v \in V$ and $\lambda \in (0, 1]$. Then

(i) *(ϕ, Φ) is reflexive if and only if φ is (T, I)-reflexive.*

(ii) *(ϕ, Φ) is symmetric if and only if φ is (T, I)-symmetric.*

(iii) *(ϕ, Φ) is transitive if and only if φ is (T, I)-transitive.*

As a result,

φ is an equivalence fuzzy relation if and only if (ϕ, Φ) is an equivalence t-fuzzy direlation on $\mathcal{F}(U)$.

Proof (i) Recall that the identity t-fuzzy direlation (ϕ_i, Φ_i) on $\mathcal{F}(U)$ is defined by

$$\phi_i(u, \lambda, v) = \begin{cases} \lambda, & \text{if } u = v \\ 0, & \text{if } u \neq v, \end{cases} \qquad \Phi_I(u, \lambda, v) = \begin{cases} \lambda, & \text{if } u = v \\ 1, & \text{if } u \neq v \end{cases}$$

for all $u, v \in U$ and $\lambda \in (0, 1]$.

Note that for all $u, v \in U$ and $\lambda \in (0, 1]$, we already have

$$\phi_i(u, \lambda, v) = u_\lambda(v) \text{ and } \Phi_I(u, \lambda, v) = u^\lambda(v)$$

where u_λ and v^λ are fuzzy point and fuzzy co-point in $\mathcal{F}(U)$, respectively.

Since (ϕ, Φ) is reflexive,

$$u_\lambda(v) \leq T(\varphi(u, v), \lambda) \text{ and } I(\varphi(u, v), \lambda) \leq u^\lambda(v).$$

That is, φ is (T, I)-reflexive. Necessity is also similar.

(ii) Let φ be (T, I)-symmetric. Then for all $u, v \in U$ and $\lambda \in (0, 1]$,

$$I(\varphi(u, v), \lambda) = I_T(\varphi(u, v), \lambda) \text{ and } T_I(\varphi(u, v), \lambda) = T(\varphi(u, v), \lambda).$$

By Theorem 7.4.6, we have

$$\Phi^{\leftarrow}(v, \lambda, u) = T_I(\varphi(u, v) \text{ and } \phi^{\leftarrow}(v, \lambda, u) = I_T(\varphi(u, v), \lambda)$$

and hence, obtain $\phi = \Phi^{\leftarrow}$ and $\Phi = \phi^{\leftarrow}$. Sufficiency can be proved using a similar way.

(iii) Let φ be (T, I)-transitive. Then by definition of (ϕ, Φ), we obtain that

$\forall u, z \in U$ and $\lambda \in (0, 1]$,

$$
\begin{aligned}
\phi(u,\lambda,z) &= T(\varphi(u,z),\lambda) \\
&\leq \bigvee\{T(\varphi(v,z),\mu) \mid v\in V,\ 0<\mu<T(\varphi(u,v),\lambda)\} \\
&= \bigvee\{\phi(v,z,\mu) \mid v\in V,\ 0<\mu<\phi(u,v,\lambda)\} \\
&= (\phi\circ\phi)(u,\lambda,z)
\end{aligned}
$$

and

$$
\begin{aligned}
(\Phi\circ\Phi)(u,\lambda,z) &= \bigwedge\{\Phi(v,\mu,z) \mid v\in V,\ \Phi(u,\lambda,\mu)<\mu\leq 1\} \\
&= \bigwedge\{I(\varphi(v,z),\mu) \mid v\in V,\ I(\varphi(u,v),\lambda)<\mu\leq 1\} \\
&\geq I(\varphi(u,z),\lambda) = \Phi(u,\lambda,z).
\end{aligned}
$$

As a result, we conclude that $\phi\circ\phi\leq\phi$ and $\Phi\leq\Phi\circ\Phi$. Necessity is clear.

7.8 t-Fuzzy Rough Set Approximations with Two Domains of Discourse

As we know, under the lattice isomorphism $\mathfrak{I}$, t-fuzzy direlations between lattices of fuzzy subsets correspond to direlations between fuzzy textures. The textural results given in Sect. 4.2 will be used for the proofs of the results due to generalized t-fuzzy approximation spaces. The proofs are easy and we generally use the lattice isomorphism $\mathfrak{I}$ between the fuzzy subset texture $(W_U, \mathcal{W}_U)$ and the fuzzy lattice $\mathcal{F}(U)$. Hence, we will randomly omit some of the proofs when they essentially have a common idea.

Let (ϕ,Φ) be a t-fuzzy direlation from $\mathcal{F}(U)$ to $\mathcal{F}(V)$. Then the quadruple (U,V,ϕ,Φ) is called *a generalized textural fuzzy approximation space* or in brief, *a generalized t-fuzzy approximation space*. A fuzzy set α can be described by the pair $(\Phi^{\to}\alpha,\phi^{\to}\alpha)$, and the sets $\Phi^{\to}\alpha$ and $\phi^{\to}\alpha$ can be regarded as the lower approximation and upper approximation of α, respectively. Similarly, a fuzzy set β can be described by the pair $(\phi^{\leftarrow}\beta,\Phi^{\leftarrow}\beta)$ where $\phi^{\leftarrow}\beta$ can be regarded as the lower approximation and $\Phi^{\leftarrow}\beta$ as the upper approximation of β. In particular, here, we show that the pairs of mappings

$$(\Phi^{\to},\phi^{\to}),\ (\phi^{\leftarrow},\Phi^{\leftarrow})$$

constitute two different interval structures.

Now we consider the following proposition.

Theorem 7.8.1 *Let (ϕ,Φ) be a t-fuzzy direlation from $\mathcal{F}(U)$ to $\mathcal{F}(V)$. The following statements hold:*

(1) (i) $\forall\alpha\in\mathcal{F}(U),\ \alpha\leq\phi^{\leftarrow}(\phi^{\to}\alpha)$.

(ii) $\forall \beta \in \mathcal{F}(V),\ \phi^{\to}(\phi^{\leftarrow}\beta) \le \beta$.

(2) (i) $\forall \beta \in \mathcal{F}(V),\ \beta \le \Phi^{\to}(\Phi^{\leftarrow}\beta)$.

(ii) $\forall \alpha \in \mathcal{F}(U),\ \Phi^{\leftarrow}(\Phi^{\to}\alpha) \le \alpha$.

Proof (i) Let us consider the corresponding direlation (r, R) from $(W_U, \mathcal{W}_U)$ to $(W_V, \mathcal{W}_V)$ where $\phi = \phi_r$ and $\Phi = \Phi_R$. By Lemma 4.2.15 (1)(ii),

$$\forall \mathfrak{I}(\alpha) \in \mathcal{W}_U,\ \mathfrak{I}(\alpha) \subseteq r^{\leftarrow}r^{\to}\mathfrak{I}(\alpha).$$

Then by Proposition 7.3.38 (i),

$$\mathfrak{I}(\alpha) \subseteq r^{\leftarrow}r^{\to}\mathfrak{I}(\alpha) = r^{\leftarrow}\mathfrak{I}(\phi^{\to}\alpha) = \mathfrak{I}(\phi^{\leftarrow}\phi^{\to}\alpha).$$

Since $\mathfrak{I}$ is a lattice isomorphism between $\mathcal{W}_U$ and $\mathcal{F}(U)$, we obtain that

$$\forall \alpha \in \mathcal{F}(U),\ \alpha \le \phi^{\leftarrow}(\phi^{\to}\alpha).$$

The proof of the rest of the items is similar and omitted.

Corollary 7.8.2 *Let (U, V, ϕ, Φ) be a generalized t-fuzzy approximation space. Further, let (r, R) be the direlation from $(W_U, \mathcal{W}_U)$ to $(W_V, \mathcal{W}_V)$ where $\phi = \phi_r$ and $\Phi = \Phi_R$. Then we have the following equivalences:*

(i) *(ϕ, Φ) is serial if and only if (r, R) is t-serial.*

(ii) *(ϕ, Φ) is inverse serial if and only if (r, R) is t-inverse serial.*

(iii) *(ϕ, Φ) is compatible if and only if (r, R) is t-compatible.*

(iv) *(ϕ, Φ) is injective if and only if (r, R) is injective.*

Proof (i) Let (ϕ, Φ) be serial. Suppose that $u, u' \in U$, $\lambda, \lambda' \in (0, 1]$ and $\overline{P}_{(u,\lambda)} \nsubseteq \overline{Q}_{(u',\lambda')}$. Then $u = u'$ and $\lambda' < \lambda$. By the assumption, for some $v \in V$,

$$\Phi(u, \lambda', v) < \phi(u, \lambda, v).$$

Let us choose a $\mu \in (0, 1])$ such that

$$\Phi(u, \lambda', v) < \mu < \phi(u, \lambda, v).$$

Then by Corollary 7.3.3,

$$r \not\subseteq \overline{Q}_{((u,\lambda),(v,\mu))} \text{ and } \overline{P}_{((u,\lambda'),(v,\mu))} \nsubseteq R.$$

This shows that (r, R) is t-serial. Now let (r, R) be t-serial. Take $u \in U$, $\lambda, \lambda' \in (0, 1]$ with $\lambda' < \lambda$. Then we have $\overline{P}_{(u,\lambda)} \nsubseteq \overline{Q}_{(u,\lambda')}$. Since (r, R) is t-serial, for some $v \in V$,

$$r \nsubseteq \overline{Q}_{((u,\lambda),(v,\mu))} \text{ and } \overline{P}_{((u,\lambda'),(v,\mu))} \nsubseteq R.$$

Therefore, again by Corollary 7.3.3,

$$\mu < \phi(u, \lambda, v) \text{ and } \mu > \Phi(u, \lambda', v).$$

This gives that $\Phi(u, \lambda', v) < \phi(u, \lambda, v)$.

(ii) Let $\overline{P}_{(v,\mu)} \not\subseteq \overline{Q}_{(v',\mu')}$ for some $\mu, \mu' \in (0, 1]$ and $v, v' \in V$. Then we have $v = v'$ and $\mu' < \mu$. By the assumption, there exists $u \in U$ and $\lambda \in (0, 1]$ such that

$$\mu' < \phi(u, \lambda, v') \text{ and } \Phi(u, \lambda, v) < \mu.$$

Then by Corollary 7.3.3, we conclude that

$$r \not\subseteq \overline{Q}_{((u,\lambda),(v',\mu'))} \text{ and } \overline{P}_{((u,\lambda),(v,\mu))} \not\subseteq R.$$

That is, (r, R) is t-inverse serial. The proof of the sufficiency is similar.

(iii) It is immediate by (i) and (ii).

(iv) It is omitted.

The pairs $(\phi^{\rightarrow}, \Phi^{\rightarrow})$ and $(\Phi^{\leftarrow}, \phi^{\leftarrow})$ constitute two different (textural) interval structures.

Theorem 7.8.3 *Let (ϕ, Φ) be a t-fuzzy direlation from $\mathcal{F}(U)$ to $\mathcal{F}(V)$ and $\{\beta_j \mid j \in J\} \subseteq \mathcal{F}(V)$.*

(1) *If (ϕ, Φ) is serial, then $\phi^{\leftarrow}\mathbf{0} = \mathbf{0}$.*

(2) (i) $\phi^{\leftarrow}\mathbf{1} = \mathbf{1}$

(ii) $\bigvee_{j\in J} \phi^{\leftarrow}\beta_j \leq \phi^{\leftarrow} \bigvee_{j\in J} \beta_j$.

(iii) $\bigwedge_{j\in J} \phi^{\leftarrow}\beta_j = \phi^{\leftarrow} \bigwedge_{j\in J} \beta_j$.

(3) *If (ϕ, Φ) is serial, then $\Phi^{\leftarrow}\mathbf{1} = \mathbf{1}$.*

(4) (i) $\Phi^{\leftarrow}\mathbf{0} = \mathbf{0}$.

(ii) $\Phi^{\leftarrow} \bigwedge_{j\in J} \beta_j \leq \bigwedge_{j\in J} \Phi^{\leftarrow}\beta_j$.

(iii) $\bigvee_{j\in J} \Phi^{\leftarrow}\beta_j = \Phi^{\leftarrow} \bigvee_{j\in J} \beta_j$.

Proof For the proof of (1) and (3), let (ϕ, Φ) be serial. Then by Corollary 7.8.2 (i), the corresponding direlation (r, R) is t-serial. Hence, by Theorem 4.4.17 (3), we have

$$r^{\leftarrow}\emptyset = \emptyset \text{ and } R^{\leftarrow}(V \times (0, 1]) = U \times (0, 1].$$

Further, note that since $\mathfrak{I}$ is a lattice isomorphism

$$\mathfrak{I}(\mathbf{0}) = \emptyset \text{ and } \mathfrak{I}(\mathbf{1}) = V \times (0, 1].$$

Hence,

$$r^{\leftarrow}\mathfrak{I}(\mathbf{0}) = \mathfrak{I}(\mathbf{0}) \text{ and } R^{\leftarrow}\mathfrak{I}(\mathbf{1}) = \mathfrak{I}(\mathbf{1}).$$

By Proposition 7.3.38 (ii), we obtain

$$\Im(\phi^{\leftarrow}\mathbf{0}) = \Im(\mathbf{0}) \text{ and } \Im(\Phi^{\leftarrow}\mathbf{1}) = \Im(\mathbf{1}),$$

that is,

$$\phi^{\leftarrow}\mathbf{0} = \mathbf{0} \text{ and } \Phi^{\leftarrow}\mathbf{1} = \mathbf{1}.$$

(2)(i) By Theorem 4.4.17 (1), we have

$$r^{\leftarrow}(V \times (0, 1]) = (U \times (0, 1],$$

and hence,

$$r^{\leftarrow}\Im(\mathbf{1}) = \Im(\mathbf{1}),$$

and so by Proposition 7.3.38 (ii), we obtain $\Im(\phi^{\leftarrow}\mathbf{1}) = \Im(\mathbf{1})$.

This implies that

$$\phi^{\leftarrow}\mathbf{1} = \mathbf{1}.$$

(2)(ii) Clearly, for all j, $\beta_j \leq \bigvee_{j \in J} \beta_j$ and by Theorem 7.3.26 (ii), we obtain

$$\phi^{\leftarrow}\beta_j \leq \phi^{\leftarrow} \bigvee_{j \in J} \beta_j,$$

and hence,

$$\bigvee_{j \in J} \phi^{\leftarrow}\beta_j \leq \phi^{\leftarrow} \bigvee_{j \in J} \beta_j.$$

(2)(iii) It was proved in Theorem 7.3.26 (i).

The proof (3) and (4) is similar. □

Theorem 7.8.4 *Let* (ϕ, Φ) *be a t-fuzzy direlation from* $\mathcal{F}(U)$ *to* $\mathcal{F}(V)$ *and* $\{\alpha_j \mid j \in J\} \subseteq \mathcal{F}(U)$.

(1) *If* (ϕ, Φ) *is inverse serial, then* $\phi^{\rightarrow}\mathbf{1} = \mathbf{1}$.

(2) (i) $\phi^{\rightarrow}\mathbf{0} = \mathbf{0}$.

(ii) $\phi^{\rightarrow} \bigwedge_{j \in J} \alpha_j \leq \bigwedge_{j \in J} \phi^{\rightarrow}\alpha_j$.

(iii) $\bigvee_{j \in J} \phi^{\rightarrow}\alpha_j = \phi^{\rightarrow} \bigvee_{j \in J} \alpha_j$.

(3) *If* (ϕ, Φ) *is inverse serial, then* $\Phi^{\rightarrow}\mathbf{0} = \mathbf{0}$.

(4) (i) $\Phi^{\rightarrow}\mathbf{1} = \mathbf{1}$.

(ii) $\bigvee_{j \in J} \Phi^{\rightarrow}\alpha_j \leq \Phi^{\rightarrow} \bigvee_{j \in J} \alpha_j$.

(iii) $\bigwedge_{j \in J} \Phi^{\rightarrow}\alpha_j = \Phi^{\rightarrow} \bigwedge_{j \in J} \alpha_j$.

Proof It is omitted.

Theorem 7.8.5 *Let* (ϕ, Φ) *be a t-fuzzy direlation from* $\mathcal{F}(U)$ *to* $\mathcal{F}(V)$. *The following statements are equivalent:*

(i) (ϕ, Φ) *is serial.*

(ii) $\phi^{\leftarrow} \circ \Phi \leq \Phi_I$ *where* Φ_I *is the identity t-fuzzy corelation on* $(\mathcal{W}_U, \mathcal{W}_U)$.

(iii) $\forall \alpha \in \mathcal{F}(U),\ \phi^{\leftarrow}(\Phi^{\rightarrow}\alpha) \leq \alpha$.

(iv) $\forall \beta \in \mathcal{F}(V),\ \phi^{\leftarrow}\beta \leq \Phi^{\leftarrow}\beta$.

(v) $\forall \alpha_1, \alpha_2 \in \mathcal{F}(U),\ \Phi^{\rightarrow}\alpha_1 \leq \phi^{\rightarrow}\alpha_2 \implies \alpha_1 \leq \alpha_2$.

Proof (i) $\implies$ (ii): Let (ϕ, Φ) be serial. Then by Corollary 7.8.2(i), (r, R) is t-serial where $\phi = \phi_r$ and $\Phi = \Phi_R$. Hence, by Theorem 4.4.11 (ii), $r^{\leftarrow} \circ R \subseteq I$. Further, by Theorem 7.3.11(ii), $\Phi_{r^{\leftarrow} \circ R} \leq \Phi_I$. By definition of composition of fuzzy corelations, $\Phi_{r^{\leftarrow}} \circ \Phi_R \leq \Phi_I$ and then $\phi_r^{\leftarrow} \circ \Phi_R \leq \Phi_I$. This implies that $\phi^{\leftarrow} \circ \Phi \leq \Phi_I$.

(ii) $\implies$ (iii): By definition of composition note that $\Phi_{r^{\leftarrow} \circ R} \leq \Phi_I$ where $\phi = \phi_r$ and $\Phi = \Phi_R$. Further, note that $\phi^{\leftarrow} = \phi_{r^{\leftarrow}}$. Now, by Theorem 7.3.11(ii), $r^{\leftarrow} \circ R \subseteq I$. Therefore, again by Theorem 4.4.11, we get $r^{\leftarrow} R^{\rightarrow}(\mathfrak{I}(\alpha)) \subseteq \mathfrak{I}(\alpha)$. Note that we also have

$$r^{\leftarrow} R^{\rightarrow}(\mathfrak{I}(\alpha)) \subseteq \mathfrak{I}(\alpha) \implies r^{\leftarrow}\mathfrak{I}(\Phi^{\rightarrow}\alpha) \subseteq \mathfrak{I}(\alpha) \implies \mathfrak{I}(\phi^{\rightarrow}\Phi^{\leftarrow}\alpha) \subseteq \mathfrak{I}(\alpha).$$

Since $\mathfrak{I}$ is a lattice isomorphism, we conclude that

$$\phi^{\leftarrow}(\Phi^{\rightarrow}\alpha) \leq \alpha.$$

(iii) $\implies$ (iv): By the assumption, for all $\alpha \in \mathcal{F}(U)$, we have $\mathfrak{I}(\phi^{\leftarrow}(\Phi^{\rightarrow}\alpha)) \subseteq \mathfrak{I}(\alpha)$ and then

$$r^{\leftarrow}(\mathfrak{I}(\Phi^{\rightarrow}\alpha)) \subseteq \mathfrak{I}(\alpha)$$

and so

$$r^{\leftarrow} R^{\rightarrow}\mathfrak{I}(\alpha) \subseteq \mathfrak{I}(\alpha).$$

Then from Theorem 4.4.11, we obtain that for all $B \in \mathcal{W}_V$, $r^{\leftarrow}B \subseteq R^{\leftarrow}B$. Hence, by Lemma 4.2.7, $r^{\leftarrow} \subseteq R^{\leftarrow}$ and so $\phi^{\leftarrow} \leq \Phi^{\leftarrow}$ where $\phi = \phi_r$ and $\Phi = \Phi_R$. This gives that

$$\phi^{\leftarrow}\beta \leq \Phi^{\leftarrow}\beta, \text{ for all } \beta \in \mathcal{F}(V).$$

(iv) $\implies$ (v): By the assumption, for all $\beta \in \mathcal{F}(V)$, the inclusion $\mathfrak{I}(\phi^{\leftarrow}\beta) \subseteq \mathfrak{I}(\Phi^{\leftarrow}\beta)$ holds. This implies that $r^{\leftarrow}\mathfrak{I}(\beta) \subseteq R^{\leftarrow}\mathfrak{I}(\beta)$. Now, by Theorem 4.4.11, we get

$$\forall \alpha_1, \alpha_2 \in \mathcal{F}(U),\ r^{\rightarrow}\mathfrak{I}(\alpha_1) \subseteq R^{\rightarrow}\mathfrak{I}(\alpha_2) \implies \mathfrak{I}(\alpha_1) \subseteq \mathfrak{I}(\alpha_2).$$

Since $\mathfrak{I}$ is a lattice isomorphism, this is, it exactly means that

$$\forall \alpha_1, \alpha_2 \in \mathcal{F}(U),\ \mathfrak{I}(\phi^{\rightarrow}\alpha_1) \subseteq \mathfrak{I}(\Phi^{\rightarrow}\alpha_2) \implies \mathfrak{I}(\alpha_1) \subseteq \mathfrak{I}(\alpha_2),$$

and this gives that

$$\forall \alpha_1, \alpha_2 \in \mathcal{F}(U),\ \phi^{\rightarrow}\alpha_1 \leq \Phi^{\rightarrow}\alpha_2 \implies \alpha_1 \leq \alpha_2.$$

(v) $\implies$ (i): Note that we have

$$\forall \mathfrak{I}(\alpha_1), \mathfrak{I}(\alpha_2) \in \mathcal{W}_U,\ r^{\rightarrow}\mathfrak{I}(\alpha_1) \subseteq R^{\rightarrow}\mathfrak{I}(\alpha_2) \implies \mathfrak{I}(\alpha_1) \subseteq \mathfrak{I}(\alpha_2).$$

From Theorem 4.4.11, (r, R) is t-serial where $\phi = \phi_r$ and $\Phi = \Phi_R$. Hence, by Corollary 7.8.2, (ϕ, Φ) is serial.

Theorem 7.8.6 *Let (ϕ, Φ) be a t-fuzzy direlation from $\mathcal{F}(U)$ to $\mathcal{F}(V)$. The following statements are equivalent:*

(i) *(ϕ, Φ) is inverse serial.*

(ii) *$\phi \circ \Phi^{\leftarrow} \leq \Phi_I$ where Φ_I is the identity t-fuzzy corelation on $(W_V, \mathcal{W}_V)$.*

(iii) $\forall \beta \in \mathcal{F}(V),\ \beta \leq \phi^{\rightarrow}(\Phi^{\leftarrow}\beta)$.

(iv) $\forall \alpha \in \mathcal{F}(U),\ \Phi^{\rightarrow}\alpha \leq \phi^{\rightarrow}\alpha$.

(v) $\forall \beta_1, \beta_2 \in \mathcal{F}(V),\ \Phi^{\leftarrow}\beta_1 \leq \phi^{\leftarrow}\beta_2 \implies \beta_1 \leq \beta_2$.

Proof It is similar to the proof of Theorem 7.8.5 and left to the reader. □

Theorem 7.8.7 *Let (ϕ, Φ) be an inverse serial t-fuzzy direlation from $\mathcal{F}(U)$ to $\mathcal{F}(V)$. Then*

(i) $\forall v_\mu \in \mathcal{F}(V),\ \Phi^{\leftarrow}v_\mu \neq \mathbf{0}$.

(ii) *If $\mathbf{0} \neq \beta \in \mathcal{F}(V)$, then $\Phi^{\leftarrow}\beta \neq \mathbf{0}$.*

(iii) *If $\mathbf{1} \neq \beta \in \mathcal{F}(V)$, then $\phi^{\leftarrow}\beta \neq \mathbf{1}$.*

Proof (i) By Theorems 7.3.38(ii) and 4.4.13, we easily observe that

$$\mathfrak{I}(\Phi^{\leftarrow}v_\mu) = R^{\leftarrow}\mathfrak{I}(v_\mu) = R^{\leftarrow}\overline{P}_{(v,\mu)} \neq \emptyset.$$

This implies that $\Phi^{\leftarrow}u_\lambda \neq \mathbf{0}$.

(ii) Again by Theorems 7.3.38(ii) and 4.4.13,

$$\mathfrak{I}(\Phi^{\leftarrow}\beta) = R^{\leftarrow}\mathfrak{I}(\beta) \neq \emptyset.$$

Since $\mathfrak{I}$ is a lattice isomorphism, we obtain $\Phi^{\leftarrow}\beta \neq \mathbf{0}$.

(iii) The proof is omitted.

Theorem 7.8.8 *Let (ϕ, Φ) be a serial t-fuzzy direlation. Then*

(i) $\forall u_\lambda \in \mathcal{F}(U),\ \Phi^{\rightarrow}u_\lambda \neq \mathbf{0}$.

(ii) *If $\mathbf{0} \neq \alpha \in \mathcal{F}(U)$, then $\Phi^{\rightarrow}\alpha \neq \mathbf{0}$.*

(iii) *If $\mathbf{1} \neq \alpha \in \mathcal{F}(U)$, then $\phi^{\rightarrow}\alpha \neq \mathbf{1}$.*

Proof It is left to the interested reader. □

Theorem 7.8.9 *Let (ϕ, Φ) be a t-fuzzy direlation from $\mathcal{F}(U)$ to $\mathcal{F}(V)$.*

(1) *If (ϕ, Φ) is serial, then for all $\alpha \in \mathcal{F}(U)$,*

$\alpha \leq \Phi^{\leftarrow}(\phi^{\rightarrow}\alpha)$.

(2) *If (ϕ, Φ) is injective, then for all $\alpha \in \mathcal{F}(U)$,*

(i) $\Phi^{\leftarrow}(\phi^{\rightarrow}\alpha) \leq \alpha$.

(ii) $\phi^{\rightarrow}\alpha \leq \Phi^{\rightarrow}\alpha$.

(iii) $\alpha \leq \phi^{\leftarrow}(\Phi^{\rightarrow}\alpha)$.

(3) *If (ϕ, Φ) is inverse serial and injective, then for all $\alpha \in \mathcal{F}(U)$,*

$\phi^{\rightarrow}\alpha = \Phi^{\rightarrow}\alpha$.

(4) *If (ϕ, Φ) is injective and serial, then for all $\alpha \in \mathcal{F}(U)$, we have the following equalities.*

(i) $\phi^{\leftarrow}(\phi^{\rightarrow}\alpha) = \alpha$.

(ii) $\Phi^{\leftarrow}(\Phi^{\rightarrow}\alpha) = \alpha$.

Proof (1) If (ϕ, Φ) serial, then (r, R) is serial where $\phi = \phi_r$ and $\Phi = \Phi_R$, by Corollary 7.8.2(i). Then by Theorem 4.4.15 (1), for all $\alpha \in \mathcal{F}(U)$,

$\Im(\alpha) \subseteq R^{\leftarrow}(r^{\rightarrow}\Im(\alpha))$.

By Theorem 7.3.38(ii),

$R^{\leftarrow}(r^{\rightarrow}\Im(\alpha)) = R^{\leftarrow}(\Im(\phi^{\rightarrow}\alpha)) = \Im(\Phi^{\leftarrow}(\phi^{\rightarrow}\alpha)) \supseteq \Im(\alpha)$.

Finally, we obtain $\alpha \leq \Phi^{\leftarrow}(\phi^{\rightarrow}\alpha)$, since $\Im$ is a lattice isomorphism.

(2)(ii) Let (ϕ, Φ) be injective. Then by Corollary 7.8.2(iv), the corresponding direlation (r, R) is also injective. Hence, by Theorem 4.4.15 (2)(ii), for all $\alpha \in \mathcal{F}(U)$, the inclusion

$r^{\rightarrow}\Im(\alpha) \subseteq R^{\rightarrow}\Im(\alpha)$.

Therefore, by Theorem 7.3.38(i),

$\Im(\phi^{\rightarrow}\alpha) \leq \Im(R^{\rightarrow}\alpha)$, that is, $\phi^{\rightarrow}\alpha \leq \Phi^{\rightarrow}\alpha$ as desired.

The rest of the proofs are left to the reader.

Dually, we have the following result.

Theorem 7.8.10 *Let (ϕ, Φ) be an inverse serial t-fuzzy direlation from $\mathcal{F}(U)$ to $\mathcal{F}(V)$.*

(1) *If (ϕ, Φ) is inverse serial, then for all $\beta \in \mathcal{F}(V)$,*

$\Phi^{\rightarrow}(\phi^{\leftarrow}\beta) \leq \beta$.

(2) *If (ϕ, Φ) is injective, then for all $\beta \in \mathcal{F}(V)$,*

(i) $\beta \leq \Phi^{\rightarrow}(\phi^{\leftarrow}\beta)$.

(ii) $\phi^{\leftarrow}\beta \leq \Phi^{\leftarrow}\beta$.

(iii) $\phi^{\rightarrow}(\Phi^{\leftarrow}\beta) \leq \beta$.

(3) *If* (ϕ, Φ) *is serial and injective, then for all* $\beta \in \mathcal{F}(V)$,

$\phi^{\leftarrow}\beta = \Phi^{\leftarrow}\beta$.

(4) *If* (ϕ, Φ) *is inverse serial and injective, then for all* $\beta \in \mathcal{F}(V)$, *we have the following equalities.*

(i) $\phi^{\rightarrow}(\phi^{\leftarrow}\beta) = \beta$.

(ii) $\Phi^{\rightarrow}(\Phi^{\leftarrow}\beta) = \beta$.

Proof It is similar to the proof of Theorem 7.8.9.

Theorem 7.8.11 *Let* (ϕ, Φ) *be a t-fuzzy direlation from* $\mathcal{F}(U)$ *to* $\mathcal{F}(V)$. *Then*

(1) $\phi^{\leftarrow}\mathbf{1} = \mathbf{1}$ *and* $\Phi^{\leftarrow}\mathbf{0} = \mathbf{0}$.

(2) *For all* $\beta \in \mathcal{F}(V)$, $\Phi^{\leftarrow}(\phi^{\rightarrow}(\phi^{\leftarrow}(\beta)) \leq \Phi^{\leftarrow}\beta$.

(3) *If* (ϕ, Φ) *is serial, then* $\phi^{\leftarrow}\mathbf{0} = \mathbf{0}$ *and* $\Phi^{\leftarrow}\mathbf{1} = \mathbf{1}$.

(4) *If* (ϕ, Φ) *is serial, then* $\Phi^{\leftarrow}\beta \leq \Phi^{\leftarrow}(\phi^{\rightarrow}(\Phi^{\leftarrow}\beta))$.

(5) *If* (ϕ, Φ) *is inverse serial, then* $\phi^{\rightarrow}\mathbf{1} = \mathbf{1}$ *and* $\Phi^{\rightarrow}\mathbf{0} = \mathbf{0}$.

Proof (4) Let (ϕ, Φ) be serial. Then (r, R) is t-serial by Corollary 7.8.2(i). Thus, for all $\beta \in \mathcal{F}(V)$, the inclusion

$$R^{\leftarrow}\mathfrak{I}(\beta) \subseteq R^{\leftarrow}(r^{\rightarrow}(R^{\leftarrow}\mathfrak{I}(\beta))$$

holds. Therefore, by Proposition 7.3.38, we get

$$\begin{aligned}\mathfrak{I}(\Phi^{\leftarrow}\beta) = R^{\leftarrow}\mathfrak{I}(\beta) \subseteq R^{\leftarrow}(r^{\rightarrow}(R^{\leftarrow}\mathfrak{I}(\beta)) &= R^{\leftarrow}(r^{\rightarrow}(\mathfrak{I}(\Phi^{\leftarrow}\beta))\\ &= R^{\leftarrow}\mathfrak{I}((\phi^{\rightarrow}(\Phi^{\leftarrow}\beta)) = \mathfrak{I}(\Phi^{\leftarrow}(\phi^{\rightarrow}\Phi^{\leftarrow}\beta)),\end{aligned}$$

and so we obtain

$$\Phi^{\leftarrow}\beta \leq \Phi^{\leftarrow}(\phi^{\rightarrow}(\Phi^{\leftarrow}\beta)),$$

since $\mathfrak{I}$ is a lattice isomorphism.

Theorem 7.8.12 *Let* (ϕ, Φ) *be a t-fuzzy direlation from* $\mathcal{F}(U)$ *to* $\mathcal{F}(V)$. *Then*

(i) (a) *If* (ϕ, Φ) *is injective, then* $\phi^{\leftarrow}(\phi^{\rightarrow}(\Phi^{\leftarrow}\beta)) \leq \Phi^{\leftarrow}\beta$.

(b) *If* $(\Phi^{\leftarrow}, \phi^{\leftarrow})$ *is injective, then* $\phi^{\leftarrow}(\phi^{\rightarrow}(\Phi^{\leftarrow}\beta)) \leq \Phi^{\leftarrow}\beta$.

(ii) *If* $(\Phi^{\leftarrow}, \phi^{\leftarrow})$ *is injective, then* $\Phi^{\leftarrow}\beta \leq \Phi^{\leftarrow}(\phi^{\rightarrow}(\Phi^{\leftarrow}\beta))$.

(a) $\Phi^{\leftarrow}(\phi^{\rightarrow}(\Phi^{\leftarrow}\beta)) = \Phi^{\leftarrow}\beta$.

(b) $\phi^{\leftarrow}(\Phi^{\rightarrow}(\phi^{\leftarrow}\beta)) = \phi^{\leftarrow}\beta$.

(c) $\Phi^{\leftarrow}(\Phi^{\rightarrow}(\phi^{\leftarrow}\beta)) = \phi^{\leftarrow}\beta$.

(d) $\phi^{\leftarrow}(\Phi^{\rightarrow}(\Phi^{\leftarrow}\beta)) = \Phi^{\leftarrow}\beta$.

(iii) *If* (ϕ, Φ) *is serial, then* $\Phi^{\leftarrow}(\phi^{\rightarrow}(\phi^{\leftarrow}\beta)) = \phi^{\leftarrow}\beta$.

Proof (ii)(b) By Theorem 4.4.18 (2)(i)(b), we have

$$r^{\leftarrow}(R^{\rightarrow}(r^{\leftarrow}\mathfrak{I}(\beta)) = r^{\leftarrow}\mathfrak{I}(\beta),$$

for all $\beta \in \mathcal{F}(V)$ where $\phi = \phi_r$ and $\Phi = \Phi_R$.

As a result,

$$\begin{aligned}\mathfrak{I}(\phi^{\leftarrow}\beta) &= r^{\leftarrow}\mathfrak{I}(\beta) = r^{\leftarrow}(R^{\rightarrow}(r^{\leftarrow}\mathfrak{I}(\beta))\\ &= r^{\leftarrow}(R^{\rightarrow}(\mathfrak{I}(\phi^{\leftarrow}\beta)) = r^{\leftarrow}((\mathfrak{I}(\Phi^{\rightarrow}\phi^{\leftarrow}\beta)) = \mathfrak{I}(\phi^{\leftarrow}\Phi^{\rightarrow}\phi^{\leftarrow}\beta).\end{aligned}$$

Since $\mathfrak{I}$ is a latice isomorphism, we obtain

$$\phi^{\leftarrow}(\Phi^{\rightarrow}(\phi^{\leftarrow}\beta)) = \phi^{\leftarrow}\beta,$$

as desired.

Theorem 7.8.13 *Let* (ϕ, Φ) *be a t-fuzzy direlation from* $\mathcal{F}(U)$ *to* $\mathcal{F}(V)$.

(i) (a) *If* $(\Phi^{\leftarrow}, \phi^{\leftarrow})$ *is injective, then* $\Phi^{\rightarrow}\alpha \leq \phi^{\rightarrow}(\phi^{\leftarrow}(\Phi^{\leftarrow}\alpha))$.

(b) *If* (ϕ, Φ) *is injective, then* $\phi^{\rightarrow}(\phi^{\leftarrow}(\Phi^{\rightarrow}\alpha)) \leq \Phi^{\rightarrow}\alpha$.

(ii) *If* (ϕ, Φ) *is injective and inverse serial, then*

(a) $\Phi^{\rightarrow}(\phi^{\leftarrow}(\Phi^{\rightarrow}\alpha)) = \Phi^{\rightarrow}\alpha$.

(b) $\phi^{\rightarrow}(\Phi^{\leftarrow}(\phi^{\rightarrow}\alpha)) = \phi^{\rightarrow}\alpha$.

(c) $\Phi^{\rightarrow}(\Phi^{\leftarrow}(\phi^{\rightarrow}\alpha)) = \phi^{\rightarrow}\alpha$.

(d) $\phi^{\rightarrow}(\Phi^{\leftarrow}(\Phi^{\rightarrow}\alpha)) = \Phi^{\rightarrow}\alpha$.

(iii) *If* (ϕ, Φ) *is inverse serial, then* $\Phi^{\rightarrow}(\phi^{\leftarrow}(\phi^{\rightarrow}\alpha)) = \phi^{\rightarrow}\alpha$.

Proof (iii) Let us take the corresponding t-inverse serial direlation (r, R). By Theorem 4.4.19 (iii), for all $\mathfrak{I}(\alpha) \in \mathcal{W}_U$, the equality

$$R^{\rightarrow}(r^{\leftarrow}(r^{\rightarrow}\mathfrak{I}(\alpha))) = r^{\rightarrow}\mathfrak{I}(\alpha)$$

holds. If we apply Proposition 7.3.38, then

$$\begin{aligned}R^{\rightarrow}(r^{\leftarrow}(r^{\rightarrow}\mathfrak{I}(\alpha))) &= R^{\rightarrow}(r^{\leftarrow}\mathfrak{I}((\phi^{\rightarrow}\alpha)))\\ &= R^{\rightarrow}(\mathfrak{I}((\phi^{\leftarrow}\phi^{\rightarrow}\alpha))) = \mathfrak{I}(\Phi^{\rightarrow}(\phi^{\leftarrow}\phi^{\rightarrow}\alpha))) = \mathfrak{I}(\phi^{\rightarrow}\alpha).\end{aligned}$$

This gives the desired equality

$$\Phi^{\rightarrow}(\phi^{\leftarrow}(\phi^{\rightarrow}\alpha)) = \phi^{\rightarrow}\alpha.$$

7.9 Definability in t-Fuzzy Approximation Spaces

Definition 7.9.1 Let (U, V, ϕ, Φ) be a generalized t-fuzzy approximation space. Then

(1) $\beta \in \mathcal{F}(V)$ is called *presectional definable* if $\phi^{\leftarrow}\beta = \Phi^{\leftarrow}\beta$.

(2) $\alpha \in \mathcal{F}(U)$ is called *sectional definable* if $\phi^{\rightarrow}\alpha = \Phi^{\rightarrow}\alpha$.

Theorem 7.9.2 *Let (U, V, ϕ, Φ) be a generalized t-fuzzy approximation space where (ϕ, Φ) is compatible.*

(1) *The following statements are equivalent:*

(i) *$\beta \in \mathcal{F}(V)$ is fuzzy presectional definable.*

(ii) *For some $\alpha \in \mathcal{F}(U)$, $\phi^{\rightarrow}\alpha = \beta$ and $\Phi^{\leftarrow}\beta = \alpha$.*

(2) *The following statements are equivalent:*

(i) *$\alpha \in \mathcal{F}(U)$ is fuzzy sectional definable.*

(ii) *For some $\beta \in \mathcal{F}(V)$, $\Phi^{\rightarrow}\alpha = \beta$ and $\phi^{\leftarrow}\beta = \alpha$.*

Proof We give the proof of (1) leaving the dual proof of (2) to the interested reader. Let $\beta \in \mathcal{F}(V)$ be presectional definable, that is, $\phi^{\leftarrow}\beta = \Phi^{\leftarrow}\beta$. Note that by Proposition 7.3.38 (ii),

$$r^{\leftarrow}\Im(\beta) = \Im(\phi^{\leftarrow}\beta) = \Im(\Phi^{\leftarrow}\beta) = R^{\leftarrow}\Im(\beta),$$

that is, $\Im(\beta)$ is presectional definable set in $\mathcal{W}_V$. Since (ϕ, Φ) is compatible, by Corollary 7.8.2, the corresponding direlation (r, R) is t-compatible. Then by Theorem 5.5.6 (i), for some $A \in \mathcal{W}_U$, we have

$$r^{\rightarrow}A = \Im(\beta) \text{ and } R^{\leftarrow}\Im(\beta) = A.$$

Hence, we get

$$r^{\rightarrow}\Im(\alpha) = \Im(\beta) \text{ and } R^{\leftarrow}\Im(\beta) = \Im(\alpha)$$

where $\Im(\alpha) = A$ for some $\alpha \in \mathcal{F}(U)$. This implies that

$$\Im(\phi^{\rightarrow}\alpha) = \Im(\beta) \text{ and } \Im(\Phi^{\leftarrow}\beta) = \Im(\alpha),$$

that is, $\phi^{\rightarrow}\alpha = \beta$ and $\Phi^{\leftarrow}\beta = \alpha$.

Now let $\phi^{\rightarrow}\alpha = \beta$ and $\Phi^{\leftarrow}\beta = \alpha$ for some $\beta \in \mathcal{F}(V)$. Clearly,

$$\Im(\phi^{\rightarrow}\alpha) = \Im(\beta) \text{ and } \Im(\Phi^{\leftarrow}\beta) = \Im(\alpha).$$

Again, by Theorem 5.5.6 (i),

$$R^{\rightarrow}\Im(\alpha) = \Im(\beta) \text{ and } R^{\leftarrow}\Im(\beta) = \Im(\alpha).$$

Therefore, $\Im(\beta)$ is presectional definable and so $R^{\leftarrow}\Im(\beta) = r^{\leftarrow}\Im(\beta)$. Finally, by Proposition 7.3.38 (ii), we conclude that $\Im(\Phi^{\leftarrow}\beta) = \Im(\phi^{\leftarrow}\beta)$, that is, $\Phi^{\leftarrow}\beta = \phi^{\leftarrow}\beta$.

Definition 7.9.3 Let (ϕ, Φ) be a t-fuzzy direlation from $\mathcal{F}(U)$ to $\mathcal{F}(V)$. Then

(i) $\beta \in \mathcal{F}(V)$ is called *fuzzy successor definable* if for some $\alpha \in \mathcal{F}(U)$,

$$\beta = \bigvee_{\alpha \not\leq u^{\lambda}} \phi^{\rightarrow} u_{\lambda}.$$

(ii) $\alpha \in \mathcal{F}(U)$ is called *fuzzy predecessor definable* if for some $\beta \in \mathcal{F}(V)$,

$$\alpha = \bigvee_{\beta \not\leq v^{\mu}} \Phi^{\leftarrow} v_{\mu}.$$

Proposition 7.9.4 *Let (U, V, ϕ, Φ) be a generalized t-fuzzy approximation space. Then we have the following equivalences:*

(1) *The following statements are equivalent:*

(i) $\beta \in \mathcal{F}(V)$ *is fuzzy successor definable.*

(ii) *For some $\alpha \in \mathcal{F}(U)$, $\beta = \phi^{\rightarrow}\alpha$.*

(2) *The following statements are equivalent:*

(i) $\alpha \in \mathcal{F}(U)$ *is fuzzy predecessor definable.*

(ii) *For some $\beta \in \mathcal{F}(V)$, $\alpha = \Phi^{\leftarrow}\beta$.*

(3) $\phi^{\rightarrow}\alpha$ *is fuzzy successor definable and* $\Phi^{\leftarrow}\beta$ *is fuzzy predecessor definable.*

Proof We give the proof of (1), leaving (2) to the reader. Let α be fuzzy predecessor definable. Then for some $\beta \in \mathcal{F}(V)$,

$$\alpha = \bigvee_{\beta \not\leq v^{\mu}} \Phi^{\leftarrow} v_{\mu}.$$

Since $\mathfrak{I}$ is a lattice isomorphism, by Proposition 7.3.38 (i),

$$\begin{aligned}\mathfrak{I}(\alpha) = \mathfrak{I}\Big(\bigvee_{\beta \not\leq v^{\mu}} \Phi^{\leftarrow} v_{\mu}\Big) &= \bigvee_{\beta \not\leq v^{\mu}} \mathfrak{I}(\Phi^{\leftarrow} v_{\mu}) \\ &= \bigvee_{\beta \not\leq v^{\mu}} R^{\leftarrow}\mathfrak{I}(v_{\mu}) = R^{\leftarrow}\Big(\bigvee_{\mathfrak{I}(\beta) \not\subseteq \overline{Q}_{(v,\mu)}} \mathfrak{I}(v_{\mu})\Big) = R^{\leftarrow}\mathfrak{I}(\beta).\end{aligned}$$

Then $\mathfrak{I}(\alpha)$ is textural predecessor definable and $\mathfrak{I}(\alpha) = R^{\leftarrow}\mathfrak{I}(\beta)$. Again, by Proposition 7.3.38 (ii),

$$\mathfrak{I}(\alpha) = R^{\leftarrow}\mathfrak{I}(\beta) = \mathfrak{I}(\Phi^{\leftarrow}\beta).$$

Finally, since $\mathfrak{I}$ is a lattice isomorphism, we obtain that $\alpha = \Phi^{\leftarrow}\beta$.

Conversely, let $\alpha = \Phi^{\leftarrow}\beta$ for some $\beta \in \mathcal{F}(V)$. Then by Lemma 1.2.4 and Corollary 7.3.28 (i),

$$\alpha = \Phi^{\leftarrow}\beta = \Phi^{\leftarrow} \bigvee_{\mu<\beta(v)} v_{\mu} = \bigvee_{\beta \nleq v^{\mu}} \Phi^{\leftarrow} v_{\mu},$$

that is, β is fuzzy predecessor definable.

(3) Note that again by Lemma 1.2.4 and Corollary 7.3.28 (i),

$$\phi^{\rightarrow}\alpha = \phi^{\rightarrow} \bigvee_{\lambda<\alpha(u)} u_{\lambda} = \bigvee_{\alpha \nleq u^{\lambda}} \phi^{\rightarrow} u_{\lambda}.$$

If we say $\beta = \phi^{\rightarrow}\alpha$, then $\phi^{\rightarrow}\alpha$ is fuzzy successor definable. Using a similar way, one can show that $\Phi^{\leftarrow}\beta$ is fuzzy predecessor definable.

We may also discuss on definabilities for generalized t-fuzzy approximation spaces with a single domain of discourse.

Theorem 7.9.5 *Let (ϕ, Φ) be a t-fuzzy direlation on $\mathcal{F}(U)$ and $\alpha \in \mathcal{F}(U)$.*

(1) *If (ϕ, Φ) is reflexive, then*

(i) $\phi^{\rightarrow}\alpha = \Phi^{\rightarrow}\alpha \implies \alpha$ *is fuzzy successor definable.*

(ii) $\phi^{\leftarrow}\alpha = \Phi^{\leftarrow}\alpha \implies \alpha$ *is fuzzy predecessor definable.*

(2) *If (ϕ, Φ) is a symmetric t-fuzzy direlation on $\mathcal{F}(U)$, then we have the following equivalences:*

(i) α *is fuzzy successor definable* $\iff \alpha$ *is fuzzy predecessor definable.*

(ii) $\phi^{\rightarrow}\alpha = \Phi^{\rightarrow}\alpha \iff \phi^{\leftarrow}\alpha = \Phi^{\leftarrow}\alpha$.

(3) *If (ϕ, Φ) is an equivalence t-fuzzy direlation on $\mathcal{F}(U)$, then the following conditions are equivalent:*

(i) α *is fuzzy successor definable.*

(ii) α *is fuzzy predecessor definable.*

(iii) $\phi^{\leftarrow}\alpha = \Phi^{\leftarrow}\alpha$.

(iv) $\phi^{\rightarrow}\alpha = \Phi^{\rightarrow}\alpha$.

Proof (1) (i) For some $\alpha \in \mathcal{F}(U)$, let $\phi^{\rightarrow}\alpha = \Phi^{\rightarrow}\alpha$. By Proposition 7.3.38 (i), we have

$$\mathfrak{I}(\phi^{\rightarrow}\alpha) - r^{\rightarrow}\mathfrak{I}(\alpha) = \mathfrak{I}(\Phi^{\rightarrow}\alpha) = R^{\rightarrow}\mathfrak{I}(\alpha),$$

where the pair (r, R) is the direlation on $\mathcal{W}_U$ and $\phi = \phi_r$ where $\Phi = \Phi_R$. Further, if (ϕ, Φ) is reflexive, then by Theorem 7.3.12, (r, R) is also a reflexive direlation on $\mathcal{W}_U$ where $\phi = \phi_r$ and $\Phi = \Phi_R$. Then by Theorem 5.5.10 (i), $\mathfrak{I}(\alpha)$ is t-successor definable. That is,

$$\mathfrak{I}(\alpha) = \bigvee_{\mathfrak{I}(\beta) \nsubseteq \overline{Q}_{(u,\lambda)}} r^{\rightarrow}\overline{P}_{(u,\lambda)},$$

for some $\mathfrak{I}(\beta) \in \mathcal{W}_U$. However, we have

$$\mathfrak{I}(\alpha) = \bigvee_{\mathfrak{I}(\beta)\nsubseteq \overline{Q}_{(u,\lambda)}} r^{\rightarrow}\overline{P}_{(u,\lambda)} = \bigvee_{\mathfrak{I}(\beta)\nsubseteq \overline{Q}_{(u,\lambda)}} r^{\rightarrow}\mathfrak{I}(u_\lambda)$$
$$= \bigvee_{\beta\nleq u^\lambda} \mathfrak{I}(\phi^{\rightarrow}u_\lambda) = \mathfrak{I}(\bigvee_{\beta\nleq u^\lambda} \phi^{\rightarrow}u_\lambda).$$

Since $\mathfrak{I}$ is a lattice isomorphism, we obtain that $\alpha = \bigvee_{\beta\nleq u^\lambda} \phi^{\rightarrow}u_\lambda$.

The proof of (ii) follows from a similar argument given in (i).

(2)(i) By Theorem 7.3.17, (r, R) is a symmetric direlation on $\mathcal{W}_U$ where $\phi = \phi_r$ and $\Phi = \Phi$. By Theorem 5.5.10 (2)(i) for all $\mathfrak{I}(\alpha) \in \mathcal{W}_U$,

$\mathfrak{I}(\alpha)$ is t-successor definable if and only if $\mathfrak{I}(\alpha)$ is t-predecessor definable.

By Proposition 5.5.8,

$$\mathfrak{I}(\alpha) = \bigvee_{\mathfrak{I}(\beta)\nsubseteq Q_{(u,\lambda)}} r^{\rightarrow}\overline{P}_{(u,\lambda)} \text{ if and only if } \mathfrak{I}(\alpha) = \bigvee_{\mathfrak{I}(\gamma)\nsubseteq Q_{(v,\mu)}} R^{\leftarrow}\overline{P}_{(v,\mu)}$$

for some $\mathfrak{I}(\beta), \mathfrak{I}(\gamma) \in \mathcal{W}_U$.

Equivalently, we also have

$$\mathfrak{I}(\alpha) = \bigvee_{\beta)\nleq u^\lambda} \mathfrak{I}(\phi^{\rightarrow}u_\lambda) \text{ if and only if } \mathfrak{I}(\alpha) = \bigvee_{\gamma)\nleq v^\mu} \mathfrak{I}(\Phi^{\leftarrow}v_\mu),$$

or

$$\mathfrak{I}(\alpha) = \mathfrak{I}(\bigvee_{\beta\nleq u^\lambda} \phi^{\rightarrow}u_\lambda) \text{ if and only if } \mathfrak{I}(\alpha) = \mathfrak{I}(\bigvee_{\gamma\nleq v^\mu} \Phi^{\leftarrow}v_\mu)$$

and finally, we obtain

$$\alpha = \bigvee_{\beta\nleq u^\lambda} \phi^{\rightarrow}u_\lambda \text{ if and only if } \alpha = \bigvee_{\gamma\nleq v^\mu} \Phi^{\leftarrow}v_\mu.$$

Therefore, we conclude that

α is fuzzy successor definable if and only if α is fuzzy predecessor definable.

(ii) Since (r, R) is symmetric, by Theorem 5.5.10 (2)(ii) for all $\mathfrak{I}(\alpha) \in \mathcal{W}_U$,

$r^{\rightarrow}\mathfrak{I}(\alpha) = R^{\rightarrow}\mathfrak{I}(\alpha)$ if and only if $r^{\leftarrow}\mathfrak{I}(\alpha) = R^{\leftarrow}\mathfrak{I}(\alpha)$,

and so, by Proposition 7.3.38,

$\mathfrak{I}(\phi^{\rightarrow}\alpha) = \mathfrak{I}(\Phi^{\rightarrow}\alpha)$ if and only if $\mathfrak{I}(\phi^{\leftarrow}\alpha) = \mathfrak{I}(\Phi^{\leftarrow}\alpha)$.

Since $\mathfrak{I}$ is a lattice isomorphism, we obtain that

$\phi^{\rightarrow}\alpha = \Phi^{\rightarrow}\alpha$ if and only if $\phi^{\leftarrow}\alpha = \Phi^{\leftarrow}\alpha$.

The equivalences given in (3) can be proved using a similar argument and hence, the proof is omitted.

7.10 Basic Properties of Fuzzy Rough Sets with Two Domains of Discourse

The vagueness and indiscernibility are the main subjects of the uncertainty of information. The combination of fuzzy sets and rough sets provide us reasonable observations for uncertainty. For instance, the clustering [18] and unsupervised feature selection [19] present successful applications in this respect (see also Sankar K. Pal et al. [20]). Recently, the studies of Kai Zhang et al. and Sankar K. Pal et al. for the fuzzy rough set systems on two universes are also notable works. In [21], Kai Zhang et al. define a new approximation operators in terms of a complicator and a t-conorm considering the approximation operators of De Cock et al. given in [11]. They use the PROMETHEE method to make decisions with the evaluation of fuzzy information.

Essentially, the fuzzy rough set models over single or two universes are widely discussed by many researchers (see, e.g., [12, 13, 15, 16, 22–29]). The starting paper of Dubois and Prade was subjected to the minimum t-norm and Kleene-Dienes S-implicator [16]. Further, Radzikowska and Kerre [26] and Wu et al. [15] presented a more general approach in terms of fuzzy implicators and t-norms. L. D'eer et al. studied the lower and upper approximations considering the minimal restrictions on the approximations [10].

Here, we argue the subject in a more general setting via the textures. Now, let (U, V, φ) be a generalized fuzzy approximation space with two domains of discourse. Further, let $T : [0, 1] \times [0, 1] \to [0, 1]$ be a left-continuous t-norm and $I : [0, 1] \times [0, 1] \to [0, 1]$ be a right-continuous implicator in the second component. Then by Theorem 7.4.2, there exists a generalized t-fuzzy approximation space (U, V, ϕ, Φ) where (ϕ, Φ) is the t-fuzzy direlation from $\mathcal{F}(U)$ to $\mathcal{F}(V)$ defined by

$$\phi(u, \lambda, v) = T(\varphi(u, v), \lambda)$$

and

$$\Phi(u, \lambda, v) = I(\varphi(u, v), \lambda)$$

for all $u \in U$, $v \in V$ and $\lambda \in (0, 1]$. Recall that by Corollary 7.3.32 and Theorem 7.4.2, for all $\alpha \in \mathcal{F}(U)$ and $v \in V$, we also have

$$(\phi^{\to}\alpha)(v) = \bigvee_{u\in U} \phi(u, \alpha(u), v) = \bigvee_{u\in U} T(\varphi(u, v), \alpha(u)) = \overline{apr}(\alpha)(v)$$

and

$$(\Phi^{\to}\alpha)(v) = \bigwedge_{u\in U} \Phi(u, \alpha(u), v) = \bigwedge_{u\in U} I(\varphi(u, v)), \alpha(u)) = \underline{apr}(\alpha)(v).$$

This link provides us a remarkable opportunity for an investigation of the basic properties of approximation operators of generalized fuzzy approximation space (U, V, φ). Note that when we discuss the equivalences of the seriality concept of the fuzzy relations, we take I as a border implicator. For almost all proofs, we use the corresponding t-fuzzy direlation (ϕ, Φ) to the fuzzy direlation $\varphi \in \mathcal{F}(U \times V)$ defined by the pairs (I, T) and (T_I, I_T) as in Theorems 7.4.2, 7.4.6 (1)(i) and (2)(i) where T is a left-continuous t-norm and I is a right-continuous implicator in the second component.

Let us consider Theorem 7.8.6 presenting the equivalent statements of the inverse seriality of a t-fuzzy relation (ϕ, Φ). If (U, V, ϕ, Φ) is a generalized t-fuzzy approximation space, then the inverse seriality of the t-fuzzy direlation (ϕ, Φ) is equivalent to the inequality

$$\forall \alpha \in \mathcal{F}(U),\ \Phi^{\rightarrow}\alpha \leq \phi^{\rightarrow}\alpha.$$

This immediately gives the following result:

Theorem 7.10.1 *Let (U, V, φ) be a generalized fuzzy approximation space with two domains of discourse. Then the following properties are equivalent.*

(i) *φ is inverse serial.*

(ii) $\forall \alpha \in \mathcal{F}(U),\ \underline{apr}(\alpha) \leq \overline{apr}(\alpha)$.

Proof (i) $\Longrightarrow$ (i): Let φ be inverse serial. Then by Theorem 7.7.9 (ii), the corresponding t-fuzzy direlation (ϕ, Φ) is inverse serial. By Theorem 7.8.6,

$$\forall \alpha \in \mathcal{F}(U),\ \Phi^{\rightarrow}\alpha \leq \phi^{\rightarrow}\alpha,$$

and so, we get $\underline{apr}(\alpha) \leq \overline{apr}(\alpha)$. Necessity is similar. □

Now, let us consider the generalized t-fuzzy approximation space $(U, V, \Phi^{\leftarrow}, \phi^{\leftarrow})$ where $(\Phi^{\leftarrow}, \phi^{\leftarrow})$ is the inverse of the t-fuzzy direlation (ϕ, Φ) given above. Then by Theorem 7.4.6, for all $\beta \in \mathcal{F}(V)$, we have the equalities

$$(\phi^{\leftarrow}\beta)(u) = \bigvee_{v \in V} \phi^{\leftarrow}(v, \beta(v), u) = \bigvee_{v \in V} T_I(\varphi(u, v), \beta(v)) = \underline{apr}'(\beta)(u),$$

$$(\Phi^{\leftarrow}\beta)(u) = \bigwedge_{v \in V} \Phi^{\leftarrow}(v, \beta(v), u), = \bigwedge_{v \in V} I_T(\varphi(u, v), \beta(v)) = \overline{apr}'(\beta)(u).$$

On the other hand, by Theorem 7.8.5, we note that the proposition

$$\forall \beta \in \mathcal{F}(V),\ \phi^{\leftarrow}\beta \leq \Phi^{\leftarrow}\beta$$

is equivalent to the seriality of (ϕ, Φ). Therefore, we have the following.

Theorem 7.10.2 *Let (U, V, φ) be a generalized fuzzy approximation space with two domains of discourse. Then the following properties are equivalent.*

(i) *φ is serial.*

(ii) $\forall \beta \in \mathcal{F}(V),\ \underline{apr}'(\beta) \leq \overline{apr}'(\beta)$.

Table 7.11 $\varphi \in \mathcal{F}(U \times V)$

φ	v_1	v_2	v_3
u_1	0	0.5	0.3
u_2	1	0	0

Proof (i) $\Longrightarrow$ (i): Let φ be inverse serial. Then by Theorem 7.7.9 (i), the corresponding t-fuzzy direlation (ϕ, Φ) is serial. Note that by Theorem 7.8.5,

$$\forall \beta \in \mathcal{F}(V),\ \phi^{\leftarrow}\beta \leq \Phi^{\leftarrow}\beta,$$

and this gives that $\underline{apr}'(\beta) \leq \overline{apr}'(\beta)$. Now, the necessity is trivial. □

In the following example, we consider a fuzzy relation φ which is not serial and $\underline{apr}(\alpha) \nleq \overline{apr}(\alpha)$.

Example 7.10.3 Let us consider the fuzzy relation $\varphi \in \mathcal{F}(U \times V)$ given in Table 7.11 where $U = \{u_1, u_2\}$ and $V = \{v_1, v_2, v_3\}$. Take the fuzzy rough set model corresponding to product t-norm and Reichenbach S-implicator. Clearly, φ is not a serial fuzzy relation. Let us consider the fuzzy set $\beta \in \mathcal{F}(V)$ where $\beta(v_1) = 0.8,\ \beta(v_2) = 0.3,\ \beta(v_3) = 0.5$. Note that

$$\begin{aligned}
\underline{apr}'(\beta)(u_1) &= \bigwedge_{v \in V} I_{T_P}(\varphi(u_1, v), \beta(v)) \\
&= I_{T_P}(\varphi(u_1, v_1), \beta(v_1)) \wedge I_{T_P}(\varphi(u_1, v_2), \beta(v_2)) \wedge I_{T_P}(\varphi(u_1, v_3), \beta(v_3)) \\
&= I_{T_P}(0, 0.8)) \wedge I_{T_P}(0.5, 0.3) \wedge I_{T_P}(\varphi(0.3, 0.5) \\
&= 1 \wedge \frac{3}{5} \wedge 1 = \frac{3}{5}.
\end{aligned}$$

Similarly, we also find that $\overline{apr}'(\beta)(u_1) = 0$, and, hence, $\underline{apr}'(\beta) \nleq \overline{apr}'(\beta)$. This is an expected result in view of the equivalence given above. On the other hand, φ is not inverse serial. Now let us take the fuzzy set $\alpha \in \mathcal{F}(U)$ where $\alpha(u_1) = 0.2$, $\alpha(u_2) = 0.3$. For $v_2 \in V$, we get $\underline{apr}(\alpha)(v_2) = 0.6$ and $\overline{apr}(\alpha)(v_2) = 0.1$. Then we have $\underline{apr}(\alpha) \nleq \overline{apr}(\alpha)$. □

Remark 7.10.4 Note that using a similar method, we easily give the results on rough set systems with two domains of discourse with respect to the pairs of fuzzy logic connectives (T, I) and (T_I, I_T) where T is a left-continuous t-norm and I is right-continuous implicator in the second component. □

By Corollary 7.3.44, we obtain the duality for the approximations.

Theorem 7.10.5 *Let (U, V, φ) be a generalized fuzzy approximation space with two domains of discourse.*

(i) $\forall \alpha \in \mathcal{F}(U),\ \underline{apr}(\alpha) = \mathbf{1} - \overline{apr}(\mathbf{1} - \alpha)$,

$\forall \alpha \in \mathcal{F}(U),\ \overline{apr}(\alpha) = \mathbf{1} - \underline{apr}(\mathbf{1} - \alpha).$

(ii) $\forall \beta \in \mathcal{F}(V),\ \overline{apr}'(\beta) = \mathbf{1} - \underline{apr}'(\mathbf{1} - \beta),$

$\forall \beta \in \mathcal{F}(V),\ \underline{apr}'(\beta) = \mathbf{1} - \overline{apr}'(\mathbf{1} - \beta).$

Proof (i) From Corollary 7.3.44 (i), for all $\alpha \in \mathcal{F}(U)$, note that

$$\underline{apr}(\alpha) = \Phi^{\rightarrow}\alpha = \mathbf{1} - \phi^{\rightarrow}(\mathbf{1} - \alpha) = \mathbf{1} - \overline{apr}(\mathbf{1} - \alpha).$$

The proof of the items (ii)–(iv) is similar.

Theorem 7.10.6 *For the fuzzy rough set systems with respect to* (T, I) *and* (T_I, I_T), *we have the following equalities:*

(i) $\forall \alpha \in \mathcal{F}(U),\ \alpha \leq \underline{apr}'(\overline{apr}(\alpha)),$

$$\overline{apr}'(\underline{apr}(\alpha)) \leq \alpha.$$

(ii) $\forall \beta \in \mathcal{F}(V),\ \beta \leq \underline{apr}(\overline{apr}'(\beta)),$

$$\overline{apr}(\underline{apr}'(\beta)) \leq \beta.$$

Proof (i) By Theorem 7.8.1 (i), we have that

$$\forall \alpha \in \mathcal{F}(U),\ \alpha \leq \phi^{\leftarrow}(\phi^{\rightarrow}\alpha).$$

If we consider the fuzzy rough systems corresponding to $(T, I$ and (T_I, I_T), we easily observe that

$$\phi^{\rightarrow}\alpha = \overline{apr}(\alpha) \text{ and so } \phi^{\leftarrow}(\overline{apr}(\alpha)) = \underline{apr}'(\overline{apr}(\alpha))$$

as desired.

If (ϕ, Φ) is a t-fuzzy direlation from $\mathcal{F}(U)$ to $\mathcal{F}(V)$, then the pair $(\Phi^{\leftarrow}, \phi^{\leftarrow})$ also constitutes an interval structure. Then in view of Theorem 7.8.4, for the systems with respect to (T, I), we obtain the following result.

Theorem 7.10.7 *Let* $\{\alpha_j \mid j \in J\} \subseteq \mathcal{F}(U)$.

(1) *If* φ *is inverse serial, then* $\overline{apr}(\mathbf{1}) = \mathbf{1}$.

(2) (i) $\overline{apr}(\mathbf{0}) = \mathbf{0}$.

(ii) $\overline{apr}(\bigwedge_{j \in J} \alpha_j) \leq \bigwedge_{j \in J} \overline{apr}(\alpha_j)$.

(iii) $\bigvee_{j \in J} \overline{apr}(\alpha_j) = \overline{apr}(\bigvee_{j \in J} \alpha_j)$.

(3) *If* φ *is inverse serial, then* $\underline{apr}(\mathbf{0}) = \mathbf{0}$.

(4) (i) $\underline{apr}(\mathbf{1}) = \mathbf{1}$.

(ii) $\bigvee_{j\in J} \underline{apr}(\alpha_j) \leq \underline{apr}(\bigvee_{j\in J} \alpha_j)$

(iii) $\bigwedge_{j\in J} \underline{apr}(\alpha_j) = \underline{apr}(\bigwedge_{j\in J} \alpha_j)$.

Proof (2) (iii) From Theorem 7.7.9 (ii), the corresponding t-fuzzy direlation (ϕ, Φ) is inverse serial. Then from Theorem 7.8.4 (2) (ii), for all $\alpha_j \in \mathcal{F}(U)$,

$$\overline{apr}(\bigwedge_{j\in J} \alpha_j) = \phi^{\rightarrow} \bigwedge_{j\in J} \alpha_j \leq \bigwedge_{j\in J} \phi^{\rightarrow}\alpha_j = \bigwedge_{j\in J} \overline{apr}(\alpha_j).$$

(4) (i) By Theorem 7.8.4 (4) (i), $\underline{apr}(\mathbf{1}) = \Phi^{\rightarrow}\mathbf{1} = \mathbf{1}$.

By Theorem 7.8.3, for the system with respect to (T_I, I_T), we also have the following properties.

Theorem 7.10.8 *Let* $\{\beta_j \mid j \in J\} \subseteq \mathcal{F}(V)$. *(1) If* φ *is serial, then* $\underline{apr}'(\mathbf{0}) = \mathbf{0}$.

(2) (i) $\underline{apr}'(\mathbf{1}) = \mathbf{1}$.

(ii) $\bigvee_{j\in J} \underline{apr}'(\beta_j) \leq \underline{apr}'(\bigvee_{j\in J}(\beta_j))$.

(iii) $\bigwedge_{j\in J} \underline{apr}'(\beta_j) = \underline{apr}'(\bigwedge_{j\in J}(\beta_j))$.

(3) *If* φ *is serial, then* $\overline{apr}'(\mathbf{1}) = \mathbf{1}$.

(4) (i) $\overline{apr}'(\mathbf{0}) = \mathbf{0}$.

(ii) $\overline{apr}'(\bigwedge_{j\in J} \beta_j) \leq \bigwedge_{j\in J} \overline{apr}'(\beta_j)$.

(iii) $\bigvee_{j\in J} \overline{apr}'(\beta_j) = \overline{apr}'(\bigvee_{j\in J} \beta_j)$.

Proof (3) From Theorem 7.7.9 (i), the corresponding t-fuzzy direlation (ϕ, Φ) is serial. Therefore, from Theorem 7.8.3 (3), we get $\overline{apr}'(\mathbf{1}) = \Phi^{\leftarrow}\mathbf{1} = \mathbf{1}$.

(4)(iii) From Theorem 7.8.3 (4)(iii), we get that

$$\bigvee_{j\in J} \overline{apr}'(\beta_j) = \bigvee_{j\in J} \Phi^{\leftarrow}\beta_j = \Phi^{\leftarrow} \bigvee_{j\in J} \beta_j = \overline{apr}'(\bigvee_{j\in J} \beta_j).$$

The above observations reveal that the seriality of the fuzzy relation is a necessity only for the equalities $\underline{apr}'(\mathbf{0}) = 0$ and $\overline{apr}'(\mathbf{1}) = 1$ which are the conditions of an interval structure. In view of Theorem 7.8.5, for the fuzzy rough set systems with respect to (T, I) and (T_I, I_T), the following conditions characterize the seriality of a fuzzy relation.

Theorem 7.10.9 *The following conditions are equivalent:*

(i) φ *is serial.*

(ii) $\forall \alpha \in \mathcal{F}(U),\ \underline{apr}'(\underline{apr}\alpha) \leq \alpha.$

(iii) $\forall \beta \in \mathcal{F}(V),\ \underline{apr}'(\beta) \leq \overline{apr}'(\beta).$

(iv) $\forall \alpha_1, \alpha_2 \in \mathcal{F}(U),\ \underline{apr}(\alpha_1) \leq \overline{apr}(\alpha_2) \implies \alpha_1 \leq \alpha_2.$

Proof (i) $\implies$ (ii) By Theorem 7.7.9, the corresponding t-fuzzy direlation is serial. Then by 7.8.5, we obtain

$$\forall \alpha \in \mathcal{F}(U),\ \phi^{\leftarrow}(\Phi^{\rightarrow}\alpha) = \underline{apr}'(\underline{apr}\alpha) \leq \alpha.$$

(ii) $\implies$ (iii) Again, by Theorem 7.7.9,

$$\forall \alpha \in \mathcal{F}(U),\ \phi^{\leftarrow}(\Phi^{\rightarrow}\alpha) \leq \alpha$$

and then we also have

$$\text{for all } \beta \in \mathcal{F}(V),\ \phi^{\leftarrow}\beta \leq \Phi^{\leftarrow}\beta.$$

This implies that

$$\underline{apr}'(\beta) \leq \overline{apr}'(\beta).$$

The implications (iii) $\implies$ (iv) and (iv) $\implies$ (i) are similar.

Theorem 7.10.10 *The following statements are equivalent:*

(i) φ *is inverse serial.*

(ii) $\forall \beta \in \mathcal{F}(V),\ \beta \leq \overline{apr}((\overline{apr}'\beta)).$

(iii) $\forall \alpha \in \mathcal{F}(U),\ \underline{apr}(\alpha) \leq \overline{apr}(\alpha).$

(iv) $\forall \beta_1, \beta_2 \in \mathcal{F}(V),\ \underline{apr}'(\beta_1) \leq \overline{apr}'(\beta_2) \implies \beta_1 \leq \beta_2.$

Proof The proof can be given using Theorem 7.8.6.

Theorem 7.10.11 *Let* φ *be an inverse serial fuzzy relation. Then*

(i) $\forall v_\mu \in \mathcal{F}(V),\ \overline{apr}'(v_\mu) \neq \mathbf{0}.$

(ii) *If* $\mathbf{0} \neq \beta \in \mathcal{F}(V)$*, then* $\overline{apr}'(\beta) \neq \mathbf{0}.$

(iii) *If* $\mathbf{1} \neq \beta \in \mathcal{F}(V)$*, then* $\underline{apr}'(\beta) \neq \mathbf{1}.$

Proof (i) Note that by Theorem 7.8.7(i), we have $\overline{apr}'(v_\mu) = \phi^{\leftarrow} v_\mu \neq \mathbf{0}$.

Theorem 7.10.12 *Let φ be a serial fuzzy relation. Then*

(i) $\forall u_\lambda \in \mathcal{F}(U)$, $\underline{apr}(u_\lambda) \neq \mathbf{0}$.

(ii) *If* $\mathbf{0} \neq \alpha \in \mathcal{F}(U)$, *then* $\underline{apr}(\alpha) \neq \mathbf{0}$.

(iii) *If* $\mathbf{1} \neq \alpha \in \mathcal{F}(U)$, *then* $\overline{apr}(\alpha) \neq \mathbf{1}$.

Proof It is left to the interested reader. □

Theorem 7.10.13 *Let φ be a fuzzy relation.*

(1) *If φ is serial, then for all $\alpha \in \mathcal{F}(U)$,*

$\alpha \leq \overline{apr}'(\underline{apr}(\alpha))$.

(2) *If φ is injective, then for all $\alpha \in \mathcal{F}(U)$,*

(i) $\overline{apr}'(\overline{apr}(\alpha)) \leq \alpha$.

(ii) $\overline{apr}(\alpha) \leq \underline{apr}(\alpha)$.

(iii) $\alpha \leq \underline{apr}'(\underline{apr}(\alpha))$.

(3) *If φ is inverse serial and injective, then for all $\alpha \in \mathcal{F}(U)$,*

$\overline{apr}(\alpha) = \underline{apr}(\alpha)$.

(4) *If φ is injective and serial, then for all $\alpha \in \mathcal{F}(U)$, we have the following equalities:*

(i) $\underline{apr}'(\overline{apr}(\alpha)) = \alpha$.

(ii) $\overline{apr}'(\underline{apr}(\alpha)) = \alpha$.

Proof In view of Theorem 7.8.9, the proof can be easily given.

Dually, we have the following result.

Theorem 7.10.14 *Let φ be a fuzzy relation.*

(1) *If φ is inverse serial, then for all $\beta \in \mathcal{F}(V)$,*

$\underline{apr}(\underline{apr}'(\beta) \leq \beta$.

(2) *If φ is injective, then for all $\beta \in \mathcal{F}(V)$,*

(i) $\beta \leq \underline{apr}(\underline{apr}'(\beta))$.

(ii) $\underline{apr}'(\beta) \leq \overline{apr}'(\beta)$.

(iii) $\overline{apr}(\overline{apr}'(\beta)) \leq \beta$.

(3) *If φ is serial and injective, then for all $\beta \in \mathcal{F}(V)$,*

$\underline{apr}'(\beta) = \overline{apr}'(\beta)$.

(4) *If φ is inverse serial and injective, then for all $\beta \in \mathcal{F}(V)$, we have the following equalities:*

(i) $\overline{apr}(\underline{apr}'(\beta)) = \beta$.

(ii) $\underline{apr}(\overline{apr}'(\beta)) = \beta$.

Proof Similar to the proof of Theorem 7.10.13.

Theorem 7.10.15 *Let φ be a fuzzy relation. Then*

(1) $\underline{apr}'(\mathbf{1}) = \mathbf{1}$ *and* $\overline{apr}'(\mathbf{0}) = \mathbf{0}$.

(2) *For all* $\beta \in \mathcal{F}(V)$, $\overline{apr}'(\overline{apr}(\underline{apr}'(\beta))) \leq \overline{apr}'(\beta)$.

(3) *If φ is serial, then* $\underline{apr}'(\mathbf{0}) = \mathbf{0}$ *and* $\overline{apr}'(\mathbf{1}) = \mathbf{1}$.

(4) *If φ is serial, then* $\overline{apr}'(\beta) \leq \underline{apr}'((\overline{apr}(\underline{apr}'(\beta)))$.

(5) *If φ is inverse serial, then* $\overline{apr}(\mathbf{1}) = \mathbf{1}$ *and* $\underline{apr}(\mathbf{0}) = \mathbf{0}$.

Proof (4) Let φ be serial. Since (ϕ, Φ) is also serial, we get

$$\overline{apr}'(\beta) = \Phi^{\leftarrow}\beta \leq \Phi^{\leftarrow}(\phi^{\rightarrow}(\Phi^{\leftarrow}\beta)) = \underline{apr}'((\overline{apr}(\underline{apr}'(\beta))).$$

□

Theorem 7.10.16 *Let (U, V, φ) be a generalized fuzzy approximation space.*

(i) *If φ is injective, then* $\underline{apr}'(\overline{apr}(\overline{apr}'(\beta))) = \overline{apr}'(\beta)$.

(ii) *If φ is injective and serial, then*

(a) $\overline{apr}'(\overline{apr}(\overline{apr}'(\beta)) = \overline{apr}'(\beta)$.

(b) $\underline{apr}'(\underline{apr}(\underline{apr}'(\beta)) = \underline{apr}'(\beta)$.

(c) $\overline{apr}'(\underline{apr}(\underline{apr}'(\beta)) = \underline{apr}'(\beta)$.

(d) $\underline{apr}'(\underline{apr}(\overline{apr}'(\beta))) = \overline{apr}'(\beta)$.

(iii) *If φ is serial, then* $\overline{apr}'(\overline{apr}(\underline{apr}'(\beta))) = \underline{apr}'(\beta)$.

Proof (i) If φ is injective, then by Theorem 7.7.14, (ϕ, Φ) is injective. Then by Theorem 7.10.16 (i), for all $\beta \in \mathcal{F}(V)$, we obtain that

$$\underline{apr}'(\overline{apr}(\overline{apr}'(\beta))) = \phi^{\leftarrow}(\phi^{\rightarrow}(\Phi^{\leftarrow}\beta)) = \Phi^{\leftarrow}\beta = \overline{apr}'(\beta).$$

(iii) By Theorem 7.10.16 (iii), we observe that

$$\overline{apr}'(\overline{apr}(\underline{apr}'(\beta))) = \Phi^{\leftarrow}(\phi^{\rightarrow}(\phi^{\leftarrow}\beta)) = \phi^{\leftarrow}\beta = \underline{apr}'(\beta).$$

Theorem 7.10.17 *Let (U, V, φ) be a generalized fuzzy approximation space.*

(i) *If φ is injective, then* $\overline{apr}(\underline{apr}'(\underline{apr}(\alpha))) = \underline{apr}(\alpha)$.

(ii) *If φ is injective and inverse serial, then*

(a) $\underline{apr}(\underline{apr}'(\underline{apr}(\alpha))) = \underline{apr}(\alpha)$.

(b) $\overline{apr}(\overline{apr}'(\overline{apr}(\alpha))) = \overline{apr}(\alpha)$.

(c) $\underline{apr}(\overline{apr}'(\overline{apr}(\alpha))) = \overline{apr}(\alpha)$.

(d) $\overline{apr}(\overline{apr}'(\underline{apr}(\alpha))) = \underline{apr}(\alpha)$.

(iii) *If φ is inverse serial, then* $\underline{apr}(\underline{apr}'(\overline{apr}(\alpha))) = \overline{apr}(\alpha)$.

Proof The proof is left to the interested reader. □

7.11 Definability in Terms of Fuzzy Logic Connectives

Definition 7.11.1 Let (U, V, φ) be a generalized fuzzy approximation space.

(1) $\alpha \in \mathcal{F}(U)$ is called (T, I)-*definable*, if $\underline{apr}(\alpha) = \overline{apr}(\alpha)$.

(2) $\beta \in \mathcal{F}(V)$ is called (T_I, I_T)-*definable*, if $\underline{apr}'(\beta) = \overline{apr}'(\beta)$.

Definability is a reasonable connection between the fuzzy rough set systems due to (T, I) and (I_T, T_I).

Theorem 7.11.2 *Let (U, V, φ) be a generalized fuzzy approximation space.*

(1) *The following statements are equivalent:*

(i) *$\beta \in \mathcal{F}(V)$ is (T_I, I_T)-definable*

(ii) $\exists \alpha \in \mathcal{F}(U)$, $\overline{apr}(\alpha) = \beta$ *and* $\overline{apr}'(\beta) = \alpha$.

(2) *The following statements are equivalent:*

(i) *$\alpha \in \mathcal{F}(U)$ is (T, I)-definable*

(ii) $\exists \beta \in \mathcal{F}(V)$, $\underline{apr}(\alpha) = \beta$ *and* $\underline{apr}'(\beta) = \alpha$.

Proof (1)(i) $\Longrightarrow$ (ii): Let β be (T_I, I_T)-*definable*. Then

$$\underline{apr}'(\beta) = \overline{apr}'(\beta)$$

where

$$(\phi^{\leftarrow} \beta)(u) = \bigvee_{v \in V} T_I(\varphi(u, v), \beta(v)) = \underline{apr}'(\beta)(u),$$

$$(\Phi^{\leftarrow} \beta)(u) = \bigwedge_{v \in V} I_T(\varphi(u, v), \beta(v)) = \overline{apr}'(\beta)(u)$$

for all $u \in U$. Therefore, β is automatically fuzzy presectional definable, that is,

$$\phi^{\leftarrow} \beta = \Phi^{\leftarrow} \beta.$$

Hence, by Theorem 7.9.2, for some $\alpha \in \mathcal{F}(U)$,

$\phi^{\rightarrow}\alpha = \overline{apr}(\alpha) = \beta$ and $\Phi^{\leftarrow}\beta = \overline{apr}'(\beta) = \alpha$.

(ii) $\Longrightarrow$ (i): If $\overline{apr}(\alpha) = \beta$ and $\overline{apr}'(\beta) = \alpha$, then $\phi^{\rightarrow}\alpha = \beta$ and $\Phi^{\leftarrow}\beta = \alpha$ and so again by Theorem 7.9.2, β is fuzzy presectional definable, namely $\phi^{\leftarrow}\beta = \Phi^{\leftarrow}\beta$. As a result we obtain $\underline{apr}'(\beta) = \overline{apr}'(\beta)$.

(2) It is omitted.

Definition 7.11.3 Let (U, V, φ) be a generalized fuzzy approximation space. Then

(i) $\beta \in \mathcal{F}(V)$ is called *T-definable* if for some $\alpha \in \mathcal{F}(U)$,

$$\beta = \bigvee_{\lambda<\alpha(u)} \overline{apr}(u_\lambda).$$

(ii) $\alpha \in \mathcal{F}(U)$ is called *I_T-definable* if for some $\beta \in \mathcal{F}(V)$,

$$\alpha = \bigvee_{\mu<\beta(v)} \overline{apr}'(v_\mu).$$

Proposition 7.11.4 *Let (U, V, φ) be a generalized fuzzy approximation space. Then we have the following equivalences.*

(1) *The following statements are equivalent:*

(i) $\alpha \in \mathcal{F}(U)$ *is T-definable.*

(ii) *For some $\beta \in \mathcal{F}(V)$, $\alpha = \overline{apr}'(\beta)$.*

(2) *The following statements are equivalent.*

(i) $\beta \in \mathcal{F}(V)$ *is I_T-definable.*

(ii) *For some $\alpha \in \mathcal{F}(U)$, $\beta = \overline{apr}(\alpha)$.*

(3) *$\overline{apr}(\alpha)$ is T-definable and $\overline{apr}'(\beta)$ is I_T-definable.*

Proof (1) (i) $\Longrightarrow$ (ii): Let α be T-definable. Then for some $\alpha \in \mathcal{F}(U)$,

$$\beta = \bigvee_{\lambda<\alpha(u)} \overline{apr}(u_\lambda) = \bigvee_{\lambda<\alpha(u)} \phi^{\rightarrow}u_\lambda.$$

Therefore, α is also fuzzy predecessor definable. Then from Proposition 7.9.4 (1), for some $\alpha \in \mathcal{F}(U)$, $\alpha = \Phi^{\leftarrow}\beta$ and so we obtain $\alpha = \overline{apr}'(\beta)$ as desired.

The proof of (ii) $\Longrightarrow$ (i) is similar.

(3) Note that from Lemma 1.2.4 and Theorem 7.10.7 (iii), we get

$$\overline{apr}(\alpha) = \overline{apr}(\bigvee_{\lambda<\alpha(u)} u_\lambda) = \bigvee_{\lambda<\alpha(u)} \overline{apr}(u_\lambda).$$

This means that $\overline{apr}(\alpha)$ is T-definable.

Corollary 7.11.5 *Let (U, V, φ) be a generalized fuzzy approximation space.*

(i) *If $\alpha \in \mathcal{F}(U)$ is (T, I)-definable, then it is also T-definable.*

(ii) *If $\beta \in \mathcal{F}(V)$ is (T_I, I_T)-definable, then it is also I_T-definable.*

Proof (i) Let $\underline{apr}(\alpha) = \overline{apr}(\alpha)$. Then $\phi^{\rightarrow}\alpha = \Phi^{\rightarrow}\alpha$, that is α is fuzzy sectional definable. Hence, from Theorem 5.5.9 (ii) fuzzy successor definable. Hence, we have

$$\alpha = \bigvee_{\alpha \nleq u^{\lambda}} \phi^{\rightarrow} u_{\lambda} = \bigvee_{\lambda < \alpha(u)} \overline{apr}(u_{\lambda}).$$

Therefore, α is T-definable.

(ii) It is similar. □

Theorem 7.11.6

(1) *If φ is inverse serial and (T, I)-injective, then for every $\alpha \in \mathcal{F}(U)$,*

α is (T, I)-definable.

(2) *If φ is serial and (T_I, I_T)-injective, then for all $\beta \in \mathcal{F}(V)$,*

β is (T_I, I_T)-definable.

Proof (1) Let φ be inverse serial and (T, I)-injective. By Corollary 7.7.7 (ii) and Theorem 7.7.14, (ϕ, Φ) is inverse serial and injective. Then by Theorem 7.8.9 (3), for all α we obtain $\phi^{\rightarrow}\alpha = \Phi^{\rightarrow}\alpha$, that is, $\underline{apr}(\alpha) = \overline{apr}(\alpha)$ as desired.

(2) The proof is omitted.

Theorem 7.11.7 *Let φ be a fuzzy direlation on $\mathcal{F}(U)$. Then for all $\alpha \in \mathcal{F}(U)$, we have the following properties:*

(1) *If φ is reflexive, then*

(i) *If α is (T, I)-definable, then α is T-definable.*

(ii) *If α is (T_I, I_T)-definable, then α is T_I-definable.*

(2) *If φ is a symmetric fuzzy relation on $\mathcal{F}(U)$, then we have the following equivalences:*

(i) *α is T-definable $\Longleftrightarrow$ α is I_T-definable.*

(ii) *α is (T, I)-definable if and only if α is (T_I, I_T)-definable.*

(3) *If φ is an equivalence fuzzy relation on $\mathcal{F}(U)$, then the following conditions are equivalent:*

(i) *α is T-definable.*

(ii) *α is I_T-definable.*

(iii) *α is (T, I)-definable.*

(iv) *α is (T_I, I_T)-definable.*

Proof (1) (i) By Theorem 7.7.21 (i), if φ is reflexive, then (ϕ, Φ) is reflexive where

$$\phi(u, \lambda, v) = T(\varphi(u, v), \lambda) \text{ and } \Phi(u, \lambda, v) = I(\varphi(u, v), \lambda)$$

for a left-continuous t-norm and a right-continuous implicator I in the second component. Let α be (T, I)-definable. Then $\underline{apr}(\alpha) = \overline{apr}(\alpha)$ and so we get $\phi^{\rightarrow}(\alpha) = \Phi^{\rightarrow}(\alpha)$. Hence, by Theorem 7.9.5 (i), α is fuzzy successor definable. Therefore, for some $\beta \in \mathcal{F}(U)$,

$$\alpha = \bigvee_{\beta \nleq u^{\lambda}} \phi^{\rightarrow} u_{\lambda} = \bigvee_{\beta \nleq u^{\lambda}} \overline{apr}(u_{\lambda}),$$

namely α is T-definable.

(ii) The proof is similar.

(2) (i) If φ is a symmetric fuzzy relation on $\mathcal{F}(U)$, then by Theorem 7.7.21 (ii), (ϕ, Φ) is symmetric. Let α be T-definable. Since (ϕ, Φ) is symmetric,

$$\alpha = \bigvee_{\beta \nleq u^{\lambda}} \phi^{\rightarrow} u_{\lambda} = \bigvee_{\beta \nleq u^{\lambda}} \Phi^{\leftarrow} u_{\lambda} = \bigvee_{\beta \nleq u^{\lambda}} \overline{apr}'(u_{\lambda}).$$

Therefore, α is I_T definable. The necessity is similar.

(ii) Let α be (T, I)-definable, that is, let $\overline{apr}(\alpha) = \underline{apr}(\alpha)$. By definition of (ϕ, Φ), we also have that $\phi^{\rightarrow}\alpha = \Phi^{\rightarrow}\alpha$. Since (ϕ, Φ) is also symmetric, we have $\Phi^{\leftarrow}\alpha = \phi^{\leftarrow}\alpha$. However, by Theorem 7.4.6, we obtain that $\overline{apr}'(\alpha) = \underline{apr}'(\alpha)$, that is, α is (T_I, I_T)-definable. The necessity can be proved by a similar argument.

(3) If φ is an equivalence fuzzy relation on $\mathcal{F}(U)$, then by Theorem 7.4.6 (iii), (ϕ, Φ) is an equivalent t-fuzzy direlation.

(i) $\Longrightarrow$ (ii): Let α be T-definable. Since (ϕ, Φ) is symmetric, in view of the proof of (2)(i), α is I_T-definable.

(ii) $\Longrightarrow$ (iii): Let α be I_T-definable. Then for some $\beta \in \mathcal{F}(U)$,

$$\alpha = \bigvee_{\beta \nleq u^{\lambda}} \overline{apr}'(u_{\lambda}) = \bigvee_{\beta \nleq u^{\lambda}} \Phi^{\leftarrow} u_{\lambda}.$$

Hence, α is fuzzy predecessor definable. Then by Theorem 7.9.5 (3),

$$\phi^{\leftarrow}\alpha = \Phi^{\leftarrow}\alpha \text{ and this gives that } \underline{apr}'\alpha = \overline{apr}'\alpha.$$

Namely, α is (T, I)-definable.

The proof of (ii) $\Longrightarrow$ (iv) and (iv) $\Longrightarrow$ (i) can be proved using a similar argument.

7.12 Revised t-Fuzzy Rough Set Models

Recall that a revised rough set model was given in [25] by Pei. In the literature, some special types of revised fuzzy rough set models were discussed. Now let us consider the operations:

$$\phi^{\rightarrow}\phi^{\leftarrow} : \mathcal{F}(V) \to \mathcal{F}(V),\ \forall \beta \in \mathcal{F}(V),\ \phi^{\rightarrow}\phi^{\leftarrow}\beta = \phi^{\rightarrow}(\phi^{\leftarrow}\beta),$$

$$\Phi^{\rightarrow}\Phi^{\leftarrow} : \mathcal{F}(V) \to \mathcal{F}(V),\ \forall \beta \in \mathcal{F}(V),\ \Phi^{\rightarrow}\Phi^{\leftarrow}\beta = \Phi^{\rightarrow}(\Phi^{\leftarrow}\beta)$$

and

$$\phi^{\leftarrow}\phi^{\rightarrow} : \mathcal{F}(U) \to \mathcal{F}(U),\ \forall \alpha \in \mathcal{F}(U),\ \phi^{\leftarrow}\phi^{\rightarrow}\alpha = \phi^{\leftarrow}(\phi^{\rightarrow}\alpha),$$

$$\Phi^{\leftarrow}\Phi^{\rightarrow} : \mathcal{F}(U) \to \mathcal{F}(U),\ \forall \alpha \in \mathcal{F}(U),\ \Phi^{\leftarrow}\Phi^{\rightarrow}\alpha = \Phi^{\leftarrow}(\Phi^{\rightarrow}\alpha).$$

Here, note that the mapping $\phi^{\rightarrow}\phi^{\leftarrow}$ is defined using the β-presection of the t-fuzzy corelation $\phi^{\leftarrow}$ and $\phi^{\leftarrow}\beta$-section of the t-fuzzy relation ϕ. Similarly, the mapping $\Phi^{\rightarrow}\Phi^{\leftarrow}$ is defined using the β-presection of the t-fuzzy relation $\Phi^{\leftarrow}$ and $\Phi^{\leftarrow}\beta$-section of the t-fuzzy corelation Φ. The mappings $\phi^{\leftarrow}\phi^{\rightarrow}$ and $\Phi^{\leftarrow}\Phi^{\rightarrow}$ may be also considered in a similar way.

Definition 7.12.1 Let (U, V, ϕ, Φ) be a generalized t-fuzzy approximation space and $\beta \in \mathcal{F}(V)$.

(i) The mappings $\phi^{\rightarrow}\phi^{\leftarrow}$ and $\Phi^{\rightarrow}\Phi^{\leftarrow}$ are called *the revised t-lower fuzzy approximation operator* and *t-upper fuzzy approximation operator* on $\mathcal{F}(V)$, respectively.

(ii) The mappings $\Phi^{\leftarrow}\Phi^{\rightarrow}$ and $\phi^{\leftarrow}\phi^{\rightarrow}$ are called *the revised t-lower fuzzy approximation operator* and *t-upper fuzzy approximation operator* on $\mathcal{F}(U)$, respectively.

Theorem 7.12.2 *Let (U, V, ϕ, Φ) be a generalized t-fuzzy approximation space. If (ϕ, Φ) is a complemented t-fuzzy direlation from $\mathcal{F}(U)$ to $\mathcal{F}(V)$, then for all $\alpha \in \mathcal{F}(U)$ and $\beta \in \mathcal{F}(V)$,*

$$\Phi^{\leftarrow}\Phi^{\rightarrow}(\mathbf{1} - \alpha) = \mathbf{1} - \phi^{\leftarrow}\phi^{\rightarrow}\alpha,$$

$$\phi^{\leftarrow}\phi^{\rightarrow}(\mathbf{1} - \alpha) = \mathbf{1} - \Phi^{\leftarrow}\Phi^{\rightarrow}\alpha.$$

$$\Phi^{\rightarrow}\Phi^{\leftarrow}(\mathbf{1} - \beta) = \mathbf{1} - \phi^{\rightarrow}\phi^{\leftarrow}\alpha,$$

$$\phi^{\rightarrow}\phi^{\leftarrow}(\mathbf{1} - \beta) = \mathbf{1} - \Phi^{\rightarrow}\Phi^{\leftarrow}\beta.$$

Proof If (ϕ, Φ) is complemented, then $\phi = \Phi'$ and $\phi' = \Phi$. Hence, by Corollary 7.3.44, we obtain

$$\begin{aligned}\Phi^{\rightarrow}\Phi^{\leftarrow}(\mathbf{1} - \beta) &= (\phi')^{\rightarrow}(\phi')^{\leftarrow}(\mathbf{1} - \beta)\\ &= (\phi')^{\rightarrow}(\mathbf{1} - \phi^{\leftarrow}(\beta)) = \mathbf{1} - \phi^{\rightarrow}(\phi^{\leftarrow}\beta).\end{aligned}$$

The proof of the second equality is dual to the first.

We have the following results.

Theorem 7.12.3 *Let (U, V, ϕ, Φ) be a generalized t-fuzzy approximation space and $\beta \in \mathcal{F}(V)$ and $\{\beta_j \mid j \in J\} \subseteq \mathcal{F}(V)$. Then we have the following properties:*

(1) $\phi^{\rightarrow}\phi^{\leftarrow}\beta \leq \beta \leq \Phi^{\rightarrow}\Phi^{\leftarrow}\beta$.

(2) $\beta_1 \leq \beta_2 \implies \phi^{\rightarrow}\phi^{\leftarrow}\beta_1 \leq \phi^{\rightarrow}\phi^{\leftarrow}\beta_2$.

(3) $\beta \leq \beta_2 \implies \Phi^{\rightarrow}\Phi^{\leftarrow}\beta_1 \leq \Phi^{\rightarrow}\Phi^{\leftarrow}\beta_2$.

(4) $\bigvee_{j\in J} \phi^{\rightarrow}\phi^{\leftarrow}\beta_j \leq \phi^{\rightarrow}\phi^{\leftarrow}\bigvee_{j\in J}\beta_j$.

(5) $\Phi^{\rightarrow}\Phi^{\leftarrow}(\bigwedge_{j\in J}\beta_j) \leq \bigwedge_{j\in J}\Phi^{\rightarrow}\Phi^{\leftarrow}\beta_j$.

(6) $\phi^{\rightarrow}\phi^{\leftarrow}(\phi^{\rightarrow}\phi^{\leftarrow}\beta) = \phi^{\rightarrow}\phi^{\leftarrow}\beta$.

(7) $\Phi^{\rightarrow}\Phi^{\leftarrow}(\Phi^{\rightarrow}\Phi^{\leftarrow}\beta) = \Phi^{\rightarrow}\Phi^{\leftarrow}\beta$.

Proof (1) By Theorem 6.3.3 (1) and Proposition 7.3.38, we get that

$$\begin{aligned}\mathfrak{I}(\phi^{\rightarrow}\phi^{\leftarrow}\beta) &= r^{\rightarrow}\mathfrak{I}(\phi^{\leftarrow}\beta) = r^{\rightarrow}r^{\leftarrow}\mathfrak{I}(\beta) \leq \mathfrak{I}(\beta)\\ &\leq R^{\rightarrow}R^{\leftarrow}\mathfrak{I}(\beta) = R^{\rightarrow}\mathfrak{I}(\Phi^{\rightarrow}\beta) = \mathfrak{I}(\Phi^{\rightarrow}\Phi^{\rightarrow}\beta).\end{aligned}$$

Since $\mathfrak{I}$ is a lattice isomorphism, we obtain that

$$\phi^{\rightarrow}\phi^{\leftarrow}\beta \leq \beta \leq \Phi^{\rightarrow}\Phi^{\leftarrow}\beta.$$

The proof of (2) and (3) is immediate from Theorem 7.3.26.

(4) From Theorem 7.8.3 (2)(ii), note that

$$\bigvee_{j\in J}\phi^{\leftarrow}\beta_j \leq \phi^{\leftarrow}\bigvee_{j\in J}\beta_j,$$

and then by Theorem 7.3.26 (i)

$$\phi^{\rightarrow}\bigvee_{j\in J}\phi^{\leftarrow}\beta_j \leq \phi^{\rightarrow}\phi^{\leftarrow}\bigvee_{j\in J}\beta_j.$$

Finally by Theorem 7.3.39 (i), we conclude that

$$\bigvee_{j\in J}\phi^{\rightarrow}\phi^{\leftarrow}\beta_j \leq \phi^{\rightarrow}\phi^{\leftarrow}\bigvee_{j\in J}\beta_j.$$

Theorem 7.12.4 *For every $\alpha \in \mathcal{F}(U)$ and $\{\alpha_j \mid j \in J\} \subseteq \mathcal{F}(U)$, the mappings $\Phi^{\leftarrow}\Phi^{\rightarrow} : \mathcal{F}(U) \to \mathcal{F}(U)$ and $\phi^{\leftarrow}\phi^{\rightarrow} : \mathcal{F}(U) \to \mathcal{F}(U)$ satisfy the following properties:*

(1) $\Phi^{\leftarrow}\Phi^{\rightarrow}\alpha \leq \alpha \leq \phi^{\leftarrow}\phi^{\rightarrow}\alpha$.

(2) $\alpha_1 \leq \alpha_2 \implies \Phi^{\leftarrow}\Phi^{\rightarrow}\alpha_1 \leq \Phi^{\leftarrow}\Phi^{\rightarrow}\alpha_2$.

(3) $\alpha_1 \leq \alpha_2 \implies \phi^{\leftarrow}\phi^{\rightarrow}\alpha_1 \leq \phi^{\leftarrow}\phi^{\rightarrow}\alpha_2$.

(4) $\bigvee_{j\in J} \Phi^{\leftarrow}\Phi^{\rightarrow}\alpha_j \leq \Phi^{\leftarrow}\Phi^{\rightarrow}\bigvee_{j\in J}\alpha_j$.

(5) $\phi^{\leftarrow}\phi^{\rightarrow}\bigwedge_{j\in J}\alpha_j \leq \bigwedge_{j\in J}\phi^{\leftarrow}\phi^{\rightarrow}\alpha_j$.

(6) $\Phi^{\leftarrow}\Phi^{\rightarrow}\alpha = \Phi^{\leftarrow}\Phi^{\rightarrow}(\Phi^{\leftarrow}\Phi^{\rightarrow}\alpha)$.

(7) $\phi^{\leftarrow}\phi^{\rightarrow}\alpha = \phi^{\leftarrow}\phi^{\rightarrow}(\phi^{\leftarrow}\phi^{\rightarrow}\alpha)$.

Proof It is omitted.

Theorem 7.12.5 *Let (U, V, ϕ, Φ) be a generalized t-fuzzy approximation space, $\beta \in \mathcal{F}(V]$ and $\{\beta_j \mid j \in J\} \subseteq \mathcal{F}(V)$.*

(1) *If (ϕ, Φ) is serial, then $\phi^{\rightarrow}\phi^{\leftarrow}\mathbf{0} = \mathbf{0}$ and $\Phi^{\rightarrow}\Phi^{\leftarrow}\mathbf{1} = \mathbf{1}$.*

(2) *If (ϕ, Φ) is inverse serial, then $\phi^{\rightarrow}\phi^{\leftarrow}\mathbf{1} = \mathbf{1}$ and $\Phi^{\rightarrow}\Phi^{\leftarrow}\mathbf{0} = \mathbf{0}$.*

(3) *Let (ϕ, Φ) be inverse serial and injective.*

(i) $\phi^{\rightarrow}\phi^{\leftarrow}\bigwedge_{j\in J}\beta_j = \bigwedge_{j\in J}\phi^{\rightarrow}\phi^{\leftarrow}\beta_j$.

(ii) $\Phi^{\rightarrow}\Phi^{\leftarrow}\bigvee_{j\in J}\beta_j = \bigvee_{j\in J}\Phi^{\rightarrow}\Phi^{\leftarrow}\beta_j$.

(4) *Let (ϕ, Φ) be serial and injective.*

(i) $\Phi^{\rightarrow}\Phi^{\leftarrow}\bigwedge_{j\in J}\beta_j = \bigwedge_{j\in J}\Phi^{\rightarrow}\Phi^{\leftarrow}\beta_j$.

(ii) $\phi^{\rightarrow}\phi^{\leftarrow}\bigvee_{j\in J}\beta_j = \bigvee_{j\in J}\phi^{\rightarrow}\phi^{\leftarrow}\beta_j$.

(iii) *Let (ϕ, Φ) be inverse serial and injective, then $\Phi^{\rightarrow}\Phi^{\leftarrow}\beta = \phi^{\rightarrow}\phi^{\leftarrow}(\Phi^{\rightarrow}\Phi^{\leftarrow}\beta)$.*

(5) *Let (ϕ, Φ) be serial and injective.*

(i) $\Phi^{\rightarrow}\Phi^{\leftarrow}(\phi^{\rightarrow}\phi^{\leftarrow}\beta) = \beta$.

(ii) $\beta = \phi^{\rightarrow}\phi^{\leftarrow}(\Phi^{\rightarrow}\Phi^{\leftarrow}\beta)$.

Proof (1) Since (ϕ, Φ) is serial from $\mathcal{F}(U)$ to $\mathcal{F}(V)$, by Corollary 7.8.2 (i), the corresponding direlation (r, R) from $(W_U, \mathcal{W}_U)$ to $W_V, \mathcal{W}_V)$ is also serial where $\phi = \phi_r$ and $\Phi = \Phi_R$. Then by Proposition 7.3.38 and Theorem 6.3.4,

$$\mathfrak{I}(\phi^{\rightarrow}\phi^{\leftarrow}\mathbf{0}) = r^{\rightarrow}\mathfrak{I}(\phi^{\leftarrow}\mathbf{0}) = r^{\rightarrow}r^{\leftarrow}\mathfrak{I}(\mathbf{0}) = \mathfrak{I}(\mathbf{0}),$$

and so since $\mathfrak{I}$ is a lattice isomorphism, we get $\phi^{\rightarrow}\phi^{\leftarrow}\mathbf{0} = \mathbf{0}$.

The proof of the second equality is similar.

(4)(i) By Theorems 7.8.10 (3) and 7.3.39 (ii), we observe that

$$\Phi^{\rightarrow}\Phi^{\leftarrow}\bigwedge_{j\in J}\beta_j - \Phi^{\rightarrow}\phi^{\leftarrow}\bigwedge_{j\in J}\beta_j = \Phi^{\rightarrow}\bigwedge_{j\in J}\phi^{\leftarrow}\beta_j = \bigwedge_{j\in J}\Phi^{\rightarrow}\Phi^{\leftarrow}\beta_j.$$

Theorem 7.12.6 *The mappings*

$$\Phi^{\leftarrow}\Phi^{\rightarrow} : \mathcal{F}(U) \to \mathcal{F}(U) \text{ and } \phi^{\leftarrow}\phi^{\rightarrow} : \mathcal{F}(U) \to \mathcal{F}(U)$$

satisfy the following properties where $\alpha \in \mathcal{F}(U)$ *and* $\{\alpha_j \mid j \in J\} \subseteq \mathcal{F}(U)$.

(1) *If* (ϕ, Φ) *is serial,* $\Phi^{\leftarrow}\Phi^{\rightarrow}\mathbf{1} = \mathbf{1}$ *and* $\phi^{\leftarrow}\phi^{\rightarrow}\mathbf{0} = \mathbf{0}$.

(2) *If* (ϕ, Φ) *is inverse serial,* $\phi^{\leftarrow}\phi^{\rightarrow}\mathbf{1} = \mathbf{1}$ *and* $\Phi^{\leftarrow}\Phi^{\rightarrow}\mathbf{0} = \mathbf{0}$.

(3) *If* (ϕ, Φ) *is inverse serial and injective, then*

(i) $\Phi^{\leftarrow}\Phi^{\rightarrow} \bigvee_{j\in J} \alpha_j = \bigvee_{j\in J} \Phi^{\leftarrow}\Phi^{\rightarrow}\alpha_j$.

(ii) $\phi^{\leftarrow}\phi^{\rightarrow} \bigwedge_{j\in J} \alpha_j = \bigwedge_{j\in J} \phi^{\leftarrow}\phi^{\rightarrow}\alpha_j$.

(4) *If* (ϕ, Φ) *is serial, and injective, then*

(i) $\phi^{\leftarrow}\phi^{\rightarrow} \bigvee_{j\in J} \alpha_j = \bigvee_{j\in J} \phi^{\leftarrow}\phi^{\rightarrow}\alpha_j$.

(ii) $\Phi^{\leftarrow}\Phi^{\rightarrow} \bigwedge_{j\in J} \alpha_j = \bigwedge_{j\in J} \Phi^{\leftarrow}\Phi^{\rightarrow}\alpha_j$.

(5) *Let If* (ϕ, Φ) *be serial and injective,*

(i) $\phi^{\leftarrow}\phi^{\rightarrow}(\Phi^{\leftarrow}\Phi^{\rightarrow}\alpha) = \alpha$.

(ii) $\Phi^{\leftarrow}\Phi^{\rightarrow}(\phi^{\leftarrow}\phi^{\rightarrow}\alpha) = \alpha$.

Remark 7.12.7 Let (ϕ, Φ) be compatible and injective.

(i), From Theorems 7.12.2, 7.12.3 (3)–(4) and 7.12.5 (2) and (3)(i)–(ii), the pair $(\phi^{\rightarrow}\phi^{\leftarrow}, \Phi^{\rightarrow}\Phi^{\leftarrow})$ constitutes an interval structure where

$$\phi^{\leftarrow}\phi^{\rightarrow} : \mathcal{F}(U) \to \mathcal{F}(U),\ \Phi^{\leftarrow}\Phi^{\rightarrow} : \mathcal{F}(U) \to \mathcal{F}(U).$$

(ii) From Theorems 7.12.2, 7.12.4 (3)–(4) and 7.12.5 (3)(i)–(ii), the pair $(\phi^{\leftarrow}\phi^{\rightarrow}, \Phi^{\leftarrow}\Phi^{\rightarrow})$ also constitutes an interval structure where

$$\phi^{\rightarrow}\phi^{\leftarrow} : \mathcal{F}(V) \to \mathcal{F}(V),\ \Phi^{\rightarrow}\Phi^{\leftarrow} : \mathcal{F}(V) \to \mathcal{F}(V).$$

For the interval structures, note that in the items (i) and (ii), the compatibility condition is necessary only for the equalities

$$\Phi^{\leftarrow}\Phi^{\rightarrow}\mathbf{1} = \mathbf{1} \text{ and } \phi^{\leftarrow}\phi^{\rightarrow}\mathbf{0} = \mathbf{0},$$

$$\phi^{\leftarrow}\phi^{\rightarrow}\mathbf{1} = \mathbf{1} \text{ and } \Phi^{\leftarrow}\Phi^{\rightarrow}\mathbf{0} = \mathbf{0}.$$

However, without any assumption, by Theorems 7.12.3 (1)–(2) and 7.12.4 (1)–(2), the revised fuzzy rough set approximations are monotonic and satisfy the conditions

$$\Phi^{\leftarrow}\Phi^{\rightarrow}\alpha \le \alpha \le \phi^{\leftarrow}\phi^{\rightarrow}\alpha,$$

$$\phi^{\rightarrow}\phi^{\leftarrow}\beta \le \beta \le \Phi^{\rightarrow}\Phi^{\leftarrow}\beta.$$

Under certain conditions, revised approximations of sets are presectional and sectional definable sets.

Theorem 7.12.8 (i) *Let* (ϕ, Φ) *be serial and* $(\Phi^{\leftarrow}, \phi^{\leftarrow})$ *be injective. Then the sets*

$$\Phi^{\rightarrow}\Phi^{\leftarrow}\beta \text{ and } \phi^{\rightarrow}\phi^{\leftarrow}\beta$$

are presectional definable.

(ii) *Let* (ϕ, Φ) *be inverse serial and* (ϕ, Φ) *be injective. Then the sets*

$$\Phi^{\leftarrow}\Phi^{\rightarrow}\beta \text{ and } \phi^{\leftarrow}\phi^{\rightarrow}\beta$$

are sectional definable.

Proof (i) By Corollary 7.8.2 (i) and (iv), the corresponding direlation (r, R) from $(W_U, \mathcal{W}_U)$ to $(W_V, \mathcal{W}_V)$ is serial and $(R^{\leftarrow}, r^{\leftarrow})$ is injective where $\phi = \phi_r$ and $\Phi = \Phi_R$. Then by Theorem 6.3.8 (i), for every $\mathfrak{J}(\beta) \in \mathcal{W}_V$, the sets

$$R^{\rightarrow}R^{\leftarrow}\mathfrak{J}(\beta) \text{ and } r^{\rightarrow}r^{\leftarrow}\mathfrak{J}(\beta)$$

are presectional definable. Therefore, by Proposition 7.3.38 and by the assumption,

$$\begin{aligned}\mathfrak{J}(\Phi^{\leftarrow}\Phi^{\rightarrow}\Phi^{\leftarrow}\beta) &= R^{\leftarrow}\mathfrak{J}(\Phi^{\rightarrow}\Phi^{\leftarrow}\beta) = R^{\leftarrow}R^{\rightarrow}\mathfrak{J}(\Phi^{\leftarrow}\beta))\\ &= R^{\leftarrow}R^{\rightarrow}R^{\leftarrow}\mathfrak{J}(\beta) = r^{\leftarrow}R^{\rightarrow}R^{\leftarrow}\mathfrak{J}(\beta)\\ &= r^{\leftarrow}R^{\rightarrow}\mathfrak{J}(\Phi^{\leftarrow}\beta) = r^{\leftarrow}\mathfrak{J}(\Phi^{\rightarrow}\Phi^{\leftarrow}\beta)\\ &= \mathfrak{J}(\phi^{\leftarrow}\Phi^{\rightarrow}\Phi^{\leftarrow}\beta),\end{aligned}$$

and so since $\mathfrak{J}$ is a lattice isomorphism, we obtain that

$$\Phi^{\leftarrow}\Phi^{\rightarrow}\Phi^{\leftarrow}\beta = \phi^{\leftarrow}\Phi^{\rightarrow}\Phi^{\leftarrow}\beta.$$

That is, $\Phi^{\rightarrow}\Phi^{\leftarrow}\beta$ is a presectional definable set. The rest of the proof is similar.

7.13 Revised Fuzzy Rough Set Models with Two Domains of Discourse

Now we consider the results given in Sect. 7.12 and in a natural way, we obtain the revised fuzzy rough set approximations in fuzzy rough set systems with two domains of discourse. Let (U, V, ϕ, Φ) be the generalized t-fuzzy approximation space where (ϕ, Φ) is the t-fuzzy direlation from $\mathcal{F}(U)$ to $\mathcal{F}(V)$ defined by

$$\phi(u, \lambda, v) = T(\varphi(u, v), \lambda)$$

and

$$\Phi(u, \lambda, v) = I(\varphi(u, v), \lambda).$$

Recall that for the section and presections of (ϕ, Φ), we have the equalities

$$\phi^{\rightarrow}\alpha(v) = \overline{apr}(\alpha)(v) = \bigwedge_{u\in U} I(\varphi(u,v),\alpha(u)),$$

$$\Phi^{\rightarrow}\alpha(v) = \underline{apr}(\alpha)(v) = \bigvee_{u\in U} T(\varphi(u,v),\alpha(u)),$$

$$\phi^{\leftarrow}\beta(u) = \underline{apr}'(\beta)(u) = \bigvee_{v\in V} T_I(\varphi(u,v),\beta(v)),$$

$$\Phi^{\leftarrow}\beta(u) = \overline{apr}'(\beta)(u) = \bigwedge_{v\in V} I_T(\varphi(u,v),\beta(v)),$$

where $\alpha \in \mathcal{F}(U)$, $\beta \in \mathcal{F}(V)$, and I and T are implicator and conjunctor, respectively.

Now, we may obtain revised fuzzy rough set approximations for two domains of discourse.

Theorem 7.13.1

(i) $\overline{apr}(\underline{apr}'(\beta))(v') = \bigvee_{u\in U} T(\varphi(u,v'), \bigwedge_{v\in V} I_T(\varphi(u,v),\beta(v)))$.

(ii) $\underline{apr}(\overline{apr}'(\beta))(v') = \bigwedge_{u\in U} I(\varphi(u,v'), \bigvee_{v\in V} T_I(\varphi(u,v),\beta(v)))$.

(iii) $\underline{apr}'(\overline{apr}(\alpha))(u') = \bigwedge_{v\in V} I_T(\varphi(u',v), \bigvee_{u\in U} T(\varphi(u,v),\alpha(u)))$.

(iv) $\overline{apr}'(\underline{apr}(\alpha))(u') = \bigvee_{v\in V} T_I(\varphi(u',v), \bigwedge_{u\in U} I(\varphi(u,v),\alpha(u)))$.

Proof Let (ϕ, Φ) be the t-fuzzy direlation given in Theorem 7.4.2. From Corollaries 7.3.32 and 7.3.33, we obtain the desired equalities as follows.

(i)

$$\begin{aligned}\overline{apr}(\underline{apr}'(\beta))(v') &= (\phi^{\rightarrow}\phi^{\leftarrow}\beta)(v')\\ &= \bigvee_{u\in U}\phi(u,\phi^{\leftarrow}\beta(u),v'))\\ &= \bigvee_{u\in U}\phi(u,\bigwedge_{v\in V}\phi^{\leftarrow}(v,\beta(v),u),v'))\\ &= \bigvee_{u\in U} T(\varphi(u,v'),\bigwedge_{v\in V} I_T(\varphi(u,v),\beta(v))).\end{aligned}$$

(ii)

$$\begin{aligned}(\underline{apr}(\overline{apr}'(\beta))(v') &= (\Phi^{\rightarrow}\Phi^{\leftarrow}\beta)(v')\\ &= \bigwedge_{u\in U}\Phi(u,\Phi^{\leftarrow}\beta(u),v'))\\ &= \bigwedge_{u\in U}\Phi(u,\bigvee_{v\in V}\Phi^{\leftarrow}(v,\beta(v),u),v'))\\ &= \bigwedge_{u\in U} I(\varphi(u,v'),\bigvee_{v\in V} T_I(\varphi(u,v),\beta(v))).\end{aligned}$$

(iii)

$$\begin{aligned}\underline{apr}'(\overline{apr}(\alpha))(u') &= (\phi^{\leftarrow}\phi^{\rightarrow}\alpha)(u')\\ &= \bigwedge_{v\in V}\phi^{\leftarrow}(v, \phi^{\rightarrow}\alpha(v), u'))\\ &= \bigwedge_{v\in V}\phi^{\leftarrow}(v, \bigvee_{u\in U}\phi(u, \alpha(u), v), u'))\\ &= \bigwedge_{v\in V} I_T(\varphi(u', v), \bigvee_{u\in U} T(\varphi(u, v), \alpha(u))).\end{aligned}$$

(iv)

$$\begin{aligned}\overline{apr}'(\underline{apr}(\alpha))(u') &= (\Phi^{\leftarrow}\Phi^{\rightarrow}\alpha)(u')\\ &= \bigvee_{v\in V}\Phi^{\leftarrow}(v, \Phi^{\rightarrow}\alpha(v), u'))\\ &= \bigvee_{v\in V}\Phi^{\leftarrow}(v, \bigwedge_{u\in U}\Phi(u, \alpha(u), v), u'))\\ &= \bigvee_{v\in V} T_I(\varphi(u', v), \bigwedge_{u\in U} I(\varphi(u, v), \alpha(u))).\end{aligned}$$

Theorem 7.13.2 (i) $\overline{apr}(\overline{apr}'(\beta))(v') = \bigvee_{u\in U} T(\varphi(u, v'), \bigvee_{v\in V} T_I(\varphi(u, v), \beta(v)))$.

(ii) $\underline{apr}(\underline{apr}'(\beta))(v') = \bigwedge_{u\in U} I(\varphi(u, v'), \bigwedge_{v\in V} I_T(\varphi(u, v), \beta(v)))$.

(iii) $\underline{apr}'(\underline{apr}(\alpha))(u') = \bigvee_{v\in V} I_T(\varphi(u', v), \bigvee_{u\in U} I(\varphi(u, v), \alpha(u)))$.

(iv) $\overline{apr}'(\overline{apr}(\alpha))(u') = \bigwedge_{v\in V} T_I(\varphi(u', v), \bigwedge_{u\in U} T(\varphi(u, v), \alpha(u)))$.

Proof The proof is left to the reader.

Martine De Cock et al. presented a useful argument explaining how the notion of approximation should be developed from Pawlak's rough set to a fuzzy rough set in rough set systems with a single domain of discourse. They emphasized the importance of the notion of equivalence class for the vague information when the fuzzy versions of rough sets is discussed. In view of this discussion, tight and loose approximations are remarkable formulations for the evaluation of membership values of the elements of domain of discourse with respect to similarity classes. Essentially, the revised fuzzy rough set approximations for two domains of discourse obtained in Theorem 7.13.1 are the natural generalizations of the tight and loose approximation operators for $U = V$.

Theorem 7.13.3

(i) $\overline{apr}(\overline{apr}(\alpha))(v) = \bigvee_{w \in U} T(\varphi(w, v), \bigvee_{u \in U} T(\varphi(u, w), \alpha(u)))$.

(ii) $\underline{apr}(\underline{apr}(\alpha))(v) = \bigwedge_{w \in U} I(\varphi(w, v), \bigwedge_{u \in U} I(\varphi(u, w), \alpha(u)))$.

(iii) $\overline{apr}(\underline{apr}(\alpha))(u) = \bigvee_{w \in U} T(\varphi(w, v), \bigwedge_{u \in U} I(\varphi(u, w), \alpha(u)))$.

(iv) $\underline{apr}(\overline{apr}(\alpha))(u) = \bigwedge_{w \in U} I(\varphi(w, v), \bigvee_{u \in U} T(\varphi(u, w), \alpha(u)))$.

Proof (i)

$$\begin{aligned}
\overline{apr}(\overline{apr}(\alpha))(v) &= \phi^{\rightarrow}(\phi^{\rightarrow}\alpha)(v) \\
&= \bigvee_{w \in U} \phi(w, \phi^{\rightarrow}\alpha(w), v)) \\
&= \bigvee_{w \in U} \phi(w, \bigvee_{u \in U} \phi(u, \alpha(u), w), v) \\
&= \bigvee_{w \in U} T(\varphi(w, v), \bigvee_{u \in U} \phi(u, \alpha(u), w)) \\
&= \bigvee_{w \in U} T(\varphi(w, v), \bigvee_{u \in U} T(\varphi(u, w), \alpha(u))).
\end{aligned}$$

(ii)

$$\begin{aligned}
\underline{apr}(\underline{apr}(\alpha))(v) &= \Phi^{\rightarrow}(\Phi^{\rightarrow}\alpha)(v) \\
&= \bigwedge_{w \in U} \Phi(w, \Phi^{\rightarrow}\alpha(w), v)) \\
&= \bigwedge_{w \in U} \Phi(w, \bigwedge_{u \in U} \Phi(u, \alpha(u), w), v) \\
&= \bigwedge_{w \in U} I(\varphi(w, v), \bigwedge_{u \in U} \Phi(u, \alpha(u), w)) \\
&= \bigwedge_{w \in U} I(\varphi(w, v), \bigwedge_{u \in U} I(\varphi(u, w), \alpha(u))).
\end{aligned}$$

(iii)

$$\begin{aligned}
\overline{apr}(\underline{apr}(\alpha))(v) &= \phi^{\rightarrow}(\Phi^{\rightarrow}\alpha)(v) \\
&= \bigvee_{w\in U} \phi(w, \Phi^{\rightarrow}\alpha(w), v)) \\
&= \bigvee_{w\in U} \phi(w, \bigwedge_{u\in U} \Phi(u, \alpha(u), w), v) \\
&= \bigvee_{w\in U} T(\varphi(w, v), \bigwedge_{u\in U} I(u, \alpha(u), w)) \\
&= \bigvee_{w\in U} T(\varphi(w, v), \bigwedge_{u\in U} I(\varphi(u, w), \alpha(u))).
\end{aligned}$$

(iv)

$$\begin{aligned}
\underline{apr}(\overline{apr}(\alpha))(v) &= \Phi^{\rightarrow}(\phi^{\rightarrow}\alpha)(v) \\
&= \bigwedge_{w\in U} \Phi(w, \phi^{\rightarrow}\alpha(w), v)) \\
&= \bigwedge_{w\in U} \Phi(w, \bigvee_{u\in U} \phi(u, \alpha(u), w), v) \\
&= \bigwedge_{w\in U} I(\varphi(w, v), \bigvee_{u\in U} \phi(u, \alpha(u), w)) \\
&= \bigwedge_{w\in U} I(\varphi(w, v), \bigvee_{u\in U} T(\varphi(u, w), \alpha(u))).
\end{aligned}$$

Let (U, φ) be a fuzzy approximation space with a single domain of discourse. Then the revised t-lower fuzzy approximation operators and t-upper fuzzy approximation operators are the fuzzy versions of the subset approximations of Yao. Indeed, for $U = V$, by Theorems 6.2.9 and 6.2.10, we have that for all $A \subseteq U$,

$$r^{\rightarrow}r^{\leftarrow}A = \underline{apr}'_{r_s}(A) \text{ and } R^{\rightarrow}R^{\leftarrow}A = \overline{apr}'_{r_s}(A)$$

and

$$R^{\leftarrow}R^{\rightarrow}A = \underline{apr}'_{r_p}(A) \text{ and } r^{\leftarrow}r^{\rightarrow}A = \overline{apr}'_{r_p}(A).$$

Then for every $\alpha \in \mathcal{F}(U)$, the sets

$\phi^{\rightarrow}\phi^{\leftarrow}\alpha,\ \Phi^{\rightarrow}\Phi^{\leftarrow}\alpha,$

$\phi^{\leftarrow}\phi^{\rightarrow}\alpha,\ \Phi^{\leftarrow}\Phi^{\rightarrow}\alpha$

can be regarded as the fuzzy versions of the subset approximations of Yao. For instance, for the first two approximations, note that

$$
\begin{aligned}
\overline{apr}(\underline{apr}'(\alpha))(v) &= (\phi^{\rightarrow}\phi^{\leftarrow}\alpha)(v) \\
&= \bigvee_{u\in U} \phi(u, \phi^{\leftarrow}\alpha(u), v)) \\
&= \bigvee_{u\in U} \phi(u, \bigwedge_{w\in U} \phi^{\leftarrow}(w, \alpha(w), u), v)) \\
&= \bigvee_{u\in U} T(\varphi(u, v), \bigwedge_{w\in U} I_T(\varphi(u, v), \alpha(w)))
\end{aligned}
$$

and

$$
\begin{aligned}
\underline{apr}(\overline{apr}'(\alpha))(v) &= \Phi^{\rightarrow}(\Phi^{\leftarrow}\alpha)(v) \\
&= \bigwedge_{w\in U} \Phi(w, \Phi^{\leftarrow}\alpha(w), v)) \\
&= \bigwedge_{w\in U} \Phi(w, \bigwedge_{u\in U} \Phi^{\leftarrow}(u, \alpha(u), w), v) \\
&= \bigwedge_{w\in U} I(\varphi(w, v), \bigwedge_{u\in U} \Phi^{\leftarrow}(u, \alpha(u), w)) \\
&= \bigwedge_{w\in U} I(\varphi(w, v), \bigwedge_{u\in U} I(\varphi(u, w), \alpha(u))).
\end{aligned}
$$

The fuzzy version of the approximations $\underline{apr}'(\overline{apr}(\alpha))$ and $\overline{apr}'(\underline{apr}(\alpha))$ can be also similarly evaluated.

The revised fuzzy approximation operators are dual.

Theorem 7.13.4 *Let (U, V, ϕ, Φ) be a generalized t-fuzzy approximation space. If (ϕ, Φ) is a complemented t-fuzzy direlation from $\mathcal{F}(U)$ to $\mathcal{F}(V)$, then for all $\alpha \in \mathcal{F}(U)$,*

$$\overline{apr}'(\underline{apr}(\mathbf{1} - \alpha)) = \mathbf{1} - \underline{apr}'(\overline{apr}(\alpha)),$$

$$\underline{apr}'(\overline{apr}(\mathbf{1} - \alpha)) = \mathbf{1} - \overline{apr}'(\underline{apr}(\alpha)),$$

and

$$\underline{apr}(\overline{apr}'(\mathbf{1} - \beta)) = \mathbf{1} - \overline{apr}(\underline{apr}'(\beta)),$$

$$\overline{apr}(\underline{apr}'(\mathbf{1} - \beta)) = \mathbf{1} - \underline{apr}(\overline{apr}'(\beta)).$$

The rest of the equalities are similar.

Proof From Theorem 7.13.4, we immediately obtain that

$$\overline{apr}'(\underline{apr}(\mathbf{1} - \alpha)) = \Phi^{\leftarrow}\Phi^{\rightarrow}(\mathbf{1} - \alpha) = \mathbf{1} - \phi^{\leftarrow}\phi^{\rightarrow}\alpha = \mathbf{1} - \underline{apr}'(\overline{apr}(\alpha)),$$

$$\underline{apr}'(\overline{apr}(\mathbf{1} - \alpha)) = \phi^{\leftarrow}\phi^{\rightarrow}(\mathbf{1} - \alpha) = \mathbf{1} - \Phi^{\leftarrow}\Phi^{\rightarrow}\alpha = \mathbf{1} - \overline{apr}'(\underline{apr}(\alpha)). \quad \square$$

Theorem 7.13.5 *Let (U, V, φ) be a generalized fuzzy approximation space and $\beta \in \mathcal{F}(V)$ and $\{\beta_j \mid j \in J\} \subseteq \mathcal{F}(V)$. Then we have the following properties.*

(1) $\overline{apr}(\underline{apr}'(\beta)) \leq \beta \leq \underline{apr}(\overline{apr}'(\beta))$.

(2) $\beta_1 \leq \beta_2 \implies \overline{apr}(\underline{apr}'(\beta_1)) \leq \overline{apr}(\underline{apr}'(\beta_2))$.

(3) $\beta_1 \leq \beta_2 \implies \underline{apr}(\overline{apr}'(\beta_1) \leq \underline{apr}(\overline{apr}'(\beta_2))$.

(4) $\bigvee_{j\in J} \overline{apr}(\underline{apr}'(\beta_j)) \leq \overline{apr}(\underline{apr}'(\bigvee_{j\in J} \beta_j))$.

(5) $\underline{apr}(\overline{apr}'(\bigwedge_{j\in J} \beta_j)) \leq \bigwedge_{j\in J} \underline{apr}(\overline{apr}'(\beta_j))$.

(6) $\overline{apr}(\underline{apr}'(\overline{apr}(\underline{apr}'\beta))) = \overline{apr}(\underline{apr}'(\beta))$.

(7) $\underline{apr}(\overline{apr}'(\underline{apr}(\overline{apr}'\beta))) = \underline{apr}(\overline{apr}'(\beta))$.

Proof (1) We know that from Theorem 7.13.5(i), if (U, V, ϕ, Φ) is the corresponding generalized fuzzy approximation space, then for all $\beta \in \mathcal{F}(V)$, we have

$$\phi^{\rightarrow}\phi^{\leftarrow}\beta \leq \beta \leq \Phi^{\rightarrow}\Phi^{\leftarrow}\beta.$$

Then we easily obtain that

$$\overline{apr}(\underline{apr}'(\beta)) = \phi^{\rightarrow}\phi^{\leftarrow}\beta \leq \beta \leq \Phi^{\rightarrow}\Phi^{\leftarrow}\beta = \underline{apr}(\overline{apr}'(\beta)).$$

(3) Let $\beta_1 \leq \beta_2$. Then from Theorem 7.13.5(iii),

$$\overline{apr}(\underline{apr}'(\beta_1)) = \Phi^{\rightarrow}\Phi^{\leftarrow}\beta_1 \leq \Phi^{\rightarrow}\Phi^{\leftarrow}\beta_2 = \underline{apr}(\overline{apr}'(\beta_2)).$$

(5) From Theorem 7.13.5(v), we observe that

$$\begin{aligned}\underline{apr}(\overline{apr}'(\bigwedge_{j\in J} \beta_j)) &= \Phi^{\rightarrow}\Phi^{\leftarrow}(\bigwedge_{j\in J} \beta_j)\\ &\leq \bigwedge_{j\in J} \Phi^{\rightarrow}\Phi^{\leftarrow}\beta_j = \bigwedge_{j\in J} \underline{apr}(\overline{apr}'(\beta_j)).\end{aligned}$$

(7) From Theorem 7.13.5(vii), we have

$$\begin{aligned}\underline{apr}(\overline{apr}'(\underline{apr}(\overline{apr}'\beta))) &= \underline{apr}(\overline{apr}'(\underline{apr}(\overline{apr}'\beta)))\\ &= \underline{apr}(\overline{apr}'(\beta)) = apr(\overline{apr}'(\beta)).\end{aligned}$$

Theorem 7.13.6 *Let (U, V, φ) be a generalized fuzzy approximation space, $\beta \in \mathcal{F}(V)$ and $\{\beta_j \mid j \in J\} \subseteq \mathcal{F}(V)$.*

(1) *If φ is serial, then $\overline{apr}(\underline{apr}'(\mathbf{0})) = \mathbf{0}$ and $\underline{apr}(\overline{apr}'(\mathbf{1})) = \mathbf{1}$.*

(2) *If φ is inverse serial, then*

$$\overline{apr}(\underline{apr}'(\mathbf{1})) = \mathbf{1} \text{ and } \underline{apr}(\overline{apr}'(\mathbf{0})) = \mathbf{0}.$$

(3) *Let φ be inverse serial and injective.*

(i) $\overline{apr}(\underline{apr}'(\bigwedge_{j\in J}\beta_j)) = \bigwedge_{j\in J}\overline{apr}(\underline{apr}'(\beta_j))$.

(ii) $\underline{apr}(\overline{apr}'(\bigvee_{j\in J}\beta_j)) = \bigvee_{j\in J}\underline{apr}(\overline{apr}'(\beta_j))$.

(4) *Let φ be serial and injective.*

(i) $\underline{apr}(\overline{apr}'(\bigwedge_{j\in J}\beta_j)) = \bigwedge_{j\in J}\underline{apr}(\overline{apr}'(\beta_j))$.

(ii) $\overline{apr}(\underline{apr}'(\bigvee_{j\in J}\beta_j) = \bigvee_{j\in J}\overline{apr}(\underline{apr}'(\beta_j))$.

(iii) *Let φ be inverse serial and injective. Then*

$\underline{apr}(\overline{apr}'(\beta)) = \overline{apr}(\underline{apr}'(\underline{apr}(\overline{apr}'(\beta)))$.

(5) *Let φ be serial and injective.*

(i) $\underline{apr}(\overline{apr}'(\overline{apr}(\underline{apr}'(\beta)))) = \beta$.

(ii) $\beta = \overline{apr}(\underline{apr}'(\underline{apr}(\overline{apr}'(\beta))))$.

Proof (1) From Theorem 7.7.7(i), (ϕ, Φ) is also serial. Then by Theorem 7.12.5(1),

$\phi^{\rightarrow}\phi^{\leftarrow}\mathbf{0} = \mathbf{0}$ and $\Phi^{\rightarrow}\Phi^{\leftarrow}\mathbf{1} = \mathbf{1}$.

Hence, we get

$\overline{apr}(\underline{apr}'(\mathbf{1})) = \phi^{\rightarrow}\phi^{\leftarrow}\mathbf{0} = \mathbf{0}$ and $\underline{apr}(\overline{apr}'(\mathbf{1})) = \Phi^{\rightarrow}\Phi^{\leftarrow}\mathbf{1} = \mathbf{1}$.

(3)(i) From Theorem 7.7.7(ii), (ϕ, Φ) is inverse serial. Further, by Theorem 7.7.14, it is also injective. Therefore, by Theorem 7.12.5 (3)(i), we conclude that

$$\overline{apr}(\underline{apr}'(\bigwedge_{j\in J}\beta_j)) = \phi^{\rightarrow}\phi^{\leftarrow}\bigwedge_{j\in J}\beta_j = \bigwedge_{j\in J}\phi^{\rightarrow}\phi^{\leftarrow}\beta_j = \bigwedge_{j\in J}\overline{apr}(\underline{apr}'(\beta_j)).$$

(4)(iii) By Theorems 7.7.7(ii) and 7.7.14, (ϕ, Φ) is also inverse and injective. Therefore, by Theorem 7.12.5 (4)(iii), we obtain

$$\Phi^{\rightarrow}\Phi^{\leftarrow}\beta = \underline{apr}(\overline{apr}'(\beta)) = \overline{apr}(\underline{apr}'(\underline{apr}(\overline{apr}'(\beta))) = \phi^{\rightarrow}\phi^{\leftarrow}(\Phi^{\rightarrow}\Phi^{\leftarrow}\beta).$$

The proof of the next two theorems is similar and omitted.

Theorem 7.13.7 *For every $\alpha \in \mathcal{F}(U)$ and $\{\alpha_j \mid j \in J\} \subseteq \mathcal{F}(U)$, the mappings*

$\underline{apr}'\overline{apr} : \mathcal{F}(U) \to \mathcal{F}(U)$ *and* $\overline{apr}'\underline{apr} : \mathcal{F}(U) \to \mathcal{F}(U)$

satisfy the following properties:

(1) $\underline{apr}'(\overline{apr}(\alpha)) \le \alpha \le \overline{apr}'(\underline{apr}(\alpha_1))$.

(2) $\alpha_1 \le \alpha_2 \implies \overline{apr}'(\underline{apr}(\alpha_1)) \le \overline{apr}'(\underline{apr}(\alpha_2))$.

3) $\alpha_1 \le \alpha_2 \implies \overline{apr}(\underline{apr}'(\alpha_1)) \le \overline{apr}(\underline{apr}'(\alpha_2))$.

(4) $\bigvee_{j\in J} \overline{apr}'(\underline{apr}(\alpha_j)) \leq \overline{apr}'(\underline{apr}(\bigvee_{j\in J}(\alpha_j))$.

(5) $\underline{apr}'(\overline{apr}(\bigwedge_{j\in J}(\alpha_j))) \leq \bigwedge_{j\in J} \underline{apr}'(\overline{apr}(\alpha_j))$.

(6) $\overline{apr}'(\underline{apr}(\alpha)) = \overline{apr}'(\underline{apr}(\overline{apr}'(\underline{apr}(\alpha)))$.

(7) $\underline{apr}'(\overline{apr}(\alpha)) = \underline{apr}'(\overline{apr}(\underline{apr}'(\overline{apr}(\alpha))))$.

Theorem 7.13.8 *The mappings*

$$\overline{apr}'\underline{apr} : \mathcal{F}(U) \to \mathcal{F}(U) \text{ and } \underline{apr}'\overline{apr} : \mathcal{F}(U) \to \mathcal{F}(U)$$

satisfy the following properties where $\alpha \in \mathcal{F}(U)$ *and* $\{\alpha_j \mid j \in J\} \subseteq \mathcal{F}(U)$.

(1) *If* φ *is serial,* $\overline{apr}'(\underline{apr}(\mathbf{1})) = \mathbf{1}$ *and* $\underline{apr}'(\overline{apr}(\mathbf{0}) = \mathbf{0}$.

(2) *If* φ *is inverse serial,* $\underline{apr}'(\overline{apr}(\mathbf{1})) = \mathbf{1}$ *and* $\overline{apr}'(\underline{apr}(\mathbf{0})) = \mathbf{0}$.

(3) *If* φ *is inverse serial and injective, then*

(i) $\overline{apr}'(\underline{apr}(\bigvee_{j\in J}(\alpha_j)) = \bigvee_{j\in J} \overline{apr}'(\underline{apr}(\alpha_j))$.

(ii) $\underline{apr}'(\overline{apr}(\bigwedge_{j\in J}(\alpha_j))) = \bigwedge_{j\in J} \underline{apr}'(\overline{apr}(\alpha_j))$.

(4) *If* φ *is serial, and injective, then*

(i) $\underline{apr}'(\overline{apr}(\bigvee_{j\in J}(\alpha_j))) = \bigvee_{j\in J} \underline{apr}'(\overline{apr}(\alpha_j))$.

(ii) $\overline{apr}'(\underline{apr}(\bigwedge_{j\in J}(\alpha_j))) = \bigwedge_{j\in J} \overline{apr}'(\underline{apr}(\alpha_j))$.

(5) *Let* φ *be serial and injective.*

(i) $\underline{apr}'(\overline{apr}(\overline{apr}'(\underline{apr}(\alpha)))) = \alpha$.

(ii) $\overline{apr}'(\underline{apr}(\underline{apr}'(\overline{apr}(\alpha)))) = \alpha$.

Remark 7.13.9 For any textural fuzzy rough approximation space (U, V, ϕ, Φ) with two domains of discourse, as is observed in Sect. 7.12, if (ϕ, Φ) is compatible and injective, then the pairs

$$(\phi^{\to}\phi^{\leftarrow}, \Phi^{\to}\Phi^{\leftarrow}) \text{ and } (\phi^{\leftarrow}\phi^{\to}, \Phi^{\leftarrow}\Phi^{\to})$$

constitute two interval structures, respectively, and hence, in view of Theorems 7.13.4, 7.13.6, 7.13.7 and 7.12.6, for the fuzzy rough approximation space (U, V, φ), the corresponding pairs

$$(\overline{apr}(\underline{apr}'), \ \underline{apr}(\overline{apr}'))$$

and

$$(\underline{apr}'(\overline{apr}),\ \overline{apr}'(\underline{apr}))$$

are also interval structures where φ is compatible and injective.

Furthermore, in view of the above theorems, we observe that without any condition, we have the following desired properties:

$$\overline{apr}(\underline{apr}'(\beta)) \leq \beta \leq \underline{apr}(\overline{apr}'(\beta)),$$

$$\beta_1 \leq \beta_2 \implies \overline{apr}(\underline{apr}'(\beta_1)) \leq \overline{apr}(\underline{apr}'(\beta_2)),$$

$$\beta_1 \leq \beta_2 \implies \underline{apr}(\overline{apr}'(\beta_1) \leq \underline{apr}(\overline{apr}'(\beta_2)),$$

$$\underline{apr}'(\overline{apr}(\alpha)) \leq \alpha \leq \overline{apr}'(\underline{apr}(\alpha_1)),$$

$$\alpha_1 \leq \alpha_2 \implies \overline{apr}'(\underline{apr}(\alpha_1)) \leq \overline{apr}'(\underline{apr}(\alpha_2)),$$

$$\alpha_1 \leq \alpha_2 \implies \overline{apr}(\underline{apr}'(\alpha_1)) \leq \overline{apr}(\underline{apr}'(\alpha_2)).$$

Example 7.13.10 (i) Take the fuzzy relation φ given in Table 7.12 and the fuzzy rough set model corresponding to minimum t-norm and Kleene-Dienes S-implicator. Clearly, φ is inverse serial. Let us take the fuzzy set $\beta \in \mathcal{F}(V)$ where $\beta(v_1) = 0.5$, $\beta(v_2) = 0.8$. We show that β satisfies Theorem 7.8.6 (iii). Note that

$$\overline{apr}(\overline{apr}'(\beta))(v_1) = \bigvee_{\lambda<0.8} (0.5 \wedge \lambda) \vee (\bigvee_{n\geq 2,\lambda<0.5} (\frac{n}{n+1} \wedge \lambda)) = 0.5,$$

$$\overline{apr}(\overline{apr}'(\beta))(v_2) = (\bigvee_{\lambda<0.8} \lambda) \vee \bigvee_{\lambda<0.5} (\frac{1}{5} \wedge \lambda) \vee (\bigvee_{n\geq 3,\lambda<0.5} 0) = 0.8.$$

As we have expected, β holds the inequality

$$\beta \leq \overline{apr}(\overline{apr}'(\beta))$$

given in Theorem 7.8.6 (iii).

Table 7.12 $\varphi \in \mathcal{F}(U \times V)$

φ	v_1	v_2
u_1	$\frac{1}{2}$	1
u_2	$\frac{2}{3}$	$\frac{1}{5}$
u_3	$\frac{3}{4}$	0
$\vdots$	$\vdots$	$\vdots$
u_n	$\frac{n}{n+1}$	0
$\vdots$	$\vdots$	$\vdots$

(ii) Now let us show that β satisfies Theorem 7.8.8 (ii) and (iii), that is, $\overline{apr}'(\beta) \neq 0$ and $\underline{apr}'(\beta) \neq 1$. By Corollary 7.3.28, we have $\overline{apr}'(\beta)(u) = \bigvee_{\mu<\beta(v)} \overline{apr}'(v_\mu)(u)$ and so we obtain

$$\overline{apr}'(\beta)(u_1) = \bigvee_{\mu<0.5} 0 \vee \bigvee_{\mu<0.8} \mu = 0.8 \neq 0.$$

On the other hand, since

$$\underline{apr}'(\beta)(u) = \bigwedge_{\beta(v)<\mu} \underline{apr}'(v^\mu)(u),$$

we have

$$\underline{apr}'(\beta)(u_2) = \bigwedge_{0.5<\mu} \mu \wedge \bigwedge_{0.8<\mu} 1 = 0.5 \neq 1.$$

(iii) Note that the revised approximations $\overline{apr}\underline{apr}'$ and $\underline{apr}\overline{apr}'$ satisfy Theorem 7.13.6 (1). Indeed, considering Theorems 7.8.3 $\overline{(2)}$(i) and $\overline{(4)}$(iii), we obtain that

$$\overline{apr}(\underline{apr}'(\mathbf{1}))(v) = \bigvee_{\lambda<1} \varphi(u, v) \wedge \lambda = 1$$

and

$$\underline{apr}(\overline{apr}'(\mathbf{0}))(v) = \bigwedge_{0<\lambda} (1 - \varphi(u, v)) \vee \lambda = 0.$$

Example 7.13.11 Let us consider the fuzzy relation φ given in Table 7.13 and the fuzzy rough set model corresponding to product t-norm and Reichenbach S-implicator. Note that φ is serial. Let us take the fuzzy set $\alpha \in \mathcal{F}(U)$ where $\alpha(u_1) = 0.2$, $\alpha(u_2) = 0.9$. By Corollary 7.3.28 (ii), we have

$$\underline{apr}'(\underline{apr}(\alpha)(u)) = \bigwedge_{(\underline{apr}(\alpha))(v)<\mu} \underline{apr}'(v^\mu)(u).$$

Note that for $n \in \mathbb{N}$, we have $\underline{apr}(\alpha)(v_n) = (\frac{1}{n})(\frac{8}{10}) + \frac{2}{10}$. Further, for $n \in \mathbb{N}$, we also have

Table 7.13 $\varphi \in \mathcal{F}(U \times V)$

φ	v_1	v_2	v_3	v_4	...	v_n	...	
u_1	0	$1 - \frac{1}{2}$	$1 - \frac{1}{3}$	$1 - \frac{1}{4}$	...	$1 - \frac{1}{n}$	...	
u_2	0	1	0	0	...	0	...	

$$\underline{apr}'(v_n^{\mu})(u_1) = \begin{cases} 1, & \text{if } 1 - \frac{1}{n} \leq \mu \\ \frac{\mu}{1-\frac{1}{n}}, & \text{if } 1 - \frac{1}{n} > \mu. \end{cases}$$

Then we obtain

$$\underline{apr}'(\underline{apr}(\alpha)(u_1)) = \bigwedge_{(\frac{1}{n})(\frac{8}{10})+\frac{2}{10}<\mu} \frac{\mu}{1-\frac{1}{n}} = \frac{2}{10} \leq \alpha(u_1).$$

It is easy to see that for all $\mu \in [0, 1]$,

$\underline{apr}'(v_1^{\mu})(u_2) = 1$,

$\underline{apr}'(v_2^{\mu})(u_2) = \begin{cases} 1, & \text{if } 1 = \mu \\ \mu, & \text{if } 1 > \mu, \end{cases}$ and

for $n \geq 3$ and for all $\mu \in [0, 1]$, $\underline{apr}'(v_n^{\mu})(u_2) = 1$.

As above, for $u_2 \in U$, we may get that

$$\begin{aligned} \underline{apr}'(\underline{apr}(\alpha))(u_2) &= \bigwedge_{1<\mu} 1 \wedge \bigwedge_{\frac{6}{10}<\mu} \mu \wedge (\bigwedge_{n\geq 3,\ (\frac{1}{n})(\frac{8}{10})+\frac{2}{10}<\mu} 1) \\ &= \frac{6}{10} \leq \alpha(u_2), \end{aligned}$$

that is, we conclude that $\underline{apr}'(\underline{apr}(\alpha)) \leq \alpha$.

(ii) Take the fuzzy set $\beta \in \mathcal{F}(V)$ where $\beta(v_{2n}) = \frac{1}{4}$ and $\beta(v_{2n+1}) = \frac{1}{3}$ where $n \in \mathbb{N}$. We show that $\underline{apr}'(\beta) \leq \overline{apr}'(\beta)$. By Corollary 7.3.28 (ii), we get $\underline{apr}'(\beta)(u) = \bigwedge_{\beta(v)<\mu} \underline{apr}'(v^{\mu}(u))$ and so $\underline{apr}'(v_{2n})^{\mu}(u_1) = \begin{cases} 1, & \text{if } 1 - \frac{1}{2n} \leq \mu \\ \frac{\mu}{1-\frac{1}{2n}}, & \text{if } 1 - \frac{1}{2n} > \mu, \end{cases}$

$$\underline{apr}'(v_{2n+1})^{\mu}(u_1) = \begin{cases} 1, & \text{if } 1 - \frac{1}{2n+1} \leq \mu \\ \frac{\mu}{1-\frac{1}{2n+1}}, & \text{if } 1 - \frac{1}{2n+1} > \mu, \end{cases}$$

$$\begin{aligned} \underline{apr}'(\beta)(u_1) &= \bigwedge_{\frac{1}{4}<\mu} (\frac{\mu}{1-\frac{1}{2n}}) \wedge \bigwedge_{\frac{1}{3}<\mu} (\frac{\mu}{1-\frac{1}{2n+1}}) \\ &= \frac{1}{4} \wedge \frac{1}{3} = \frac{1}{4}. \end{aligned}$$

$$\underline{apr}'(\beta)(u_2) = \bigwedge_{\frac{1}{3}<\mu} 1 \wedge \bigwedge_{\frac{1}{4}<\mu} \mu = \frac{1}{4}.$$

By Corollary 7.3.28 (ii), we get

$$\overline{apr}'(\beta)(u) = \bigvee_{\mu<\beta(v)} \overline{apr}'(v_\mu(u)).$$

Since

$$\overline{apr}'(v_{2n})_\mu(u_1) = \begin{cases} 0, & \text{if } \mu \le \frac{1}{2n} \\ \frac{\mu-\frac{1}{2n}}{1-\frac{1}{2n}}, & \text{if } \mu > \frac{1}{2n}, \end{cases}$$

$$\overline{apr}'(v_{2n+1})_\mu(u_1) = \begin{cases} 0, & \text{if } \mu \le \frac{1}{2n+1} \\ \frac{\mu-\frac{1}{2n+1}}{1-\frac{1}{2n+1}}, & \text{if } \mu > \frac{1}{2n+1}, \end{cases}$$

we obtain

$$\begin{aligned}\overline{apr}'(\beta)(u_1) &= \bigvee_{\mu<\frac{1}{4}} \Big(\frac{\mu-\frac{1}{2n}}{1-\frac{1}{2n}}\Big) \vee \bigvee_{\mu<\frac{1}{3}} \Big(\frac{\mu-\frac{1}{2n+1}}{1-\frac{1}{2n+1}}\Big) \\ &= \frac{1}{4} \vee \frac{1}{3} = \frac{1}{3}.\end{aligned}$$

Further, for all $\mu \in [0, 1]$, since

$$\overline{apr}'(v_1^\mu)(u_2) = 0$$

and

$$\overline{apr}'(v_2^\mu)(u_2) = \begin{cases} 0, & \text{if } 1-\mu \ge 1 \\ \mu, & \text{if } 1-\mu < 1, \end{cases}$$

we conclude that

$$\overline{apr}'(\beta)(u_2) = \bigvee_{\mu<\frac{1}{3}} 0 \vee \bigvee_{\mu<\frac{1}{4}} \mu = \frac{1}{4}.$$

As a result, we get $\underline{apr}'(\beta) \le \overline{apr}'(\beta)$.

(iii) As we have proved in Theorem 7.13.6 (1), for serial fuzzy relation φ, in view of Corollary 7.3.28 (i), we see that

$$\begin{aligned}\overline{apr}(\underline{apr}'(\mathbf{0}))(v) &= \bigvee_{\lambda<\underline{apr}'(\mathbf{0})(u)} \overline{apr}(u_\lambda)(v) \\ &= \bigvee_{\lambda<0} \overline{apr}(u_\lambda)(v) = 0,\end{aligned}$$

$$\begin{aligned}\underline{apr}(\overline{apr}'(\mathbf{1}))(v) &= \bigwedge_{\overline{apr}'(\mathbf{1})(u)<\lambda} \underline{apr}(u^\lambda)(v) \\ &= \bigwedge_{1<\lambda} \underline{apr}(u^\lambda)(v) = 1.\end{aligned}$$

The approximation operators of the systems corresponding to pairs (T, I) and (T_I, I_T) do not change the information with respect to approximations of the subsets of the domains of discourse U and V.

Theorem 7.13.12 *Let φ be a fuzzy relation. Then for all $\alpha \in \mathcal{F}(U)$ and $\beta \in \mathcal{F}(V)$, we have the following propositions:*

(i) $\overline{apr}(\underline{apr}'(\overline{apr}(\alpha))) = \overline{apr}(\alpha)$.

(ii) $\underline{apr}(\overline{apr}'(\underline{apr}(\alpha))) = \underline{apr}(\alpha)$.

(iii) $\underline{apr}'(\overline{apr}(\underline{apr}'(\beta))) = \underline{apr}'(\beta)$.

(iv) $\overline{apr}'(\underline{apr}(\overline{apr}'\beta)) = \overline{apr}'(\beta)$.

Proof (i) From Theorem 7.3.41 (i), we have that

$$\forall \alpha \in \mathcal{F}(U),\ \phi^{\rightarrow}(\phi^{\leftarrow}(\phi^{\rightarrow}\alpha)) = \phi^{\rightarrow}\alpha.$$

Therefore, we conclude that

$$\overline{apr}(\underline{apr}'(\overline{apr}(\alpha))) = \phi^{\rightarrow}(\phi^{\leftarrow}(\phi^{\rightarrow}\alpha)) = \phi^{\rightarrow}\alpha = \overline{apr}(\alpha).$$

(iii) From Theorem 7.3.41 (iii),

$$\forall \beta \in \mathcal{F}(V),\ \phi^{\leftarrow}(\phi^{\rightarrow}(\phi^{\leftarrow}\beta)) = \underline{apr}'(\overline{apr}(\underline{apr}'(\beta))) = \phi^{\leftarrow}\beta = \underline{apr}'(\beta).$$

Theorem 7.13.13 *Let $\varphi \in \mathcal{F}(U \times V)$ be serial and (T_I, I_T)-injective. Then*

(i) *For all $\alpha \in \mathcal{F}(V)$, the sets*

$$\overline{apr}'(\underline{apr}(\alpha)),\ \underline{apr}'(\overline{apr}(\alpha))$$

are (T, I)-definable.

(ii) *For all $\beta \in \mathcal{F}(U)$, the sets*

$$\overline{apr}(\underline{apr}'(\beta)),\ \underline{apr}(\overline{apr}'(\beta))$$

are (T_I, I_T)-definable.

Proof We give the proof of (ii) leaving (i) to the interested reader. From Corollary 7.7.7, the corresponding t-fuzzy direlation (ϕ, Φ) is serial and from Theorem 7.7.14 (ii), the corresponding t-fuzzy direlation $(\Phi^{\leftarrow}, \phi^{\leftarrow})$ is injective, and so by Theorem 7.12.8, we have

$$\phi^{\leftarrow}(\Phi^{\rightarrow}(\Phi^{\leftarrow}\beta)) = \Phi^{\leftarrow}\Phi^{\rightarrow}(\Phi^{\leftarrow}\beta), \text{ and}$$

$$\Phi^{\leftarrow}(\phi^{\rightarrow}(\phi^{\leftarrow}\beta)) = \phi^{\leftarrow}\phi^{\rightarrow}(\phi^{\leftarrow}\beta).$$

Then we conclude that

$$\begin{aligned}\underline{apr}'(\underline{apr}(\overline{apr}'(\beta))) &= \phi^{\leftarrow}(\Phi^{\leftarrow}(\Phi^{\leftarrow}\beta)) \\ &= \Phi^{\leftarrow}\Phi^{\rightarrow}(\Phi^{\leftarrow}\beta) = \overline{apr}'(\underline{apr}(\overline{apr}'(\beta)))\end{aligned}$$

and

$$\overline{apr}'(\overline{apr}(\underline{apr}'(\beta))) = \Phi^{\leftarrow}(\phi^{\rightarrow}(\phi^{\leftarrow}\beta))$$
$$= \phi^{\leftarrow}\phi^{\rightarrow}(\phi^{\leftarrow}\beta) = \underline{apr}'(\overline{apr}(\underline{apr}'(\alpha))).$$

Corollary 7.13.14 *Let (U, V, φ) be a generalized fuzzy approximation space.*

(1) *If φ is (T, I)-injective and serial, then for all $\beta \in \mathcal{F}(V)$ the sets*

$\underline{apr}'(\beta)$ *and* $\overline{apr}'(\beta)$

are definable with respect to revised approximation on $\mathcal{F}(U)$.

(2) *If φ is (T_I, I_T)-injective and inverse serial, then for all $\alpha \in \mathcal{F}(U)$, the sets*

$\underline{apr}(\alpha)$ *and* $\overline{apr}(\alpha)$

are definable with respect to revised approximation on $\mathcal{F}(V)$.

Proof We give the proof of (1) leaving the dual proof (2) to the reader. We know that by Theorems 7.7.9 and 7.7.14, the corresponding t-fuzzy direlations (ϕ, Φ) is also serial and injective. Then we may consider the equalities given in Theorem 7.8.9 (4)(i)–(ii)(c). if we take $\alpha = \phi^{\leftarrow}\beta$, then we obtain

$$\phi^{\leftarrow}\phi^{\rightarrow}(\phi^{\leftarrow}\beta) = \phi^{\leftarrow}\beta = \Phi^{\leftarrow}\Phi^{\rightarrow}(\phi^{\leftarrow}\beta)$$

and so, we get

$$\underline{apr}'(\overline{apr}(\underline{apr}'(\beta))) = \overline{apr}'(\underline{apr}(\underline{apr}'(\beta))).$$

That is, the set $\underline{apr}'(\beta)$ is definable with respect to revised approximation on $\mathcal{F}(V)$. For the definability of $\overline{apr}'(\beta)$, it is enough to get $\alpha = \Phi^{\leftarrow}\beta$.

References

1. B. De Baets, J.C. Fodor, Residual operators of uninorms. Soft Comput. **3**, 89–100 (1999)
2. M. Baczynski, B. Jayaram, *Fuzzy Implications, Study in Fuzziness and Soft Computing*, vol. 231 (Springer, Berlin, 2008)
3. J.C. Fodor, M. Roubens, *Fuzzy Preference Modelling and Multicriteria Decision Support* (Kluwer, Dordrecht, 1994)
4. S. Gottwald, *A Treatise on Many-Valued Logics* (Research Studies Press, Baldock, 2001)
5. Y. Narukawa, M. Taha, V. Torra, On the definition of probabilistic metric spaces by means of fuzzy measures. Fuzzy Sets Syst. Article in Press. https://doi.org/10.1016/j.fss.2023.108528
6. B. Schweizer, A. Sklar, *Probabilistic Metric Spaces* (Elsevier, North-Holland, 1983)
7. J.A. Goguen, L-fuzzy sets. J. Math. Anal. Appl. **18**, 145–174 (1967)
8. R. Bělohlávek, *Fuzzy Relational Systems: Foundations and Principles* (Kluwer, New York, 2002)
9. I.U. Tiryaki, L.M. Brown, Relations and correlations between lattices of fuzzy subsets. Hacet. J. Math. Stat. **34**(S), 69–89 (2005)

10. L. D'eer, N. Verbiest, C. Cornelis, L. Godo, Implicator-conjunctor based models of fuzzy rough sets: definitions and properties, in *RSFDGrC, LNAI*, vol. 8170 ed. by D. Ciucci, M. Inuiguchi, Y. Yao, D. Slezak, G. Wang (Springer, Heidelberg, 2013), pp. 169–179
11. M. De Cock, C. Cornelis, E. Kerre, Fuzzy rough sets: the forgotten step. IEEE Trans. Fuzzy Syst. **15**(1), 121–130 (2007)
12. Q. Hu, L. Zhang, D. Chen, W. Pedrycz, D. Yu, Gaussian kernel based fuzzy rough sets: model, uncertainty measures and applications. Int. J. Approx. Reason. **51**, 453–471 (2010)
13. J. Mi, W. Zhang, An axiomatic characterization of a fuzzy generalization of rough sets. Inf. Sci. **160**, 235–249 (2004)
14. D. Yeung, D. Chen, E. Tsang, J. Lee, W. Xizhao, On the generalization of fuzzy rough sets. IEEE Trans. Fuzzy Syst. **13**(3), 343–361 (2005)
15. W.-Z. Wu, Y. Leung, J. Mi, On characterizations of (I, T)-fuzzy rough approximation operators. Fuzzy Sets Syst. **154**, 76–102 (2005)
16. D. Dubois, H. Prade, Rough fuzzy sets and fuzzy rough sets. Int. J. Gen. Syst. **17**(23), 191–209 (1990)
17. K. Gödel, Zum intuitionistischen aussagenkalkul. Auzeiger der Akademie der Wissenschaften in Wien, Mathematisch, Naturwissenschaftliche Klasse **69**, 65–66 (1932)
18. A. Ganivada, S.K. Pal, A novel fuzzy rough granular neural network for classification. Int. J. Comput. Intell. Syst. **4**(5), 1042–1051 (2011)
19. P. Mitra, C.A. Murthy, S.K. Pal, Unsupervised feature selection using feature similarity. IEEE Trans. Pattern Anal. Mach. Intell. **24**(4), 301–312 (2002)
20. S. Pal, S.S. Ray, A. Ganivada, *Granular Neural Networks, Pattern Recognition and Bioinformatics*. Studies in computational intelligence, vol. 712 (Springer, Heidelberg, 2017)
21. K. Zhang, J. Zhan, W.Z. Wu, Novel fuzzy rough set models and corresponding applications to multi-criteria decision-making. Fuzzy Sets Syst. **383**, 92–126 (2020)
22. C. Cornelis, M. De Cock, A.M. Radzikowska, Fuzzy rough sets: from theory into practice, in *Handbook of Granular Computing*, ed. by W. Pedrycz, A. Skowron, V. Kreinovich (Wiley, Chichester, 2008)
23. J. Mi, Y. Leung, H. Zhao, T. Feng, Generalized fuzzy rough sets determined by a triangular norm. Inf. Sci. **178**, 3203–3213 (2008)
24. N. Morsi, M. Yakout, Axiomatics for fuzzy rough set. Fuzzy Sets Syst. **100**, 327–342 (1998)
25. D. Pei, T. Fan, On generalized fuzzy rough sets. Int. J. Gen. Syst. **38**(3), 255–271 (2009)
26. A. Radzikowska, E. Kerre, Characterisation of main classes of fuzzy relations using fuzzy modal operators. Fuzzy Sets Syst. **152**, 223–247 (2005)
27. D. Yeung, D. Chen, E. Tsang, J. Lee, W. Xizhao, On the generalization of fuzzy rough sets. IEEE Trans. Fuzzy Syst. **13**(3), 343–361 (2005)
28. W.-Z. Wu, J. Mi, W. Zhang, Generalized fuzzy rough sets. Inf. Sci. **151**, 263–282 (2003)
29. W.-Z. Wu, W. Zhang, Constructive and axiomatic approaches of fuzzy approximation operators. Inf. Sci. **159**, 233–254 (2000)

Index

M. Diker, *Texture Spaces*, Studies in Fuzziness and Soft Computing 411,
https://doi.org/10.1007/978-3-031-39748-6

T

U